Praise for the *Voice of the Infinite in the Small*

With incredible sensitivity and astonishing insight, Lauck has tackled a subject many of us would prefer to tackle with a can of repellant. What "bugs" us so about insects? Drawing upon myth, anecdote, history and research, Lauck weaves a compelling tale of the soul and psychology beneath our savage and unrelenting war on the nations of six-and eight-leggeds. Kinship with ALL Life? Even creepy crawlies? Lauck says "Yes," and says it with well-informed conviction. A jewel of a book—a classic.

—Susan Chernak McElroy, author of *Animals as Teachers and Healers*

Lauck, an environmental educator, writes with infectious enthusiasm about everyday matters that may well determine the fate of the earth, namely, our relationship to its most numerous inhabitants: the insects. Lauck is a storyteller, too, and in her effort to replace our underlying adversarial myths about our relation to insects, she relates tales from cultures around the world in which insects are helpers, heroes, teachers, even gods. Sections of detailed material on the lives of insects are surprising and fascinating. ...this is important material, informative and highly readable, a good resource book for teachers, parents and citizens, as practical as can be, personally and politically.

—*Publishers Weekly*

Not only is this a tremendously insightful and thought provoking book but it is essential reading for anyone who is willing to explore the profound human connection with all life forms, regardless of how great or small, for only within the understanding of this relationship can we comprehend our own place in the scheme of life. In *The Voice of the Infinite in the Small*, Joanne Lauck summons us to view a far larger window.

—Bill Schul, author of *Life Song: In Harmony With All Creation*

After reading this book, you're going to want to kiss a fly, hug a cockroach and take an ant out to dinner.

—Machaelle Small Wright, author of *Behaving As If the God in All Life Mattered*

At last, finally, a book that treats insects as fellow passengers (and beloved mythmakers) on the ride we call life on earth. It took courage and farsightedness for Joanne Lauck to speak out for a new relationship with the six- and eight-leggeds who travel among us. That she has done so with such clarity, means we shall all learn from her vision.

—Jim Nollman, author of *Spiritual Ecology*

Insects are the great shapeshifters. Not only do they serve a vital function within all of nature, but they can teach us much about shapeshifting our own lives and spirits. *The Voice of the Infinite in the Small* is a wonderful resource for anyone wishing to open their hearts to the wealth of knowledge and spirit available to us through the insect world. Whether scientifically-focused or mystically-minded, this book will inspire an enriching metamorphosis within your perceptions about this amazing kingdom!

—Ted Andrews, author of *Animal-Speak* and *Animal-Wise*

It has always seemed strange to me that our modern religions and philosophies have no place in them for those kingdoms of life that so vastly outnumber us. *The Voice of the Infinite in the Small* restores to the insect kingdoms their rightful place among us. I was touched beyond words by the timeliness of this important work and by the beauty of the personal stories included within its pages. I will be recommending this book to my clients for many years to come.

—Sharon Callahan, interspecies communicator and author of *AnaFlora: Flower Essences for Animals*

At last there is a book that confronts our war against insects directly. *The Voice of the Infinite in the Small* is a timely book that helps the reader step out of their human perspective and cultivate a feeling connection for creeping creatures. An inviting read, Joanne Lauck pioneers a new vision for harmonious coexistence with this aspect of Nature that is at once practical and illuminating. I would recommend it.

—John Seed, Director of the Rainforest Information Centre in Australia and co-author of *Thinking Like a Mountain*

Those of you who read my book *Buffalo Woman Comes Singing* know that one of my most powerful shamanic teachers opened me to the world of insects as messengers of spirit and Earth at many levels. Now we are gifted with Joanne's wonderful book, which directly supports these small messengers, and in doing so awakens us to another level of the embrace which is spoken by many of us in the words, All My Relations. Using the insects as very clear models, she invites us to listen to the wisdom of all life around us as we move forward in creating solutions to the crises we have created—solutions which work for everyone and everything in the circle of life for seven generations. This amazing work can change your mind and heart forever in a very good way.

—Brooke Medicine Eagle, American Native teacher, Earthkeeper, healer, singer, ceremonial leader, and catalyst for positive change

Filled with fascinating facts and psychological insights, *The Voice of the Infinite in the Small* uproots the outworn attitude that insects are merely pests, instilling readers with new appreciation for the amazing arthropods and astonishing arachnids who populate our planet.

—Gary Kowalski, author, *The Souls of Animals*

Truly marvelous. Enthralling. A book full of wisdom. I feel as though I've been initiated into a purer, rarefied realm of awareness, where it IS possible to celebrate even the most painful events and to feel the Big Connection more often than not.

Joanne Lauck has worked, with obvious devotion and inspiration, to turn the great body of knowledge she has garnered into viable solutions to deep problems—ways of improving the way we see and behave toward all creatures, methods we can employ in our daily lives. This makes the knowledge truly valuable, giving it the potential of becoming wisdom we can all hold within us. Lauck has made many beautiful links between her own ideas and insights and those of others whose time has also come. She is a great webspinner, practicing what she preaches.

—Gwynn Popovac, artist, author of *Conversations With Bugs*

We stand at a critical juncture in our human consciousness, when we must choose whether to continue the old patterns of disregard, disrespect, and ill-considered destruction of other life forms that share our planet, or embrace instead, reverence for all life. The *Voice of the Infinite in the Small* stands forward as an essential tool that will help each of us to extend such reverence to those of the insect realms that none may be excluded from our unconditional acceptance and compassion. Joanne Lauck is a most skillful teacher and guide. Her presentation is fascinating, a pleasure to read, and will absolutely change any lingering negative feelings toward insects one may have.

—Rita Reynolds, author of *Blessing the Bridge* and editor of *laJoie* magazine.

Lauck has researched her topic well. Her persuasive and stimulating discussion touches on science, mythology, literature, and film. Her book is a timely challenge, a message that great opportunities for healing sometimes come in small, six-legged packages.

—*NAPRA ReView*

The Voice of the Infinite in the Small is not only beautiful, but it is informative and profound. Joanne Lauck has taken the innermost desire within all of us to feel the reality of our Oneness on a deep level and shown us a clear path toward embracing not just what is easy to love and to be in awe of, but what has awakened our deepest fears. This book is a breakthrough in awareness for human beings. It is a challenging and spirited call to a greater wisdom. I, for one, am very grateful for the work. It blesses us all. It brings back that childhood special sense of wonder of the incredible richness and diversity of insects in a way I can't explain and I will be recommending this to our readers and making it available through our bookstore service.

—Kurt Lauren de Boer, editor of *Earthlight* magazine

Insects have been made humanity's scapegoat. They share the planet with us. If we are committed to fully conscious living, we have a responsibility to reexamine our hatred for bugs. This captivating book is an excellent tool for that purpose. I am redefining my relationship with the insect kingdom.

—Michael Peter Langevin, publisher, *Magical Blend* magazine

The Voice of the Infinite in the Small

The Voice of the Infinite in the Small

Revisioning the Insect-Human Connection

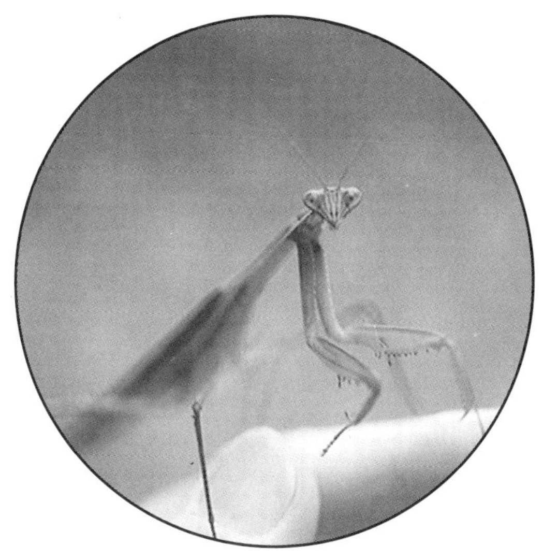

By
Joanne Elizabeth Lauck

Swan•Raven & Co.
P.O. Box 190
Mill Spring, NC 28756

Library of Congress Cataloging-in-Publication Data
Joanne E. Lauck
The Voice of the Infinite in the Small: Revisioning the Human-Insect Connection
by
Joanne E. Lauck p. cm
Includes bibliographical references and index.
ISBN 0-926524-49-6
1. Insects.
2. Insects--Psychological aspects.
3. Insects-Religious aspects.
4. Human-animal relationships.
5. Nature - Effect of human beings on.
I. Title
QL463.L28 1998
595.7--DC21
98-7162
CIP

Cover and book design by imagination@work
San Jose, California

Volume V - Ecology
The New Millennium Library

Address all inquiries to:
Swan•Raven & Co.
an imprint of Blue Water Publishing, Inc.
P. O. Box 190
Mill Spring, NC 28756
U.S.A.

Blue Water Publishing, Inc., is committed
to using recycled or tree-free paper.

Printed in the United States of America

For the Insects

and

In appreciation and honor of
Sir Laurens van der Post
and a Bushman named Xabbo

You should try to hear the name the Holy One has for things,
We name everything according to the number of legs it has;
The other one names things according to what they have inside.

—Jelaluddin Rumi
version by Robert Bly

The Voice of the Infinite in the Small

Acknowledgments

Although many people have supported me directly and indirectly as I worked on this book, my deepest gratitude and love goes to my parents—Emery and Naomi Lauck. To my father for passing on his deep love of the natural world and encouraging me to always have a vision, and to my mother for her love and the unfailing support that she has provided in large and small ways—including proofreading the final manuscript. I also thank my stepchildren Ashley, Andy, and Matthew for bringing the fire of transformation and renewal into my life and my nieces Andrea, Ellen, Beth, Sarah, and Laura who have been the inspiration for my work with children in the schools and the ones who shared their insect dreams and stories and went on bug hunts with me.

I also wish to extend my love and appreciation to my sisters Linda Buecken and Cheryl Thomas for their encouragement and for sharing my love of the "other ones." And to my brother and business partner Tom Lauck for his patience and willingness to tutor me, not only in the art of computing, but in the art of moving out into the world with clarity, integrity, and purpose. Throughout the process of living and writing this book, Tom's sense of humor saved many a day when I let my concerns about the book darken my mood. I am also indebted to him for his willingness to carry more than his share of the workload so I could meet the book's numerous deadlines.

Special love and deep gratitude goes as well to Marcia Lauck, who has been friend, sister-in-law, and mentor for over twenty years. Marcia, an author herself, is the "universal dreamer" whose dream I include at the end of the chapter on insects in dreams. As the catalyst for my self-explorations and my guide, Marcia has freely shared her vision, knowledge, and heart with me over the years. In fact, it was she who first understood and taught me that my deep feelings for other species were not the eccentricity of a personality out of sync with cultural expectations, but the expression of an inner blueprint that would eventually lead me to begin writing and teaching about these "other nations of consciousness." Her influence in my life and my gratitude extend far beyond what I can express here.

I also want to acknowledge and give thanks to my dear friend Trisha Lamb Feuerstein. Trisha and her husband, Georg Feuerstein, have supported this project from the moment they first heard about it. Trisha has been sending me insect articles and information over the last four or five years and many of the stories included in the book are her discoveries. Trisha has also given her time and energy in providing editorial assistance, and her suggestions have greatly improved the text.

I wish to extend appreciation as well to my publishers Brian Crissey and Pam Meyer for their belief in the importance of this subject and for their willingness to learn from the material (I suspect that Brian is being tutored in the language of the sting and Pam in the creative medicine of the spider!). Thanks also to my neighbor Pam Van Dyck, an insect enthusiast for years (the gods must have been laughing when we became neighbors eight years ago, unaware of our common interest). And to other friends, old and new, who have provided encouragement and enthusiasm along the way and "fresh eyes" during the editing process, including Gwynn Popovac, the insect artist you will meet in the chapter on affinity, my cousin James Lauck who brought his teacher's eye to the manuscript, my Brazilian sister Silvia Jorge, as well as Deborah Koff-Chapin, Hariana Chilstrom, Sharon Callahan, Rita Reynolds, Phylis Rollins, Darlene Dressler, Diana Rys, Sara Willow, Harry Hart-Browne, Jean Collins, and Patricia Earl.

And this page of acknowledgments would not be complete without thanking Brian Goddard, the young man featured in the stories about my "Thinking Like A Bug" class. Brian, who is now in high school, became my assistant in the years that followed his initial introduction to insects and his memorable encounter with Cedar the cockroach, and I have never had a better helper.

Thanks and appreciation as well to Booksin Elementary teachers Peggy Tanger and Lu Anne Behringer for inviting me into their classrooms each year to teach their children how to "think like a bug."

And last, but not least, I want to thank the many creatures who have shared their wisdom with me, Mantis under whose watchful presence this work was created and moves out into the world, and Leaf, child of my heart, teacher, and the being in fur who started it all.

Preface

One night when he was six years old, writer Daniel Quinn had a dream about a beetle. In his dream he was walking home late at night when he found the sidewalk blocked by a fallen tree. A large black beetle scurried along the trunk. He shrank back, terrified of insects in general and worried that the beetle might blame him for cutting down the tree and destroying its home. But the beetle, who had "an aura of great wisdom and authority," told him not to be afraid, that it just wanted to talk to him.

The insect told Quinn that the community of life—that is, all the nonhuman species (who lived in a forest next to the sidewalk and were now at its edge watching the interchange)—needed Quinn's help. They wanted him to leave the sidewalk and enter the forest. "We'll all be there, waiting for you," said the beetle, but "it will mean almost giving up your life ...[and] becoming one of us, ...and we need to tell you the secret of our lives."[1]

Young Quinn stepped off the sidewalk and woke up. He immediately burst into tears. When his mother tried to comfort him, reassuring him that it was just a bad dream, he told her that she didn't understand, that he was crying because the dream was so beautiful. Some forty years later in *Providence: The Story of a Fifty-Year Quest*, Quinn described the dream as the call of destiny.

Quinn knew from that early age that someday he would enter the forest where the other ones lived and help these beings. Twenty-five years later he used the framework of this dream in his novel *Ishmael*. Instead of a beetle, however, a gorilla confronts the narrator, offering to reveal secrets unknown to humans and inviting him to embark on a journey of discovery.

Around the same time that Quinn had his beetle dream, I was tending a wagonful of dirt and worms in my backyard in Michigan. At four years old, a longing for connection with other species was already a strong and steady impulse within me.

From my current perspective it seems that, like Quinn, a whole generation of people born around the same time were wired to step off the sidewalk. Each of us found our beetle equivalents, and each of us were compelled by intense life-shaping communions with other forms of life to push beyond the accepted cultural and scientific assumptions about them. My own life experiences, like those of many of my contemporaries, found affirmation in the perennial wisdom of indigenous cultures that believed we were never alone—that we were immersed in a sentient world—a "forest of eyes."[2]

In the midst of global crises—caused by paradigms that justify our separation from nature and rationalize our unconscionable destruction of the biosphere—more people are beginning to ask critical questions. They also are learning how to listen to and learn from the wisdom encoded in the natural world—a wisdom that has informed and guided indigenous peoples the world over. What has been commonly overlooked in today's society, however, for reasons this book details, is the wisdom manifested in and by creeping creatures.

This work on revisioning our connection with insects (and I use the term "insect" as a convenience to refer to all six-, eight-, and multi-legged creeping and flying creatures) has grown out of the contributions of many others—the majority of whom I have referred to in the text. The relatively new paradigms of ecopsychology and deep ecology, for example, have provided the psychological foundation for a new relationship with other species. I have merely applied those principles to insects. And Edward Wilson's concept of "biophilia," which in essence refers to our innate love of or affinity for life, not only confirms our need for positive relationships with other species, it implies there are consequences when we ignore this need and exclude so many species.

It is the work of the late Laurens van der Post, however, to whom I am particularly indebted. He illuminated in his books and lectures the "first" transcendent, creative pattern within each of us (which he saw in its purest and most mature form in the Stone Age race of hunters in Africa called Bushmen), explaining why an insect—the praying mantis—was chosen as the Bushmen's highest representation of value and meaning or "the voice of the infinite in the small."[3] Van der Post believed—and it is an idea that has greatly influenced my thinking—that the restoration of the despised and profoundly rejected wilderness or instinctual self inside us (our foundation in spirit) is critical to our surviving the challenges of this time.

One of my objectives in writing *The Voice of the Infinite in the Small* was to reveal just how and why we have become so antagonistic toward insects, while at the same time showing the results of our attitude. Although my first inclination was not to detail the abuses, I decided it was necessary. Exposing the effects of this habit of enmity will strip the largely unconscious forces that fuel it of their influence over us. It also has been my experience that bringing awareness to the widespread belief in the insect as enemy is necessary before we can stop the onslaught of pesticides and move toward real cooperation and/or communion with these creatures.

By revealing current beliefs and practices and the adversarial context through which we have learned to interpret the behavior of insects, I hope not to solve every issue, but to set the relationship between insects and people on its right course. A new context that assumes kinship and seeks correspondences between our worlds and harmonious coexistence will position us to see and to hear the voice of the infinite in its many-legged guises—and provide the motivation to heal the rift. And I have little doubt that this new orientation will lead to more people rejoicing in insects. Rejoicing is an appropriate and natural response when we take delight in someone or something—whether insect or other. The stories I have included of people interacting positively with insects also demonstrate that compassion and empathy for insects require only a willingness to merge the boundaries that separate and distinguish one from another and participate imaginatively in the other's worldview.

In the course of researching this book, many people informed me cheerfully that they hate insects, and just as many shared their favorite bug story, happy to have an attentive listener. Others were simply put off by the subject, their absence of curiosity about insects noticeable and congruent with their belief that insects have nothing important to teach us. At one of my first conferences, for instance, my speaker badge prompted a conference attendee sharing my lunch table to ask me what I would be speaking about. I said, "Insects." She leaned toward me eagerly and said, "Incest?" I said, "No, insects." She was taken aback as though I had rebuffed her, and managed only a muffled "Oh" before she hastily excused herself from the table.

A friend in Florida stopped writing after learning of my intent to write a book that championed all insects, even Florida's cockroaches and mosquitoes. So did another correspondent who was working on exposing pesticide abuses.

A puzzled relative wanted to know why I wasn't writing about something more worthwhile, like abused children. I explained to him that I was trying to address the issues underlying all forms of abuse. I believe that this work on revisioning the insect-human connection includes all the basic education and all the psychological principles and lines of reform needed to help us end abuse and bring about harmonious relations between all beings. If we learn to respond compassionately to insects, for instance, we will have deepened our capacity to respond compassionately to all species, including our fellow human beings. We will also have developed a discriminating eye toward the kinds of propaganda that feed our fears, and we'll have understood that we have the capacity to reinvent ourselves as a compassionate, cooperating species.

After much consideration I have chosen not to include the scientific names of insects in the text, except in a very few circumstances where they are needed for clarification. In doing so, I hope to appeal to the average person who is largely unfamiliar with insects and uninterested in delving into the classification system that entomologists find so useful. I have in many instances, however, used other criteria that laypeople find easy to remember—like grouping together species that fly and bite fiercely or those with a penchant for blood.

I am not an entomologist, as readers will quickly determine, although I share a common purpose with the many entomologists who have devoted considerable energy and resources to changing the public's view of insects; to that end we are in agreement. My approach, however, emerges from my understanding of the human psyche, from a belief that the microcosm reflects the macrocosm, and from an abiding interest in the healing potential inherent in our relationship to other species.

Joanne Lauck
March 1998

Introduction

In his great vision that he had in 1872 when he was nine years old, the Lakota Indian Black Elk experienced a moment when he saw the entire universe dancing together to the song of the stallion in the heavens: the leaves on the trees, the grasses, the waters in the creeks and in the rivers, the four leggeds and the two leggeds and the wingeds of the air "all danced together to the music of the stallion's song." Throughout his work Black Elk insists on this great unity of the entire world of the living. He tells us that "One should pay attention to even the smallest of crawling creatures for these too may have a valuable lesson to teach us, and even the smallest ant may wish to communicate to a man."

In an even more intensive manner Joanne Lauck is dealing here with the intimate bonding of humans with the insects. To appreciate the relation of humans to these other modes of living beings does not come easily for us in the Western cultural tradition. That we are so lacking in both our intellectual and our emotional appreciation of the insects is part of the retarded cultural development of the peoples of Western civilization. We have intense emotional bonds with the dog and cat. Yet we have a completely utilitarian view of the animals that we raise for food under rather horrendous conditions. The wild animals that we confine in zoos lead desolate, distorted lives, their beauty gone, their emotions frustrated. We seem to think that we are doing good to animals to bring them into our own life context in a patronizing manner. We do not like association with the wolf, or the snake, or the rat, or the scorpion in their wild environment.

We even associate our own vices with the animals: gluttony with the pig, deceit with the snake, viciousness with the wolf, malevolence with the mosquito. We avoid these animals, especially the insects, as beings that somehow should be destroyed as far as we are able, unless we could make them serve some obvious human need. We have little awareness that by losing any of the animal species we are losing splendid and intimate modes of divine presence. By destroying some insect species we could even completely upset the pattern of survival upon which we ourselves depend.

We need to be reminded constantly of such basic realities as we pass through these critical years of transition into the twenty-first century. In our earlier years of tribal existence, when we were awakening to our human form of consciousness, we had a deeper understanding of our unity with other life-forms. We could speak of the various forms of life as "all our relations." These other life-forms were our guardians, our ancestors, our teachers, our healers. We set up totem poles to indicate this relationship, we carved masks and kachinas, painted our bodies and our dwellings, and enacted rituals and ceremonies to indicate this relationship. When we needed assistance we went to these other members of our great family.

As Chang Tsai, an official in the Chinese administration in the twelfth century A.D., stated in the inscription that hung on the west wall of his office: "Heaven is my father and Earth is my mother, and even such a small creature as I finds an intimate place in their midst. Therefore that which extends throughout the universe I consider as my body and that which directs the universe I consider as my nature. All people are my brothers and sisters and all things are my companions." We find such thoughts attractive, until we think of insects. Immediately we become ambivalent. Insects, in terms of species, of individuals, and in terms of sheer volume of living matter, outnumber and outweigh all other forms of animal life combined. Since we do not know why they exist in such abundance, we project onto them our own desire for dominance and then react to them with fear. What we fail to realize is that they exist in such large numbers and such mass because they are necessary for the functioning of the Earth and for the survival of all other species.

We are most conscious of such insects as flies, fleas, roaches, and mosquitoes that may afflict us personally at times but which deserve a better understanding, an understanding that *The Voice of the Infinite in the Small* seeks to provide. Added to these are the insects that devastate our crops: the weevil, the potato bug, and the insects that damage grain crops. We need to understand how to limit the damage they do, but not by saturating our fields with pesticides and herbicides. These only increase the difficulty. After all these years we are still suffering the same amount of losses to our crops as we were before we began use of such harsh repression measures. We need to enter into that system of mutual limitation that nature has designed so that no one species or group of species can overwhelm the other species.

We might also reflect more fully on those insects that bestow benefits on us, the honeybee and the silkworm that contribute products that we find delightful and most helpful. And insects such as the praying mantis protect us from other insects that if not checked would do greater harm to the vegetation. Still others nourish the birds that adorn our world so beautifully in their form and color and who delight us so in their singing. Most needed are those insects who do the immense service of pollinating the flowers. The Earth that we know would be unimaginable without this pollination. Most important also are those insects who recycle our bodily waste and other decaying material. They cleanse the Earth while renewing its fertility.

Yet in the human world a good part of the chemical business consists of producing various sprays to rid our houses of insects of any form and pesticides to be spread over cultivated areas. First used on crops on a large scale in the 1860s, these pesticides are now applied by airplanes that carry out a procedure known as "dusting" the fields. One of the difficulties with such pesticides and herbicides is that the insects have an amazing genius for transmutation whereby they not only survive the toxins used against them, but on occasion they also turn the pesticide itself into nourishment.

There are a few instances in which the chemical sprays used on crops have been successful over a long period, but, in general, the insects survive better than the crops. Only a fraction of one percent of the sprays ever gets to its target organism. Meanwhile the pesticides destroy an immense number of insects that protect crops from other insects that do the harm. A third, even more significant, disadvantage is that pesticides end up in our own bodies since they saturate the food itself. This is the case with the notorious DDT, which is still being used in some parts of the world and is still turning up in the animal forms that humans use for food.

The real question that needs discussion, however, is the inner psychic-spiritual attitude that we adopt toward insects. However irritating they can be at times, the insects belong to the same organic social order that we do. They are integral members of our own life community. We cannot live without them. To speak of a comprehensive compassion toward all insects is indeed challenging. Yet we cannot refuse the right to existence to any aspect of the universe. Every being came from the same source as every other being. Every being has its own unique yet needed role in relation to all the others. As soon as we reject any part of the universe we have upset the order of things.

We must indeed protect ourselves against the harmful dimension of other modes of being, yet we easily develop a psychic imbalance through our irrational and unfounded fears and cause more damage than we avoid. We begin to eliminate insects from any positive role in the integral functioning of the planet. Once we begin this process of rejecting, once we place ourselves at war with the insects, then it is not easy to determine just how far we will go in applying this process of species suppression.

The sting of the insects is a language we need to understand. If we turn from this generalized attitude of antagonism to one of discernment, we discover a world of beauty, of skill, of communication, a world of genius for adaptation beyond anything that we could ourselves ever be able to accomplish.

We are missing fully half of nature when we eliminate insects from our world of interest. Considering the beauty of color and form and song added to our world and the increase in intelligence offered to us, it is utterly objectionable to impose our antagonism on the insect components of the Great Community of the Earth. Each of these tiny insects is, by definition, an animated being, a being with an anima, a soul; not a human soul indeed, but an insect soul, a thing of marvelous beauty expressing some aspect of the divine.

In establishing our basic attitude toward the world around us we might simply reflect on the awakening of consciousness in our earliest years. As soon as we awaken to consciousness, the universe comes to us, while we go out to the universe. This intimate presence of the universe to itself in each being is the deep excitement of existence. The word "universe," *uni-versa* in the Latin, indicates the turning of the grand diversity of things back toward their unity. I mention this tendency here because the purpose of this book on insects is simply to indicate the intimate presence that exists between ourselves and the insects. The immediate corollary is that we and the insects depend on each other in some profound manner.

This was the primordial insight of the Taoists of ancient China: the movement of the Tao is "to turn." After differentiating, all things turn back to that primordial unity where each is fulfilled in the others. To go far is to come near. Such is the basic law of existence. We are at the moment of turning. The time has come for humans and insects to turn toward each other. It is our way to wisdom, the source of our healing, our guidance into the twenty-first century.

Thomas Berry
March 1998

Part I

Redrawing the Circle

1

Coming Home

We are all culturally hypnotized from birth.
—Willis Harman

In Lewis Carroll's *Through the Looking Glass*, a gnat inquires of Alice, "What sort of insects do you rejoice in where you come from?" Puzzled by the question, Alice replies, "I don't rejoice in insects at all."[1] In these lines, an ancient world that acknowledged and celebrated all life-forms meets our modern world which has little regard for insects* and related creatures who look and act so differently from ourselves. In the West we don't rejoice in insects, our interactions with them fraught with anxiety and mistrust. If not actively hostile, we are at best, like Alice, ambivalent, our feelings colored by habitual enmity and fear.

If we rejoiced in insects, we would take delight in their presence. It is a state already invoked by particular animals. In fact, our battles with insects take place against a backdrop of increasing sensitivity to other creatures. It seems that the rapid decline of species throughout the world has, at last, sounded a chord in the collective psyche, arousing people to celebrate their presence and act on their behalf. Yet, only specialists act on behalf of threatened and endangered insect species. Most of us feel that a few less insects in the world would be a good thing.

As the number of people seeking personal contact with wild animals increases—including contact with dolphins, wolves, elephants, lions, and even gorillas—stories of these human-animal friendships are making the news. The accounts feed a deep longing in us for connection with other species, but we have drawn our circle of concern and appreciation too small when we exclude insects, believing that our desire for connection doesn't extend to this kingdom. Many insects remain outside our attention, dismissed and discarded as insignificant. Others, the objects of our fears and prejudices, are condemned to death on sight as we interpret their behavior as hostile or invasive, or judge their appearance as unacceptable.

* As a convenience I use the term *insect*, usually reserved for six-legged invertebrates, to refer to all six-, eight-, and multi-legged land-dwelling, creeping, and flying creatures commonly considered bugs or creepy crawlers.

When I first began to investigate the insect-human connection, I was looking for stories of cooperation and kinship. What I found was animosity. It was everywhere I looked—in the news, pop culture, scientific journals, children's books, and even on the World Wide Web. I was aware of our culture's lack of regard toward insects, but I assumed its position on insects was merely an extension of a general lack of regard for nonhumans. I was wrong. The level of fear and hatred that governs typical responses to insects is far greater than our negative feelings toward most other species in the animal kingdom.

Although a growing number of specialists eagerly cultivate insects as a resource, few individuals enter their world with empathy and a desire to understand and appreciate them. In the general literature on insects, observations that might lead to rejoicing are skewed by the language of war and images of insects as enemies. It is apparent that the myth of our separation from nature has found a disturbing foothold in our interactions with the members of this vast Lilliputian world. Equally disturbing is the fact that despite evidence of the far-reaching and dangerous effects of killing insects with toxic chemicals, we are still using insecticides in alarming quantities.

Attempts to explain our negative feelings typically focus on the insects: their independence, their strange appearance and behaviors, and their tendency to appear in great numbers. But I wondered how much our assumptions color what we see when we look at these creatures. I also wondered how much a lack of knowledge about insects and the absence of a context that would support our kinship with them influence how we interpret our encounters with them.

The Belief in Insect as Adversary

Large and small battles against insects occur every day in Western industrialized society. A widespread assumption of insect as adversary is operating as fact, routinely shaping our experiences with real insects. Looming across the landscape, this shadowed view of insects has tangled roots embedded in the dominant mechanistic view of the natural world and our bid to master it. From this view and with our support have risen the regulatory agencies and big business conglomerates whose orientation revolves around eradicating or controlling insects. Propaganda generated by these institutions has followed, simplifying the complexities of our relationship with insects and feeding us hostile images that perpetuate a militaristic stance toward thousands of species. The taproot, however, is our fear.

Allegiance to Fear. "I didn't know what it was so I killed it"[2] is what biologist Ronald Rood heard over and over again when he questioned people who brought him the dead bodies of insects or other unfamiliar creatures to identify. The curiosity natural in a nonthreatening situation was absent. They had killed it—just to be safe—to crush the fear the creature's presence invoked.

A large part of our fear revolves around being bitten. Bug-zappers, those glowing fixtures in suburban yards, keep those fears at bay, killing billions of insects each year. Research on electric traps, however, reveals they do more harm than good and give us little more than psychological comfort and the illusion of controlling biting bugs. Every sizzle brings satisfaction. One entomologist contends that we buy these killing devices for amusement and suggests selling them in the home-entertainment section of stores. A recent study on their effectiveness showed that less than one-fourth of one percent of the insects zapped were ones with a penchant for blood; half were nonbiting aquatic insects that feed fish, frogs, birds, and bats, and another fourteen percent were insect predators.

The estimated seventy-one billion nontarget insects killed each year by these devices, however, does not generate much concern for a human society with a perception of insects as adversaries. In fact, little influences our preset ideas. When our experiences do not substantiate our fears, the human need for psychological congruency prompts us to stockpile experiences that support our belief in insects as the enemy and dismiss contradictory evidence. Convinced we are justified and acting realistically, we are quick to defend our aggressive stance. The possibilities that arise for a different, more positive relationship are rarely entertained.

Images of Kinship. When I first started looking at insect-related material, I sought images of kinship, communion, and cooperation. I discovered these images in certain religions, philosophies, and cultures that felt differently about insects and included them in their circle of community. In tribal societies, for instance, the strategies for dealing with biting insects and the discomfort of insect bites did not negate their feelings of kinship with these species. Consider a story, condensed below, called "The Old Woman Who Was Kind to Insects"[3] which was told around a fire in West Greenland during a plague of mosquitoes.

> One winter an old woman was left behind by her family and tribe who had to follow what winter prey was available. She was so old that she could barely chew anymore, and all her family left her to eat was a few insects. The woman looked at the insects and said: "I'm not going to eat these poor creatures. I am old and perhaps they are young. Perhaps even a few are children. I'd rather die first."
>
> Soon after a fox entered her hut. It leaped up and started to bite her. The old woman thought, "Well I'm really dead now." But the fox acted strangely and bit her all over her body, as if it were taking off her clothing. Soon all her skin fell away, revealing a new younger skin underneath, for the grateful insects had instructed their friend the fox to rid her of her old skin.
>
> The next summer when her family returned to the camp, they didn't find the woman or her bones. She had gone to live with the insects. It is said that she married a little blowfly of whom she had grown quite fond.

This story of compassion toward insects and its surprising reward gains added importance because it was told during a time when the air was thick with mosquitoes. No hero's tale of battle against mosquito foes, it seeds the listener's imagination with images of respect and kinship, teaching the transformative power of compassion. It also underscores this tribe's belief in the bond between their people and all other species, even when the behavior of another species is painful or inconvenient.

In Western culture we have moved away from this bond, thinking ourselves separate from other species. But denying our links hasn't dissolved them. As the science that supported the separation gives way to the understanding of our profound interdependence with all species, only the fear-driven habits of enmity—which largely determine our present responses to insects—hold us to a false path. In the following pages, we will look at this habit closely as it relates to insects, identifying its roots and its influence over our perception and imagination. By taking a hard look at the current state of affairs in regard to insects, we will help strip this habit of hostility and mistrust of its covert power and clear a path to the multifaceted transformative power of the insect realm.

Creating a Hostile World

There are real reasons why we mistrust and fear insects. Few have much to do with the actual insect—a statement to take on faith for now. Most involve misperceptions and are tied to a multitude of beliefs about who we are in relationship to the rest of the Earth community.

Three hundred years ago when we desacralized nature and adopted a mechanistic model of the world (a collective decision explored in recent years through a great number of lenses) we transferred our trust to science and technology. We also turned into an absolute belief the simple assumption that what is strange or unknown may be dangerous (a cautionary stance in previous cultures). Then we added to it, attributing an evil intent to the strange and unknown. It is understandable then that we view the often bizarre looking, multi-legged creatures suspiciously and arm ourselves against them.

By drawing our boundaries of self and community too small, we have created a world outside those arbitrary and narrow limits that frightens us. Imbuing the unfamiliar and strange with malevolence has transformed the once-sacred Earth community into an environment populated by monsters. It has also exaggerated and distorted whatever survival instincts—whatever healthy fear—we had evolved as a species to keep us cautious and appropriately alert.

Modern culture's habitual hostility toward insects, the product of both the assumption of insect as adversary and the belief in the malevolence and danger of the unfamiliar, operates throughout the industrialized world. Its widespread acceptance as a realistic response makes its considerable influence invisible. The stain of this prejudice, however, is clearly evident in most insect-related items, events, movies, stories, and news reports generated in the West.

Our Assumptions at Work in Movies. The manner in which our assumptions about insects are expressed in the culture is not subtle. Our beliefs surround us. What is subtle is the way they perpetuate our misperceptions and encourage a combative attitude toward all creeping creatures.

In movies, we see our beliefs paraded before us in uninhibited display. Although we may discount their significance, these imaginative expressions of our culture reflect the mainstream's core assumptions and fears about insects and, entertainment aside, help to maintain our enmity.

Movies in the last hundred years, for instance, have consistently portrayed insects as hungry for power and human flesh. In many films, common insects assume immense proportions, often after some lab accident or catastrophic natural event. Their appetite increases in proportion to their size. When they inevitably seek human prey, a hero-scientist must outwit them to save the terrorized human community.

In *The Deadly Mantis*, a typical science-fiction thriller, a giant praying mantis, frozen during its prehistoric life, thaws out and begins to hunt humans for food. "A thousand tons of horror from a million years ago..." devastates the Arctic before being driven south by the United States military. The insect arrives in Washington, D.C., leaving a path of death and destruction in its wake. Finally, a beleaguered, but determined, military task force under the direction of a scientist traps and kills it.

New Films, Old Fears. Filmmakers express the culture's concerns about certain species by weaving their particular survival strategy into the plot. In the 1957 film *The Beginning of the End*, flesh-hungry giant locusts ravage the Midwest. When they swarm on Chicago, a resourceful scientist, intent on saving the besieged city, implements a strategy based on his knowledge of locust sexual behavior. He records their mating calls and broadcasts the recording from a boat on Lake Michigan. They rush into the lake pursuing the sound and drown. Tricked by their insatiable instinct to procreate, the locusts prove no match for the hero-scientist.

Forty years later the same underlying fear is expressed in the 1996 movie *Independence Day*. In this immensely popular film, aliens, described as swarming locusts, ruthlessly attack the Earth, trying to exterminate the human species.

The next year *Men in Black* made its debut. In this rather comic sci-fi movie the ruthless extraterrestrial alien is now a cockroach, hated and feared by humans and other extraterrestrials alike. No matter that the cockroach villain has teeth like a shark and a body that resembles a lizard, the real cockroaches that drop out of its sleeve throughout the movie are enough to pair it with the loathing people feel toward this particular insect.

Scientific Discoveries Add Credibility. To provide variations on a standard theme, filmmakers keep their antennae tuned to scientific discoveries about insects, incorporating them into their scripts to make them more plausible. New findings either create a new insect problem or provide the heroic scientists with a critical advantage over insect monstrosities ravaging the human community.

The use of wasp enzymes and bee serum in the cosmetic industry led to a 1960 film called *Wasp Woman* in which a cosmetic queen given wasp enzymes to stay young turns into a killer wasp. And in *Bug*, a 1975 movie, a scientist crosses a common cockroach with a new insect species that feeds on carbon left from a recent earthquake, and their offspring is a flesh-eating insect. Another movie, perhaps a sequel, titled *The Nest*, also features cockroaches that crave human flesh.

The 1997 sci-fi thriller *Mimic* incorporates gene-altering biotechnology with the remarkable ability of many insects to camouflage themselves. The story begins when genetically altered cockroaches, inadvertently released in New York City, multiply and grow, camouflaging themselves as human beings to escape detection and eradication.

And a recent psychological thriller called *The Silence of the Lambs* relies on techniques from the field of forensic entomology by which the presence of insect larvae on a corpse helps determine the time and place of a murder and whether the body was moved after death. In this macabre film, a serial killer and amateur lepidopterist (one who studies butterflies and moths) places a cocoon of the death head moth in the bodies of his victims.

Short Stories Reflect Our Beliefs. Like movies, books, and magazine articles also follow the culture's beliefs faithfully. In T. S. Eliot's short story "The Cocktail Party," the missionary heroine is staked near an anthill and tortured to death by swarming ants. Ants are also the culprits in William Patrick's science thriller novel *Spirals* in which they carry a virus from a sealed lab to the outside world. In "Itching For Action," a short story by Charles Garofalo, fleas take revenge on a man who poisons dogs and cats.

Thomas M. Disch's story "The Roaches," written in 1965, incorporates layers of negative beliefs. The heroine Marcia is a lonely young woman who hates cockroaches and spends every evening killing them in her apartment. The story's effectiveness depends on readers who are repelled by the look and mobility of cockroaches and horrified by the thought of thousands living behind their walls. The story begins:

> Miss Marcia Kensell had a perfect horror of cockroaches.... She couldn't see one without wanting to scream.... It was horrible, unspeakably horrible, to think of them nesting in the walls, under the linoleum, only waiting for the light to be turned off, and then.... No, it was best not to think about it.[4]

The intensity of Marcia's hatred and her single-minded pursuit of them create a link between herself and the insects. One day she discovers that they understand and obey her commands. Once aware of her powers over them, she sends them to kill boisterous and erotic neighbors who offend her:

> When the insects return, the deadly deed done, they seem to regard her calmly. Marcia believes she can read their one repetitious thought: "We love you we love you we love you we love you." To her amazement she answers, "I love you too.... Come to me, all of you. Come to me."[5]

The story ends as roaches from every corner of the city hear her and crawl toward their mistress. The story's message—that loathing can be an expression of hidden sympathies for the thing we loathe—is a subject we will touch on when we examine affinity. The idea that cockroaches will kill on command reflects the common belief in the robotic mentality of insects and is itself a kind of infantile fantasy of power over others.

Comic Books Reflect the Culture's View. Comic books are another medium that expresses the culture's beliefs about insects in exaggerated fashion. A comic strip called "Winged Death Came Out of the Night" plays out our fears with a generic bug who combines size, a fetid odor, thorny claws, and drooling and piercing mandibles (jaws) to engage the widest range of fears in the reader. This nightmarish anomaly comes smashing against a screen door like a torpedo, startling a man alone in a cabin.

Each night the bug grows bigger, hurling itself against the door in unexplained fury. The man sleeps and dreams that the bug is human-size and bursts through the screen. In the dream when he threatens to kill it, the bug says scornfully, "You can't kill me, man! You can't keep me from getting you! My intelligence is greater than yours! We, the lowly bugs shall rule the earth."[6]

The dream becomes reality the next day. The now human-size bug bursts through the screen:

> He felt the slash of the dripping mandibles...the piercing ripping grip of the thorned claws. This was no nightmare. This was real...He screamed and...he died! Then there was only the heat and silence, except for the soft rustle and the wet sticky sound of chewing![7]

This is our stuff. It is our belief in the insect as adversary. It is also our belief in the malevolence of the unknown and unfamiliar, shaped by discoveries from those who study insects, expressed by writers and artists, and accepted by men and women of good heart and fine intention who unconsciously live out the cultural consensus concerning insects. We imagine in some dark recess of our mind that given the chance insects, who innately possess an evil intent, will want to take over the world and dominate us.

This tendency to attribute to insects our desire for power over others is also seen in articles on alien abductions and extraterrestrial life. The author of an article called "Big Bugs From Outer Space," for example, describes the similarities between "the bugs we squash and the alien that could squash us."[8] Recalling the "flying bee hives" reported in the 1950s and other sightings and accounts of abductions by insectoid aliens, he painstakingly describes what these creatures could do to the humans they abduct.

But is the will to dominate another species in the insect, or is it in the human personality? Our beliefs about insects and our readiness to think the worst of them color our experiences with real insects and transform the natural world into an alien landscape populated by robotic and malevolent specks of life. Anxiety levels elevated, we become increasingly isolated, pitted against the imagined ill-will of the Earth community.

Organic Controls Keep the Insect as Enemy

After Rachel Carson's exposé on insecticides, *Silent Spring*, hit the bookstores in the sixties, the growing concern about using toxic chemicals to fight insects increased. The new awareness kept intact the belief in insect as adversary, focusing on the need to replace pesticides with organic controls. We still wanted them dead, we just didn't want to kill ourselves in the process—or poison creatures we liked. New products flooded the market. Advertising kept its aggressive tone. Headlines from organic gardening magazines read like military news: "Be Ready for the Annual Beetle Invasion."

The World Wide Web, a contemporary means of transmitting cultural beliefs through a network of computers and users, followed suit. The web site "Fighting Pests Without Pesticides" delivers a message of old-fashioned hostility, disseminated through high-tech wizardry that is anything but subtle: "We all know that ants, flies, beetles, and roaches are ugly, disgusting, gross, and plain old nasty, but they are also annoying and destructive."[9] The site contains many different ways of killing insects, most of them, like the electronic bug-zappers, indiscriminate about which insects are killed.

Mutant Species Resist Insecticides. As reports circulated about the dangers of insecticides, scientists also reported that many target insects were developing a resistance to specific poisons, and mutant species were evolving. That information went into our already fear-ridden imaginations and stirred things up.

A new series of revenge-of-nature films expressed our agitation about mutating insects and the effect of toxic chemicals on the environment. In *Kingdom of the Spiders*, for instance, tarantulas mutate after coming into contact with a new crop-dusting spray and seek humans for food. And in a 1978 movie, *The Bees*, a mutant strain of bees fight exploitation by a large American corporation. The bees devastate North America until the United Nations agrees to take better care of the environment.

The Language of War

Only a few steps away from movies, short stories, and comic books are editorials and news reports about insects that employ military terms and hostile descriptions without trying to disguise the assumption of insect as enemy. As noted before, any assumption with widespread acceptance becomes invisible. And once hidden from the light of reason, it can wield considerable power without calling attention to itself. Since we see our enmity as a realistic response to the reality of these creatures, our news reflects that operating premise and few of us question why the language of war punctuates most insect-related news. Its congruency with our belief that they are the enemy safeguards it from examination.

A newspaper reporter under the spell of the cultural bias, for example, writes about a "vicious" fly biting innocent people unfortunate enough to live in the fly's territory. According to this news item, the fly is a "savage animal" who rampages across the county looking for people to attack. Another news item, about a company that manufactures biological insecticides, describes an "insidious new weapon against humanity's ancient enemy—the lowly cockroach." Still another newspaper headline reads "Swat Teams Rev Up Campaign Against Larvae," and two magazine briefs on insects are called "Battling a New Intruder" and "Moth Wars."

Messages of Hatred. Insect-related advertising on television and in newspapers and magazines, a tool of the industries that make their money killing insects, reminds us at every turn that insects lie waiting for a chance to hurt us or our property, or to spread filth and disease. A pleasant-looking woman on a recent television commercial for an insecticide tells the viewers in a confidential tone, "They deserve to die." Few of us will question this statement. Fewer still will identify it as an expression of hatred or realize the well documented and real psychological and physiological consequences of hating anything.

We simply don't take issue when insects are called aggressive, vicious, or malevolent. In fact, we are more likely to object and dismiss as sentimentality descriptions of insect behavior using favorable characteristics, with the exception of the industrious bee and ant. Worse, cultural pressure shames those who might naturally respond compassionately to individual insects. Offenders are ridiculed and dismissed perfunctorily by those who would uphold the culture's condemnation of insects and our right to ridicule and exterminate what we find offensive.

Ironically, the tendency to anthropomorphize or attribute human characteristics to other species, a taboo of science, is often accepted without comment in scientific circles when the traits employed are negative and congruent with the culture's fears and opinions. In a recent book, for example, ethnologist-primatologist Frans de Waal points out that current scientific literature routinely depicts other species as "suckers," "grudgers," and "cheaters" whose members act "spitefully," "greedily," and "murderously."[10] When, however, a species displays altruism, friendship, or curiosity, scientists are quick to use negative or dehumanized neutral language to prevent accusations of naiveté, romanticism, or ridicule by their colleagues. It seems that no one is automatically immune to the influence of the collective view simply by the weight of his or her academic credentials.

Celebrating Our Hostility

Our enmity toward insects also appears in community activities. In fact, we often celebrate our hostility, staging events that express it collectively. An archery club on the West Coast holds an annual bug shoot. Archers from around the state gather to shoot at more than forty replicas of worms, caterpillars, snails, scorpions, beetles, butterflies, and bees. Although the public might protest if they used models of dogs and cats, no one protests that the targets are creeping creatures.

Mosquitoes are a favorite target for group hostility, especially in tropical regions where they thrive. Each year a small town in Texas, for example, hosts a mosquito festival as a tourist lure. At their 11th Annual Great Texas Mosquito Festival, however, the eighty-four species of mosquitoes native to the state were noticeably absent. During the festival the year before, townspeople had passed out tiny swatters to participants to keep the mosquitoes away and planned on doing it again—only this time there were not enough insects around to kill. The mayor was visibly and vocally unhappy about the absence of the normally abundant mosquitoes—but he made the best of it. In the opening ceremony, he showed the crowd a jar with a live, captive mosquito in it. Then he opened the jar and killed it ceremoniously with a four-foot-long, "Texas size" swatter as the crowd cheered.

Contamination of Educational and Cultural Events. Attempts to balance the prejudice toward insects and promote an appreciation for them are often contaminated by the unexamined, but active, forces operating within us. An example is the popular hands-on exhibition on insects currently touring the country's museums and featuring six huge, robotic insects. It could have been called Backyard Wonders, but was named Backyard Monsters by its creators.

A local art gallery in the San Francisco Bay area that prides itself on its innovative exhibitions featured a collection of mixed media art devoted to insects. One "artist" killed flies and mounted their dead bodies on paper so that they spelled "fly paper." Prejudice passed for creativity, and the audience was appreciatively amused.

Another artist on the East Coast, known for his portraits of black men, drew the man who killed Huey P. Newton with a praying mantis over his face. The artist either ignored or was unaware of the complexities of the praying mantis and nature's checks and balances on its population when he chose this creature to suggest how the black race is destroying itself.

When art reflects the limitations of the popular culture, it becomes little more than a propaganda device, reflecting not vision, but our blind spots and narrow interpretation of life. In Paris, for instance, a Chinese artist put spiders, snakes, scorpions, and toads together in a small cage so they could battle and eat each other in an exhibition at a contemporary art center. The artist insists that his work, recently restricted to photographic display only, is valuable as a statement about the condition of our human world.

A contemporary artist in Britain, whose work was featured recently on the World Wide Web, uses live insects in his exhibits to invoke correspondences in observers. In one exhibit, for example, houseflies are hatched and killed by an electric fly-zapper. The artist says he hopes that an observer will become uncomfortable when they realize they are like the hapless flies. In another exhibit, the artist has made holes in a display cabinet, and when insects enter they are killed by an electrical device. The canvas, which is permanently wet, also acts as a deathtrap for other live creatures imprisoned for the exhibit.

Teaching Prejudice

The message we give our children is understandably in line with our beliefs. We train their imaginations, instilling hostile images that uphold the cultural stereotypes of other creatures. This type of indoctrination also prevents undue curiosity from interfering with the official view—especially if the insect has been categorized as a pest.

Our control/management orientation toward the natural world, taught from grade school on, emphasizes that every plant and animal must fit our human-centered agenda to be of value on the Earth. Having divided all species into beneficial ones or pests, we are quick to ask "What good is it?" and "How can it further our plans?" The answer given determines how we feel about that insect, and those we judge as pests are stripped of all rights and subjected to ridicule and extermination.

Training the Hostile Imagination. Our teaching materials naturally reflect the biases we have come to view as reality. An Eastern World Wide Web site geared toward children joined a new science center and calls itself "The Yuckiest Site on the Internet."[11] A news article about the site says it presents photos of "humankind's worst enemy: the cockroach." In its on-line learning center, Rodney the Cockroach offers the inside story on cockroaches, facts stained by our pest-management orientation and the accepted view of cockroaches as loathsome creatures.

Teaching hatred under the guise of education is effective. The site's message board includes an entry from a young visitor who shares his "fun" roach-killing methods: "Turn a fan vertical and drop one in [splat]."[12] Lest we think this unusual, a major technology institute teamed up with an insecticide company to sponsor a cockroach-killing exhibit for children. Billed as an educational event, it was well attended.

A "find out about science" book for young children follows the same line of thinking. Illustrations on the opening page of this primer show adults and children killing different kinds of insects in a variety of ways:

> Some insects are pests. People don't like them. They swat flies. They slap mosquitoes. They squash cockroaches. Of all the insects, cockroaches are the ones people dislike most. People try to kill cockroaches when they see them. But cockroaches are tough. They are hard to get rid of.[13]

In the same book, a second illustration shows a boy's high-top sneaker about to come crashing down on a cockroach to illustrate the function of the cockroach's cerci—two appendages that help alert the insect to danger. The author invites the child reader to imagine what would happen if they tried to step on a cockroach.

One of the underlying messages of books like that one is that human beings don't like certain insects and that it is okay to kill what we don't like. By the age of ten, the information will be embedded in the child's mind. As adults, the odds are that they will not remember why they hate them. Their ingrained response will feel natural, and they will believe that they have no points of kinship with cockroaches or any creeping creatures introduced to them in this manner.

Justifying the Rift. These same negatively conditioned individuals will be likely to accept the popular theory that people dislike and fear insects because they look so different from what they are accustomed to seeing. They may also accept the theory that insect fear as it exists in modern society is a protective mechanism, a genetic holdover from a time when humans were still learning which insects were dangerous. They may not even consider the hypothesis that their responses are exaggerated and distorted by simple conditioning and the lack of a context that would encourage kinship and appreciation. And news articles laced with combative language and warnings about insects combined with exposure to popular films and advertising from the billion-dollar pesticide industry will support their beliefs, keeping their imaginations harnessed to the cultural consensus on these creatures.

The Weapons of War

The official view of insects encourages a dangerously high use of toxic chemicals thirty-plus years after Rachel Carson's eloquent appeal and well-documented conclusions about their hazards. A 1996 news article reported that acre for acre, United States homeowners apply more pesticides (which includes herbicides, insecticides, fungicides, and rodenticides) to their yards and homes than farmers do to their fields. If we consider the widespread dependency on chemicals in agriculture, this statistic is even more disturbing.

Children and the elderly are particularly sensitive to these synthetic chemicals, a fact that has not, until recently, been taken into consideration by the Environmental Protection Agency (EPA) when it sets tolerance levels for toxic ingredients. One study, reported by Food & Water, a nonprofit national grassroots organization against pesticide use, concluded that the average child in the United States consumes more pesticide residues by their first birthday than the EPA considers safe for an entire lifetime.

Research has linked pesticides and their by-products to cancer, birth defects, hormonal disruptions, and genetic alterations. Synergistic effects have not been thoroughly evaluated, but new evidence shows that individual pesticides, known to disrupt the endocrine system, can increase one thousand times when mixed together. Many studies have also demonstrated that pesticides lower reproductive rates in birds, threatening the survival of species like the peregrine falcon. Newer studies link reproductive problems in humans to pesticides.

In the environment, pesticides destroy beneficial soil organisms, which in turn escalates soil erosion and nutrient loss. Pesticides seep into groundwater, streams, and lakes, contaminating our drinking water and damaging aquatic ecosystems. Pesticide use also leads to secondary insect infestations that necessitate further applications, putting us on a deadly pesticide treadmill.

Biological Pest Control and Genetic Engineering. One solution to pesticides that safeguards our belief in insect as adversary and allows us to continue current agricultural single-crop practices is called biological pest control—using insects to kill other insects. We have imported insects from around the world to prey on local insect species we want to control or eliminate—setting aside the complexities and interrelationship between plants and insects in the target habitat. John McLaughlin of the United States Department of Agriculture admits biological pest control "is a bit of a crapshoot, so you don't release just one species; you try to introduce several."[14] Sometimes it appears to have worked. Other times it definitely has not. Sometimes the imported species has switched hats—becoming the enemy and preying on and threatening to exterminate species that were not even supposed to be part of the game.

Genetically engineered plants have also entered the battlefield. Engineered to resist specific insects, their effectiveness has been shown to decrease over time as insects mutate to survive. Then another gene is added to address that resistance. Applying one fix on top of another gone awry is standard operating procedure despite the fact that hundreds of cases of resistance by mutation have been recorded. In 1997, for example, a study indicated cotton-eating moth larvae and other insects were developing a resistance to a natural insecticide produced by a genetically engineered variety of cotton. These findings may also apply to corn and potato plants genetically altered to produce the same insecticide.

Upping the ante, scientists have recently succeeded in making genetically engineered insects to carry on the war. A public interest group angry over their intended release has called it "ecological dynamite." One scientist, in an attempt to assuage concern, told reporters, "We can use genes the same way we use insecticides." But scientists, even those in the field of applied entomology, do not know for certain what they are unleashing in the environment. Their understanding of insect behavior and ecology is far from complete. They are also under considerable pressure to act, because their research is funded by regulatory agencies that demand efficient short-term solutions. More importantly, most of these well-meaning individuals lack the psychological maturity to unplug the war machine and harness genetic technological prowess in service of the nonhostile imagination. Only then might an appropriate response to a complex problem be found.

The late mythologist Joseph Campbell says the popular culture never rises above issues of power, and it deals with this theme in all its infinite variations. It is in this mode, then, that we are caught between opposites: either we kill the insects, or we are defeated by them. We rarely see a third possibility. We rarely put down our weapons long enough to consider the effect we might have if we entered their world with empathy and compassion—like the old woman who was kind to insects. Perhaps we underestimate the powers of Providence that would suddenly appear if we could align ourselves with the Earth and its small creatures. It is time to try.

Revisioning Our Connection to Insects

Thomas Berry writes in *The Dream of the Earth* that "we are returning to our native place after a long absence, meeting once again with our kin in the earth community."[15] The journey back to our native place—to the place where we might meet our insect kin—does not begin with a lecture or textbook on insects. It begins with uprooting the beliefs and fears about insects that we hold individually and collectively.

It would be a simple matter to revision our connection to insects if all it involved were the right facts about an insect presented in the right way. But remembering how to rejoice in insects has less to do with insect biology and behavior than it does with untangling the threads of self-deception, misperception, and fear that prevent us from including insects in our circle of community.

Depth psychologist James Hillman maintains that the insect problem is in our heads. He says that unless we develop an inner ecology to stop us from acting out our fears and dark fantasies, we will become increasingly isolated and continue to infect, pollute, and poison the world in an attempt to eradicate what we perceive as evil. "When we imagine that they are killing us, we kill them first, and overkill them. So if we don't remove the bug problem in the psyche...we go on with insecticides until we've actually poisoned ourselves out of existence."[16]

Transforming Ourselves. To unplug from the perception of insect as adversary and radically alter the way we perceive and relate to insects, we must be willing to change ourselves. The required transformation promises to rattle our most cherished ideas, but will serve us in the end. Anytime we let go of assumptions we blindly adopted as children and prejudices we retain as adults, we move under the influence of a powerful catalyst for insight and growth in directions unforeseen.

Our task includes clearing the lens through which we see another and becoming more psychologically sophisticated. We must learn, for instance, how enemies, personal and collective, are created in the human psyche. And we must be willing to wrestle with the complex issues inherent in our relationship to species we have judged as economic pests and disease carriers.

We must also bring a sharp and discriminating eye to the propaganda around us that encourages us to battle insects. It's big business to wage war on insects. As Rachel Carson pointed out, when a for-profit organization is a "sustaining associate" of a scientific society, whose voice are we hearing when they speak about a chemical's harmlessness?

The *Wall Street Journal* has estimated that insecticides and extermination services are a $3.5 billion annual business. Advertising in service of these industries strips away the ambiguities inherent in our relationship with insects and plays on our emotions, telling us, "The only good bug is a dead bug." When we take our instructions from these types of organizations, we stay on the battlefield, expending great amounts of energy fighting insect phantoms created by advertising or insect pests created by those invested in current nonsustainable agricultural practices.

A further challenge involves scrutinizing our notions of sickness and health and who causes what disease and then rerouting research dollars into preventative measures. Traditional medicine, built on a militaristic foundation (that battles disease instead of fostering health), has fueled the heroic attempts to curb disease by eradicating the insect carrier. Yet, insects and microbes are mutating, and diseases like tuberculosis and malaria are increasingly drug-resistant and on the rise. Worse, chemical intervention has stripped the indigenous people of any hard-won immunity developed over generations from living in these insect-thick tropical regions. It's clear we need a different conceptual base underlying medicine—one that adds more complexity to our formulations and treatment strategies.

Finding a New Context. After rooting out the beliefs that set up the war with insects, we will be ready for a new context to help us translate our interactions with them. The context sets the stage and determines whether we enter a battlefield, amusement park, or temple when we meet an insect. The right context will help us relinquish the role of hero in our modern good-versus-evil drama and discover something heroic and of great merit in changing our customary stance. Besides giving us a way to think about encounters in which we are bitten or stung by an insect, it will also provide us with guidelines for how to respond.

Stories are the vehicle through which any cultural worldview or commonly held beliefs are disseminated and upheld in the population. From the story we tell ourselves about who we are and how the world and its creatures work, an environmental ethic, sound or unsound, emerges.

A new life-serving story that weaves new scientific discoveries and ancient truths about our interdependence with other species will be the ground from which an appropriate ethic about insects will emerge. Indigenous practices and stories like "The Old Woman Who Was Kind to Insects" are a response to an understanding of that bond. Studying the ways of tribal societies may help us find a contemporary expression aligned with the truth of our profound interconnectedness. Then we can teach the children.

We might begin by adopting the perspective of indigenous people who believed each species had value and operated in both physical and spiritual worlds. Seeing in the insect a mode of divinity would position our spirits correctly for rejoicing in this aspect of creation. But lest we think our task easy, our new context must also address the species science has linked to illness. How are we to rejoice in insects who have the power to inflict pain, disease, and even death?

Coming Home

Having briefly entered and marked the territory—the existing bias against insects, the beliefs that have created our habits of enmity, the forms that uphold it, and the nature of the change that will free us for rejoicing—we are ready to embark on the journey back to our native place. By transforming ourselves, we can forego our fugitive status and come home, finally, to an Earth community that awaits our return.

2

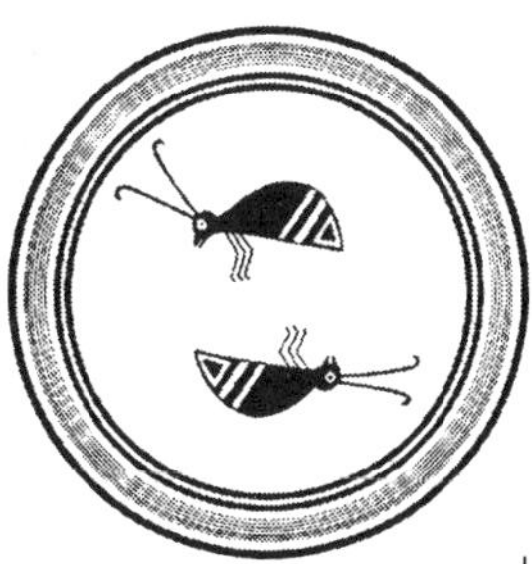

Clearing the Lens

In the beginning we create the enemy.
—Sam Keen

What prevents us from rejoicing in insects also prevents us from seeing them. We see, instead, an image shaped by our beliefs and by our fears. Perhaps we assume our eyes operate like competent machines, reflecting the world back to us without distortion, or that seeing is an objective phenomenon, a detached, efficient, and rational ability. The insect presents itself to us, and, if we have our eyes open, we see it. But studies in human perception have proven that seeing is undependable, inconsistent, and caught up in unconscious desires and fears. Our ability to perceive something follows our expectations.

What we hear and remember about insects also depends on psychological factors. When we study insect biology and behavior, facts bend themselves around our beliefs. Inner circuits built on the myth of our separation from insects, intermingled with notions about insects as robotic and malevolent, control what we let in, what we recall, and what conclusions we draw. In this unexamined state, the drama and diversity of the real insect kingdom is virtually inaccessible. Our inner insect images obscure the view.

Beginner's Mind. It isn't easy to see an insect or any creature without having preconceived ideas about it. When we do, however, we open ourselves to the most vital and significant facts arising out of the moment. It's called "beginner's mind," a state of dynamic receptiveness that meditators seek. Zen masters teach that this state of mind is the space or attitude from which all wisdom arises, and it takes practice to achieve it.

Since we are most often praised for having answers, we usually have ideas and beliefs about everything. It is a way to ward off the unknown. But we lose a deep feeling connection to other creatures when we trade the mystery of their existence—which should increase our own sense of wonder and aliveness—for a neat bundle of classifications and summary explanations. We also lose our ability to see accurately and respond appropriately when we attribute dark motives to another and create enemies.

Enemy-Making

Enemy-making depends on a defense mechanism in the human psyche called projection. Projections interfere with our capacity to see objectively and relate humanely to another, be it insect or person. When a projection is operating, we attribute qualities, traits, and motives to another. Their identity becomes our creation, in line with our own psychological needs and beliefs, which are often far removed from the other's true nature.

The Human Shadow. Clearing our perceptual lens of projections is the first step toward seeing anyone—human or nonhuman—as they exist outside of our fears, needs, and opinions. Projections reside in what psychologists call the "shadow." It's a part of the personality that contains qualities repressed or made unconscious so other traits (ones the person deems good and acceptable) can be emphasized and developed. Jungian psychologist Edward C. Whitmont says the shadow is the urge in every person to find someone to blame and attack in order to vindicate oneself and be justified; it is "the archetypal experience of the enemy..."[1]

The creation of a shadow is a universal phenomenon in people, a pattern formed as a child's ego or center of awareness is formed. This developmental process always entails a clash between the individuality of the child and the opinions of the culture and parents. Each child learns its culture's acceptable ways of looking and acting, modified only by the parent's values. All other qualities contrary to the cultural and familial views of what is seen as good and appropriate are relegated to the child's unconscious shadow. But repressing qualities does not eliminate them. They continue to function unsupervised and unchecked, projected out into the world onto others.

Group Shadow. Groups, like individuals, have a shadow that contains repressed qualities incongruent with how the members of the group like to think of themselves. The shadow of Western culture is expressed collectively as the Enemy, the personification of evil typically seen in minority nationalities, races, and religions. Mythologically, our culture's shadow is seen in characters like *Star Wars'* Darth Vadar, the devil, tempters, and the dark brother or sister of a pair.

Both the personal and collective shadow help us avoid looking at our own hostilities, because they are always projected out onto someone else. Propaganda is an integral component of preserving the shadow projections toward a particular target. It simplifies the issues and glosses over contrary evidence. The collective shadow's influence compels us to find public enemies, replacing one nation with another, one leader with another, and one species with another, as power shifts and economic and political alliances change.

The fact that we project our rejected qualities onto another doesn't automatically mean that the other is innocent of these projections, but the intensity of our fear, repulsion, or dislike is a barometer of the strength of the repressed energies within us—not a response to the traits as they actually exist in another. In fact, we can't really know how much is in the other until we suspend our beliefs and study the source of our projections. When we have done our personal work and understood our complicity, we will have defused the hot buttons that cause us to react with self-righteousness and militancy. We will also have freed ourselves to see another accurately and choose an appropriate response.

Insects as a Hook for Our Projections. In *Faces of the Enemy*, Sam Keen's insightful book on the psychology of enmity, he says we are driven to fabricate an enemy, a scapegoat to carry our enmity. First we create a target or public enemy from the hostility that resides in our unconscious shadow. He sees the wars that follow as compulsive rituals in which we continually attempt to kill those parts of ourselves we despise and deny having.

Although Keen is referring to the faces of human enemies, the psychological realities apply to any situation in which we deny our hostility, and project it out into the world. Insects are an easy hook for our projections because of their strange appearances and habits.

Each generation has learned from the previous one to project a good measure of hostility onto insects and then react to them as the enemy—alien and malevolent. Children watch their parents respond to the creatures around them and imitate their reactions. The learning process is simple and effective. It can start with a comment similar to one I recently overheard in a local craft store. While standing in an aisle amidst an assortment of plastic crickets and ladybugs created to add interest to floral arrangements, a woman with a toddler in hand walked by me and pointed to the plastic insects. She said, "Oh, Jeffrey, look, bugs! Ick!" "Ick," echoed the child, the lesson learned without protest.

Once we learn what is acceptable and what is not, other psychological mechanisms called selective perception and selective recall impeccably uphold our projections into adulthood. They filter what we let in through our senses. When the subjects are insects, we remember only those facts and behaviors of insects that support the images we have already endorsed.

Pairing Insects with Other Enemies. Occasionally the faces of our enemies, human and insect, merge. In 1994, when the Haldeman diaries implicated Richard Nixon in the Watergate cover-up, *San Jose Mercury News* cartoonist Scott Willis drew a cockroach with Nixon's head, the association between Nixon's behavior and our belief in the contemptibility of this insect readily understood.

During Operation Desert Storm, political cartoonist D. B. Johnson drew a fly with Iraqi President Saddam Hussein's head and a hand with a fly swatter ready to annihilate the fly-man. No one protested, although as we will see in a later chapter, the fly has many more redeeming qualities than have been demonstrated by this man or the policymakers on this side of the Atlantic reluctant to be held accountable for the effects of the chemical warfare they authorized.

Projections prevent us from seeing our actions and motivations clearly. Psychologist James Hillman, who pointed out that the insect problem is in our heads, says that our fears about insects that lead to our automatic extermination reaction—that is, fears about their multiplicity, monstrosity, autonomy, and parasitic propensity—are actually *our* attributes and the methods *we* employ to eradicate insects. In fact, he points out, our prolific use of multiple deadly chemicals has infested every bit of soil, water, and air, and every household and institution in the United States. This use of insecticides displays an autonomous or out-of-control quality (evidenced by the felt helplessness of the individual to stop the outpouring by large faceless institutions) that we also project on insects. It seems, observes Hillman, that we are mimicking the "enemy," having become the deadly ones who infest places and operate by automatic reflex.

The Quest for an Insect-Free World

To add further complexity to the picture, the contents of our personal shadows have joined with the economic incentives of big-business agriculture to support what Rachel Carson called "the crusade to create a chemically sterile, insect-free world..."[2] It is this crusade that is fueled by what Carson saw as a "fanatic zeal on the part of many specialists and most...control agencies."[3] It is likely, however, that this zeal was not so much a conscious and collective rallying of forces as it was an unconscious gathering of images and beliefs from many projections.

The propaganda that feeds our personal and cultural projections of insects as the enemy and supports continued pesticide use—beyond safety and reason—is largely generated by special-interest groups. These include regulatory entomologists and decision-makers invested in the status quo and employed by powerful industries and government regulation agencies. With the implicit consent of an unaware public, they make policy and garner support for the eradication of insects using whatever method is deemed necessary. And typical of decisions and goals fueled by shadow projections are the ill-conceived strategies of eradication based on brute strength and ruthlessly applied power in the form of toxic chemicals.

The very notion that characteristics in others that we find objectionable could be pointing at ourselves is an insight with the potential to rattle not only our individual self-image but our identity as a nation. If we are to take back our power and apply pressure to stop the daily onslaught of chemicals into the environment, our ways of doing business and those in power who benefit from them will have to change.

As Vice President Gore wrote in his introduction to the 1994 reprinting of *Silent Spring*, "Cleaning up politics is essential to cleaning up pollution."[4] The necessary upheaval in all spheres of traditional influence linked to agriculture and toxic chemical manufacture and use, however painful and costly at the onset, could move our technologically advanced society toward a psychological maturity that so far eludes us.

Reinventing Ourselves as Friendly Human Beings. As individuals, if we could comprehend the psychological truth of projection, we might take a second look at insects and others who disturb us. Simply by understanding how we create the image of insect as enemy, for instance, we begin the process of retrieving the projection.

The goal is nothing less than to radically change ourselves, to create, as Keen suggests, "A new human being—*Homo amicus* ('friendly human')—who is animated by kindness, has a friendly psyche, and a politics of compassion."[5] It is this friendly and compassionate potential inside each of us that already knows how to rejoice in insects. It is our long-forgotten wilderness self—intuitive, instinctual, and natural—waiting for acknowledgment.

For as long as our projections rule our vision, dictating what we see and experience, we will defend our view of insect as enemy. By moving our focus from the creature to the eyes that see it and the mind that interprets it, we open the door to the complexities of those species against whom we fight with such ferocity. We also bring back charity and clarity and dispel the free-floating anxiety and vague fears that pool beneath our awareness and urge us to war.

A Feeling for the Organism

Spiritual philosopher and naturalist Henry David Thoreau believed that looking at something and seeing it were two different acts and that only the act of seeing involved understanding. According to Thoreau, when something was truly seen, it unlocked a new faculty of the individual's psyche or soul. Another naturalist, Annie Dillard, observed that only when you love something or have spent the time and energy to become knowledgeable about it, can you see its reality.

Dillard's insight is echoed by geneticist and Nobel Prize laureate Barbara McClintock, who conducted her research by getting to know each plant of corn intimately. McClintock's deep reverence for nature and capacity for union with whatever she studied (what French ethnologist Lucien Lévy-Brühl saw in aboriginal people and called "mystical participation"[6]) enhanced her vision and led to her discovery of gene transposition in corn plants.

Harvard entomologist and evolutionary biologist Edward O. Wilson's vast knowledge of ants—his specialty—led him beyond insects to the formulation of a sophisticated theory on the nature of life and community. He believes when people add knowledge about other species to a natural tendency to love and affiliate with them, their capacity to care about them deepens and expands.

So, to look at insects rightly, we must have, as McClintock would assert, "a feeling for the organism"[7] comprised of empathy, appreciation, and enough knowledge to respond appropriately when we meet them.

Neither loving toward nor knowledgeable about insects, we have accepted the common view that some creatures are worthless. It is part of the story we tell ourselves, part of the myth of our separation from the natural world that ranks life in a hierarchy of perceived value and places humans at the top and insects at the bottom.

It is a story, however, that is changing. New research demonstrates that we are tied to insects in large and small ways.

The Underpinning of the Insect-Human Connection

The biological sciences have revealed specific links to the insect community that provide the physical underpinnings of the new relationship we seek. Besides sharing the same basic DNA (a surprising connection revealed in genetic research), we need insects to survive. Our destinies are tied to the small creatures of the Earth.

Edward Wilson calls insects and other land-dwelling arthropods the "cornerstone of life on Earth."[8] He believes we could lose creatures like the wolf, bear, and hawk and still survive. Not so if we lose insects. If they or other groups of major small organisms like bacteria or fungi were to disappear, it would be catastrophic. We would follow them into extinction within a few short months, unable to live on the Earth without their recycling, harvesting, and pollinating services.

Chemical Sages of Nature. Our physical survival may also depend in part on the chemical prowess of the insect kingdom. According to recent estimates from studies in South American and Borneo rain forest canopies, ten to fifty million insect species inhabit the earth—most of them undiscovered. Insect champions, typically entomologists, field biologists, beekeepers, and other specialists who study them, aware of the importance of insects, inform us that any one of the million new species awaiting identification could offer scientific information, new products, and pharmaceuticals of great value. It was in the gland of an obscure water beetle, for instance, that a Heidelberg chemist named Schildknecht discovered substantial quantities of cortisone, a substance used in the synthesis of certain medicines.

More recently, Cornell entomologist Thomas Eisner found potential heart and nerve drugs in insects. Others have discovered in the wings of a butterfly and in the common yellow jacket's venom compounds that may prove useful in cancer treatment. Researchers are even exploring bloodsuckers like mosquitoes and ticks because they make a potent anticoagulant chemical that keeps blood from clotting. Genetic engineers can now reproduce this substance in large quantities and pharmaceutical companies hope it will be highly effective for heart-disease patients.

Insects as Food. Some specialists see insects as a valuable food source and the solution to the world's food shortages. Insects have substantial nutritive value, often containing significant amounts of protein, vitamins, minerals, and important amino acids. Ironically, before insecticides and our prejudices reached Third World countries, most indigenous people already included insects in their diet. In the West, some twist of mind lets us eat with relish multi-legged creatures that live in the water and reject the land-loving ones. But like it or not, we eat them anyway. They are present in all of our processed foods, and we may even receive a significant fraction of our daily protein from this inevitable intake of insects.

Feelings as the Gateway. Our physical dependence on insects for survival, the treasure house of chemical knowledge they embody, and their potential as food is only a part of the substructure that supports relating to insects differently. It is territory that, by itself, does not rouse us to conserve, much less rejoice. Even the use of insects to develop new medicines does not engender the level of support necessary to save threatened species. Most of us live our lives far away from the chambers where those discoveries are made. Intellectually, we might know that the cure for cancer or AIDS may be hidden in a dragonfly or a beetle, but the revelation dims before our shadowed opinions and disappears when we encounter that insect in our homes.

A petition that relies on numbers and common sense is obviously not enough. Rachel Carson wrote in *A Sense of Wonder* that "it is not half so important to know as to feel."[9] She says that if facts are the seeds of knowledge and wisdom, then emotions and sense impressions are the fertile soil. Only after our emotions are aroused do we want to know something about the creature or object that stirred us.

We need to cultivate a feeling for insects that is not based on our needs and desires. Feelings fuel action. If insects are part of our circle of community, then we'll care about them and act appropriately when we meet them, or when they are threatened. Positive feelings are the gateway to positive experiences, appropriate responses, and peaceful coexistence.

Psychological Links to Insects

Although we typically justify our lack of feeling for insects by citing our lack of structural similarity (although scientists tell us that people share key genes with insects), there are stronger links than physical and genetic equipment to support a feeling connection to them.

Those who've delved deeply into human nature and the natural world—scientists, mystics, psychologists, naturalists, and poets—all tell us that human beings are psychologically closer to other species than we have acknowledged in modern times. Evidence of our deep emotional links to the natural world and its inhabitants appears repeatedly across the many disciplines that have examined this connection phenomenon and its benefits to the human species.

Benefits of Being in Nature. Recent studies in the field of human-animal interactions, for instance, demonstrate that just being in the natural world invokes within us a sense of completeness. Without its presence, when we reside in a purely human-oriented environment, we suffer from emotional loss and unrealized potential. We even lose a measurable degree of vitality and general health.

Many of us intuitively sense this psychological dependence when we are outside in an area populated with wildlife. When we have become dulled by our daily human-centered routines, encounters with other species heighten our awareness and return us to an energetic state of being. Outdoors, insects sometimes surprise and arouse us. As teachers of the natural world, they continually invite us closer to inspect and reflect on their ways. Were we more open, they could awaken us out of our complacency, returning us to an alert, watchful state of being.

It is not surprising that after spending time in wilderness areas, people report feeling renewed and infused with energy. We have only to look at our overcrowded national parks, deluged with people seeking this renewal, to see evidence of the potent draw of the natural world.

Our sense of self and community is also intertwined with other species. Philosopher Howard Thurman said that the more people become aware of the affinity between their consciousness and the consciousness of the many forms around them, the more they feel their own sense of self enlarged and their lives enriched by a sense of belonging to a great mystery. We are diminished as individuals and a society when we exclude species from our circle of community because of fears which, when magnified by our collective assumptions, allow us to justify short-term economic goals.

Perhaps we have not included insects in our circle of community because we are reluctant to accord them consciousness. We have only begun to recognize our own consciousness and its influence on our lives. That some kind of universal consciousness is at work at every level of reality, informing and directing it, is a radical idea for those steeped in a mechanistic model of the universe and an idea only recently advanced in this age. It is directly opposed to the central myth of Western science that maintains our nature is only the by-product of random evolutionary events and natural selection. Forward-thinking individuals across many disciplines say that the eventual dissemination of this idea into mainstream society promises to drive revolutionary change in every area of our lives, including our interactions with other species.

Biophilia. Our psychological ties to insects, part of a greater web that connects us to all other species, includes another radical idea—an inborn capacity to love insects. Edward Wilson's "biophilia hypothesis,"[10] a proposition gaining recognition and serious consideration in the biological sciences, maintains that people have an innate emotional affiliation for other living organisms. It implies that our association with insects and other species plays a role in our mental health. How we translate the affiliation then, allowing, blocking, or distorting its expression, merits our attention.

The healthy expression of a biophilic affiliation to insects might include rejoicing. If blocked or distorted by the prejudice of the culture, however, we might respond instead—and to our detriment—with malice, fear, and ridicule.

Our Need for Other Species. Ecopsychology, a synthesis of ecology and psychology, assumes a sympathetic bond between every individual and the world, as ancient shamanic traditions have always practiced. The late pioneer ecopsychologist Paul Shepard took the premise further by arguing that within each of us there is a deep-seated need for other species—recognized or not. The need revolves around the unique manner in which animal images and forms shape our personalities, identities, and social consciousness. According to Shepard, other species, including insects, play a large role in the growth and development of the qualities we most value in the human psyche.

Environmental psychologist James Swan believes that our core identity is comprised of an "inner zoo"[11]—natural images that resonate with each aspect of the natural world. Perhaps the recent development of insect zoos and butterfly insectariums in cities across the world has been prompted by impulses arising from our inner insect kingdom. And if so, what happens to our core identity when the corresponding aspects in the natural world no longer exist?

Consider that if a need for other species and a longing for connection is encoded in our psyche, and if our interior is populated with images of the creatures of the world, the persistent exclusion of so many species from our circle of concern and community means we are expending great amounts of energy excluding parts of ourselves. It also means that what we do to insects we do to our inner menagerie. When we destroy an aspect of the natural world, or when a species becomes extinct, we lose a piece of ourselves. Ecological restoration is at its essence a restoration of ourselves, a retrieval and revival of the disowned parts of self.

The Primitive Self

At the center of the human psyche is a natural, intuitive, and instinctive self at home in the world. The late Laurens van der Post, champion of the South African Bushmen, called this fundamental identity our "primitive" or "wilderness" self. He saw it standing "in rags and tatters, the lowest of the low in the hierarchy of the contemporary mind."[12] No mere alarmist, van der Post believed the resurrection of this wilderness person was absolutely vital to our continuation as a species.

Within this natural self is an equally alienated insect aspect that we have, individually and as a culture, demonized and banished to what myths call the underworld and what psychologists call the unconscious. Retrieving and reconnecting to this aspect of our core identity is an appropriate and necessary task for us as we tackle the problems facing us.

Judging from the number of insect species and their importance to the ecosystem, it would seem that our wilderness self has a great fondness for creeping creatures. Perhaps it builds our complex responses to events from basic qualities they embody. If so, the number and importance of insects in the world means that our current beliefs and the animosity created by them is a false overlay on an extended identity that encircles all life-forms.

This extended identity is the focus of an emerging paradigm called "deep ecology." Deep ecologists view the self and the world as a seamless whole. By "greening" the individual self, they cast our identities over a living, self-regulating world. This world, which biochemist James Lovelock called "Gaia," is a living system, like ourselves, that exhibits the behavior and intelligence of a single living organism. As we relate to Gaia and think of ourselves as part of the collective intelligence of all life-forms, we free ourselves from the conventional idea of the self as a finite entity contained by our skin. And we rediscover ourselves as part of every ecosystem, every creature on Earth. We might dare to ask then who is this fly aspect or this cockroach inside us? And more importantly, how can we hold and heal these misunderstood, mistreated, and hated aspects of ourselves?

Journey to Wholeness. By knowing that the primitive self and its insect aspect exist, we start the retrieval process. Within us is a longing for wholeness—a longing for our true nature—and an impulse to move toward it. Psychologists tell us we can only inhabit that nature more fully through knowledge and reclamation of the missing parts. It is a journey that Carl Jung called "individuation." Each step we make toward our essence and each initiation we experience is reflected in our outer lives, which then become more coherent with our fundamental nature. Having strayed from our instinctual human wholeness, the wilderness self, we actualize it only by repeated acts of choice. Consciously choosing to cultivate a feeling for insects, we take a giant step toward healing this aspect of our psyche that has been battered by our hatred and warfare against these species.

It may comfort some and disturb others to realize we may already have guides back to our native place and helpers to redeem our lost aspects of self. There are insect species zealous in their efforts to remain close to us. When we abandoned our primitive selves, moving away from the natural world and into artificial living environments, certain insects followed us, moving into the cracks and crevices we couldn't plug. Maybe it is these insect species, the ones that insist on sharing our living space and that carry the heaviest burden of our ill will, who are the messengers of our insect self. How many then are we in conflict with? How many parts of ourselves have we condemned to hell, and at what point do we see their appearance not as something unpleasant to deal with, but as a remarkable opportunity to stop the war and make peace?

If we can stay open to this possibility and give our hearts and imaginations free rein, there is immense power for positive change in our current struggles with certain species. We might pray that our job is as straightforward as relinquishing our current stance toward them and clearing the lens through which we see them. Seeing for the first time, we become beginners, and all possibilities for understanding and compassion are once again present in the moment.

3

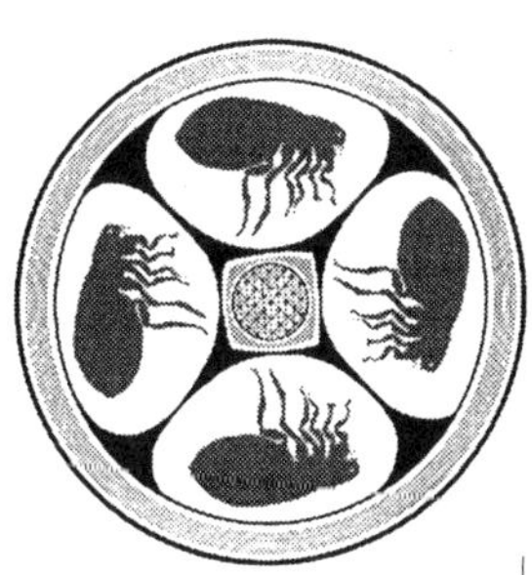

Mirrors of Identity

> One should pay attention to even the smallest crawling creature for these too may have a valuable lesson to teach us.
> —Black Elk

Native people observed other species closely, looking for insight and help in meeting their own life challenges. Unhampered by the hierarchical organization that positions one species above or below another, they had great freedom to learn from every species. Different situations demanded different skills. Since each species was endowed with its own unique strengths and strategies, they knew they could benefit from studying the ways of Ant* in situations requiring patience and attention to detail and use Eagle's vision when circumstances changed and they needed a far-ranging perspective.

Native people also acknowledged the invisible forces at work behind physical form, recognizing that everything emanated from a hidden spiritual realm. Native ceremonies created pathways between the physical world and the spiritual world to enlist the cooperation of the forces operating from this spiritual realm.

The Hopi tribe of southwestern United States and neighboring Pueblo tribes like the Zuñi, Shoshone, and Tiwa communicated routinely with the spirit essence of everything that existed in the physical world, calling them "tihu" or "kachinas." By invoking these spiritual entities through paint, symbols, actions, costumes, and rituals, these people encouraged interaction between the two worlds.

Insects had considerable influence and occupied as important a position as other animals in the tribe's pantheon of spirits. Each insect had some power or attribute associated with it that made it an essential kachina. Throughout the territory kachinas like *Kokopelli* or Assassin Fly, Butterfly Girl and Man (*Poli Mana*, and *Poli Taka*), Hornet (*Tatangaya*), Cricket (*Susöpa*) and Scorpion

---

* Capitalization of a common insect designates that it is the representative for its species or the overlighting spirit.

(*Puchkofmok' Taka*) were represented more often on altars and in ceremonies than the animal kachinas that we might presume to have more influence.

Buffalo Bug. The tribes of southwestern United States were not unique in their view of insects. Tribal people revered life in all its forms, so they naturally accorded insects value and respect and looked to them for information. One of the insects the Oglala Lakota honored was the "buffalo bug." This insect was sensitive to the vibrations of the Earth, and Oglala scouts would seek it out and study its position to determine the location of the nearest buffalo herd. The Lakota also celebrated a tiny wood grub and depended on its boring abilities to create the opening in their ceremonial pipe stems.

Honoring life in all its manifestations was a central orientation for indigenous people. The tribe acknowledged through prayers and rituals the lives of other species given so that the tribe might live. An insect might be killed inadvertently because of its small size, or gathered and eaten with due respect and gratitude, but there were understandings about the interdependence of all life that discouraged insects or other creatures from being unnecessarily harmed. To do so would place the perpetrator outside the covenant with the natural world that was at the center of tribal society and spiritual practices.

Creatures as Guides

Indigenous people routinely studied the local terrain and its animals, looking for spiritual guidance that would help them meet the trials inherent in everyday life. Sometimes the specific attribute of a creature was sought. For example, the protection of the butterfly, which exemplified elusiveness in some tribes, might be invoked before battles. Warriors would paint their bodies with butterfly symbols to assume the insect's power and gain help in dodging the arrows and bullets of their enemies.

It was also a tradition in many such cultures for individuals to take as a spiritual guide the first creature that caught his or her eye when starting the day. They understood that whoever commanded their attention—insect or otherwise—had sought them out, knowing inexplicably that its particular guidance was needed on some level.

Likewise, an expectant mother in some tribal societies would look for the creature or aspect of nature—whether rainbow, beaver, or ant—that presented itself to her during her pregnancy. She would know intuitively that its spirit had sought her out, wanting to communicate its connection to the unborn child.

When a creature appeared in a dream or vision, it was acknowledged as a reflection of the Creator as well as an intermediary link to what was sacred in the world. The dream was investigated and interpreted within this context, and its message was puzzled out—and if called for, acted upon.

Underlying all such behavior was the intuitive knowledge that an encounter with any species was not haphazard or accidental, but an intentional meeting carrying immense healing potential and gifts for the recipient. If you approached the creature with genuine humility and intuition, you could be gifted in unforeseen ways and healed on the level you needed to be healed.

Animal Medicine. Most tribes believed that each species had an overlighting spirit who, residing in the spiritual world behind matter, animated the physical form of each individual earthbound creature and infused them with the power unique to its species—called its "medicine." This behind-the-scenes entity could also orchestrate contact between one of its creatures and a human being.

Since each creature manifested its medicine through its appearance and behavior, for the perceptive individual the creature was the curriculum. In its behavior and in its unique way of being were lessons for living life well. Poets, naturalists, and mystics also recognized the wisdom inherent in the natural world. The Christian mystic Meister Eckhart said, "Every single creature is full of God and is a book about God."[1] Viewing the insect as a divine textbook and a part of our human curriculum provides us with a lifelong course of study and contemplation leading toward health and wholeness. Simply paying attention to those that command our attention and applying their strategies for living to our circumstances, we begin the learning process.

Insects as Messengers

Most indigenous cultures thought that the simply constructed creeping creatures responded more readily to the creative impulse of the divine than other creatures. In fact their simplicity was believed to have been carefully designed so that they might be ready emissaries of these powers. That insects are messengers is a truth known to spiritual teachers of every tradition, yet it is wholly unknown to most lay people. For example, when a bee flew into the facial hair of Dawn Boy, the spiritual teacher of psychologist, poet, and author Brooke Medicine Eagle, she laughed. But Dawn Boy lectured her sternly, telling her that the bee's action was a message from his spiritual elder, who spoke regularly to him through the insect world.

Later when Brooke Medicine Eagle began to pay attention to insects, she noticed that a specific kind of insect would appear just before one of her spiritual elders dropped by for a visit. Repeated experiences convinced her that the insects were, as Dawn Boy had indicated, messengers from those of the Dawn Star (or Christ) lineage of light.

Meaningful Coincidences. Often insects arrive at a critical juncture in our lives, creating an event that weaves our inner and outer world together in what we call a meaningful coincidence. Interestingly, a beetle first brought the phenomenon of meaningful coincidences to the attention of Carl Jung, the pioneer of dream psychology. When this insect made its appearance, Jung was listening to his client describe a dream she had about the sacred Egyptian scarab beetle. As Jung listened to the dream, his back to a closed window, he heard a slight tapping sound. Turning around, he saw a flying insect knocking against the window glass from the outside. He opened the window and the beetle flew in. Jung caught it and held it gently in his hand, studying it. He recognized the beetle as a species as close to the golden scarab beetle as one could find in that geographical region.

In Egyptian mythology the beetle is a potent symbol of rebirth, and its appearance in the dream of Jung's client was a significant event in her therapy. Jung reports that until that day, she had made no progress. She clung to a rigid view of the world that only an irrational, acausal event—like the beetle's unlikely appearance—could break. And it did. After that session, she began to change. Jung named the coincidence "synchronicity," and his investigation of the phenomenon led him to conclude that these occurrences emerge from another order outside our conscious control.

The interweavings of the outer world with our subjective one—a truth traditional science is prone to ignore—are frequently experienced as a form of internal and psychological help. Inner and outer worlds come together in a manner that speaks specifically to us, and we sense the Divine at work. Some numinous power has reached out and touched us. These correspondences, recognized by ancient matriarchal cultures and aboriginal tribes, tell us that all is connected. Feelings, intentions, thoughts, dreams, and intuitions are tied to outer activity and the physical world of biology and physics, and insects are frequently the messengers that alert us to these connections.

Insects as Creators

In many cultures insects were viewed not only as messengers, but also as creators. Since autonomy was generally considered a behavior of the gods, the autonomous vitality of insects as well as their strange appearances gave them considerable mystique and interest. In tribal mythologies, the local insect species, including creatures like spiders and scorpions, played an assortment of roles, sometimes acting as primary Creator and other times assisting in some aspect of creation.

Every culture, indigenous or not, has a creation story. Creation stories answer questions like: Where do things and creatures come from? and Why do they look and act like they do? A Malagasy creation myth traces the descent of one tribe from a moth, and in the Dong Ba culture silkworms, flies, bees, and other insects played key roles in the world's creation. In a Sumatran creation myth, Butterfly laid three eggs from which the first three people were born, and Butterfly is a form of the Creator as well as a totem for the Pima tribes of southwestern United States. The Andamanese islanders believed the first man emerged from a bamboo stick and formed the first woman out of clay from an anthill. And for the South African Bushmen, Mantis (praying mantis) was the primary creator and representative of God, the "voice of the infinite in the small."[2]

Spiders were also cast in the role of Creator. In the mythology of the tribes inhabiting Nauru Island in the South Pacific, for instance, the world was created by Areop-Enap, or Ancient Spider. Spider Woman was also an early creatrix for the Hopi, and in other Native American traditions Spider Man's web connected Heaven with Earth. In Navajo myths Spider Woman joined Spider Man to instruct the newly created Earth people in the art of weaving, while Cornbeetle Girl gave these people their voices.

Navajo Insect People. In the complex Navajo creation story, the surface of the Earth is the Fifth World. The first four worlds lie stacked beneath it, one on top of the other, with a specific color dominating each world and substituting for light. Some versions tell of the first world as red and inhabited by the Insect People. In this account, the Insect People moved from world to world, evolving, until in the fourth world they took on human form.

In another version, the Navajo's First World was in darkness, inhabited by nine "People"—six kinds of ants and three kinds of beetles. All spoke the same language. This was a world before stones, vegetation, or light. But within it were impulses toward activity and the desire for upward movement. As the ants and beetles began to explore and climb upward, more worlds and characters joined them, but in the myth no differentiation was made between insects, animals, and humans. In both versions, creation was a conscious activity from within, not an external act by a greater being.

As unfamiliar as this creation story may appear, what modern biology believes about the shared origins of life echoes aspects of the Navajo tale and promises to revise our ideas about our origins.

The New Story. The work of pioneer biologist Lynn Margulis suggests that far from leaving microorganisms or bacteria (which we commonly call bugs) behind on an evolutionary ladder, human beings are both surrounded by and composed of them. Evidence is mounting, says Margulis, for the idea that "life did not take over the globe by combat, but by continual cooperation, strong interaction, and mutual dependence among life-forms."[3] In this view, remarkably similar to the Navajo belief of activity originating from within a primary world, all life-forms from the paramecium to the human are meticulously organized, sophisticated aggregates of evolving microbial life.

In story form, imagine how these scientific insights might be conveyed to our children to instill a sense of awareness and connection to bacteria:

> In the beginning were bugs, creeping creatures of all kinds that together provided the building blocks for all other life-forms. They were hard workers and great inventors. For the first two billion years of life on Earth, they transformed the planet's surface and atmosphere, creating all life's essential chemical systems and processes like fermentation, photosynthesis, and

> oxygen breathing. Since these ancient bugs shared a single gene pool, they were in constant contact with each other and could use genes from the entire kingdom when they needed to adapt rapidly to changing environmental conditions. They had a great zest for life, and, propelled by a spirit of adventure, they combined their bodies with other bacteria and formed alliances that were often permanent and resulted in another life-form. Each life-form told its story through its unique form and behavior. Eventually entire societies of bacteria joined and became like states, and the states became like nations, and one of the nations was the People, the ones who, according to Australian aboriginal legend, could read the living text of all the other creatures and tell their stories.

And if Margulis's work isn't enough to link today's discoveries with the intuitive knowledge in the ancient Navajo creation story, the discovery of new forms of microbial life deep inside the Earth has gotten many scientists speculating about a hidden biosphere beneath the Earth's surface whose total number of organisms may rival or exceed that on the Earth's surface. Add the recent evidence of possible microbial life in a frozen ocean on Europa, one of Jupiter's sixteen moons, and the new story unfolding takes an unexpected turn in a direction we have only begun to consider.

Different Eyes

Scientific breakthroughs across a spectrum of disciplines are bringing us closer to the intuitive wisdom of native cultures and away from a desacralized view of nature. The language is different, but the common underlying truth of our interdependence and the need for cooperation and respect is unmistakable.

The practices of tribal people offer us lessons in interdependence. Adopting their worldview provides a clear perspective on what is out of balance in modern culture. An overemphasis on economics and expediency before other values, for instance, has exacted a heavy toll, creating an unparalleled Earth crisis and sending thousands of species into extinction.

Native cultures also provide us with evidence that we are not necessarily by nature a hostile species. We can unlearn our habits of hatred. Many examples exist of nonhostile human cultures, including the stone-age race of Bushmen in Africa and more culturally sophisticated tribes like the Hodi of south-central Venezuela, which operates on the principles of generosity and noncompetition. In fact, the name Hodi means "all of you like me," and each member looks after themselves by looking after one another.

Chewong Med Mesign. The Chewong people are another nonhostile people who for eons have inhabited the tropical rain forests of peninsular Malaysia. The Chewong believe that every species inherently deserves human respect and that each possesses *med mesign*,[4] a concept that translates roughly into "different eyes." The way each species perceives the world is considered legitimate.

For example, the tiger worldview, the fruit bat worldview, and the beetle worldview are all different but valid views.

Chewong stories about local creatures explain the med mesign of different species. The tribe's children are taught that the intent and behavior of any species, even when it is threatening or disconcerting to humans, arises out of its worldview. This insight encourages them to bring compassion and understanding to every encounter.

From a belief in the value of each species springs implicit rules governing ethical behavior toward other species. The guidelines grant great freedom for each species to pursue its own peculiar strategy for survival. Acceptable behavior for human beings—what is deemed good—includes the need to accord respect to other species, regardless of size or appearance. And there are taboos against ridiculing or laughing at another creature.

If we overlay the Chewong worldview on our culture, however, we find it does not fit. What is acceptable behavior today is based on a cultural consensus about creatures that denies them the right to pursue their life with dignity and respect. Picking and choosing whom we value, we let this hierarchy determine our feelings and behavior. The belief that we are at the top of the heap and entitled to exploit other species and do with them anything we deem fit means that those below us live only with our permission. And the ones on the bottom rungs are stripped unceremoniously of any rights, including the right to pursue their lives without undue human intervention.

The stories we tell ourselves reflect this limited understanding of life and our broken relationship to nature. It does not take much psychological sophistication to see that many advertising tales, for example, are created and disseminated to convince us to buy and consume a product. Lacking the integrative vision of authentic stories, their messages do little to expand our sense of self or connect us to the rest of the Earth community. True to their origins, they reflect an organizing myth that places economic logic and values above other values. And since their primary goal is to boost sales, vital information is commonly omitted.

Consider a company that promotes an insecticide that will dramatically increase crop yields for five years by destroying a particular insect. It has abused the story form if its advertisements neglect to tell the consumer that this same insecticide will also destroy all the organisms in the soil that keep it fertile, condemn natural predators that feed on the target insect to death by starvation, increase the likelihood of secondary insect infestations, and create a dependence on the product.

Native Stories versus Modern Tales

Stories always mirror the cultural beliefs from which they emerge. The contrast between native insect stories and those created by individuals living in the industrialized age is understandably great. In a British short story called "Caterpillars," by E. F. Benson, for example, the narrator of the tale encounters a pyramid of immense luminous caterpillars, each a foot or more in length with rows of crab-like pincers and gaping mouths without faces. Mysteriously, they notice him, and when they do, they drop to the floor and move toward him. Their numbers and anonymous vitality lend weight to the man's conviction that they are possessed of a malevolent supernatural mind that lets them choose a target at will.

In contrast, when native stories include fear, destructive acts, or even death, the context is different. The reason that an animal pursues a person is spelled out. Often the main character is punished for harming another species or for acting in a disrespectful manner. For instance, in a story from the Sandy Lake Cree tribe of Ontario, who lived in an area thick with mosquitoes, a Cree warrior is angry about being bitten by mosquitoes. He scoops up a handful of the insects one summer day, keeping them captive until winter. On a bitterly cold morning, he dumps them in the snow and watches them die. The following summer the mosquitoes pursue and bite him relentlessly until he dies from loss of blood. The message of the story is clear: Vindictive behavior toward another species will not be tolerated.

Transmitting Information Through Biting. In tribal cultures, a biting or stinging insect was not necessarily something to avoid. It was believed that knowledge or attributes could be transmitted from other species to human beings through an act of biting or stinging.

A contemporary Eskimo stone-cut print called "Mosquito Dream"[5] depicts an ancient scene in which five larger-than-life mosquitoes are instructing an Eskimo shaman-initiate by biting him. They surround the man, biting his neck, lower back, and back of the knee—penetrating three areas of great vulnerability. Biting creates the connection between the two species, perhaps by a mingling of fluids, and allows the mosquito to impart knowledge, or its "medicine," to the man.

For shamans to attain their vocation as healer, seer, and visionary, they had to experience wounding, death, and rebirth. Animals—including biting and stinging insects—were typically a vehicle for acquainting novices with suffering. Experiences that we might consider merely misfortune—sickness, decrepitude, and pain—would strengthen the shaman and prepare the shaman-initiate for his or her actual mission.

If we return to Benson's caterpillar story, we find no initiation, no final transformation, and no transmission of attributes, information, or power. The caterpillars wield no spiritual power. They are the personification of insatiable appetite and amoral intent, bringing only a horrible death. We are led to believe that the hero does not provoke the attack but is instead an innocent victim of the mindless hunger of the caterpillars. We finish the story feeling, understandably, that we too are vulnerable to unswerving, demonic-like forces that prey on the innocent.

The Legend of the Queen Mosquito. The Sandy Lake Cree tribe has another story that explains the bond between the Cree and the mosquito, and accounts for the swift and unforgiving punishment of the warrior who acted cruelly toward mosquitoes in the previous story.

The hero of this legend is a Cree warrior called Mama-gee-sic who is forever down on his luck. One day, he goes into the forest and is beset by hungry mosquitoes. Unable to protect himself, he finally submits, removing his shirt and lying down, prepared to die, as the swarm covers his body. When he is close to death, he hears a loud humming voice saying, "Let this unfortunate man alone. He has had trouble enough for one man." Hearing these words, all the mosquitoes fly away. When Mama-gee-sic opens his eyes, he sees a huge mosquito hovering over him. It descends, touching him gently with her wings. An amazed Mama-gee-sic finds his wounds healed instantly and his strength returned. From that day forward he is successful in his endeavors. The Queen mosquito had become his protector. The Cree people end the story by telling their children that the Queen mosquito never harms them. "She has befriended the tribe, and for that reason Cree hunters respect the mosquito."[6]

Mosquitoes as the Enemy. We teach our children a different story, a view that develops the hostile imagination necessary to reinforce and uphold the culture's view of mosquitoes. The first few lines in a children's book sets up the battle against mosquitoes this way: "Nobody likes mosquitoes. A mosquito bite hurts. You have probably been bitten by mosquitoes many times."[7]

Once instilled, mosquito hatred takes many forms, like the Texas Mosquito Festival mentioned in Chapter 1. Contemporary children with this kind of education have, understandably, become disconnected from the Mama-gee-sic lineage. If we do not provide them a context in which to understand affinity and investigate mosquitoes heartfully, how will they ever know if the Queen mosquito is their protector?

Indigenous tribes like the Cree acknowledged animal (and plant) guardians and totems. These ties to specific local creatures (like the mosquito) distinguished and defined people. Acknowledging the link was also a practical way to promote harmonious coexistence with those who inhabited the same territory. Although remnants of our animal totems are seen in the names of our athletic teams and some drinking establishments, they have lost their psychic roots and with them their power to influence and direct the contemporary individual or the group—or provide a context in which to work with adversity.

Opportunities Disguised as Misfortune

The notion that adversity (like being overcome by a swarm of mosquitoes) can be transformed into opportunity (such as being befriended by the overlighting spirit) is an ancient idea. Many Eastern religions teach that adversity provides an opportunity to deepen one's faith and cultivate certain qualities. In both Buddhism and shamanism, when pain and distress are confronted directly, they are transformed into manifestations of wisdom. In this way weaknesses become strengths and the source of compassion for others.

A story, condensed below, of how adversity becomes wisdom is told about Gyelsay Togmay Sangpo, the author of *The Thirty-Seven Practices of Bodhisattva:*

> When Togmay Sangpo was at a monastery and studying Mahayana texts on training the mind, he looked after a destitute man covered with lice. One night the man wasn't in his usual place behind the monastery and Togmay Sangpo finally found him huddled in a hollow by the side of the road. When he asked him why he was there, the man said that people were so sickened by him, they'd driven him away. Moved to tears Togmay Sangpo took the man back to his room, fed him, and gave him a clean sheepskin coat. When morning came, he sent him on his way with the new coat and other necessities.
>
> But now he had another problem—what to do with the lice in the man's old coat? He knew if he threw it away, all the lice would die—an unacceptable option. He put on the coat, determined to nourish them with his own blood. Soon after he began to look and feel awful and could no longer attend classes. Some of his classmates tried to persuade him not to continue to feed the lice, but Togmay Sangpo was not to be swayed from the course of action he had chosen. He told them that in past lives he had lost one body after another in meaningless ways and that this time he would make a gift of his body.
>
> After seventeen days, the lice began to die naturally and no new ones were born. He picked out the dead lice saying mantras over them. Then he ground up their bodies, mixing them with clay and making votive candles out of them for the lice's benefit (a practice usually done with the bones of those people revered and cherished for their kindness).[8]

Togmay Sangpo's classmates learned from his example that any adverse circumstance could be turned into a condition conducive to spiritual development.

Today we combat all adversity with heroic measures. We have lost (or rejected) the context that once taught people how to transform misfortune into blessing. The author of a juvenile book on lice introduces the louse as a creature "almost universally despised by both human and beast,"[9] and "at its best…a nasty little pest that seems to serve no good purpose."[10] We don't object because we think that the author's statement is accurate. In the West, head and body lice appall people. In fact, the presence of head lice is often seen as a cause for shame (although a lice specialist says that lice prefer clean hair).

Few of us wonder if there is another way to think about these creatures, not knowing that native people accepted the presence of lice with aplomb. A story in the Navajo tradition of why humans have body and head lice is an example of how native mythology creates the positive connection that prevents members of the tribe from taking an adversarial stance. In the time when things were still being created, Monsterslayer threatened to kill Louse. Louse pleaded with him to let him live, but Monsterslayer wanted to know why he should. Louse explained that if he is killed, humans will be lonesome and will not have anyone to keep them company. So Monsterslayer relented and let him live.

Native people also had an unusual (for us) strategy for dealing with lice—they ate them. In fact this practice appears in tribes from Alaska to South America and in Asian, African and Australian aboriginal societies. Many native people believed that eating lice was healthy (and it probably is) and picked off and ate lice from family members and friends as an act of courtesy, favor, and friendship. In Papua and New Guinea, for instance, tribal people ate their own lice and deloused members of the opposite sex as an act of intimacy. The Yurok Indians of western North America also ate their own body lice, thinking that they could absorb some of the character of a departed soul that had temporarily assumed the form of a louse.

Although we are repulsed at this environmental and human-friendly strategy for dealing with lice, our preferred method of chemical warfare—especially against head lice—has not only done little to arrest the presence of lice, but it has also created "superlice"—pesticide-resistant creatures who intend to stay close to us.

Without instruction and openness to all species—yes, even lice and mosquitoes—we miss the lessons of how to work effectively with adversity and even transform it into opportunity or wisdom. We also short-circuit the links that could lend us strength and assurance. Few of us today ever discover the insect species that fits the unique contours of our psyche—the kind of insect that, mirroring some vital aspect of self, has the ability to expand our self-identity, and the ones from whom we might particularly learn and benefit by observing their unique ways of being in the world.

Affinities

Each of us has innate affinities for particular species, activated by the sight of them. We respond with wonder, or, if the creature is misunderstood, or condemned by the culture, with fear and loathing.

Entomology and pest eradication and control are the primary tracks in Western culture available for someone with a strong affinity for insects, and neither career focuses on rejoicing in insects. In fact, entomology in North America is so linked to control and eradication that often the only goal of gathering knowledge about insects is to kill them more efficiently. In Frank Graham, Jr.'s book on biological control *The Dragon Hunters*, an entomologist says about his colleagues, "These people loathe insects…their life is a crusade against them."[11]

Whether true for the majority in applied entomology or not, many other professional and amateur entomologists are fascinated with insects. The rest of us, however, find little evidence of an innate emotional affiliation with creeping creatures, especially to the over ten thousand species classified as pests. These unfortunate beings have an appetite or appearance that we disapprove of, or their behavior costs us money, makes us uncomfortable, or is just plain inconvenient.

Unlike the Sandy Cree tribe who came to terms with the mosquitoes in their habitat, we have closed the door on affinity with any species we consider a pest. As a result, those with a special inner connection or affinity for mosquitoes and other pest species are cut off from an aspect of their essential selves.

Traces of Affinity. In the popular culture, we see evidence of affinity or special connections with certain species in comic heroes like the Astonishing Ant Man, Spiderman and Spiderwoman, and comic villains like the Beetle and the Fly. Each of these characters takes on the power of the species for which they are named. But the original event that gave them insect attributes does not stem from conscious affinity with the creature, nor is the insect given any credit or respect for its innate abilities.

When insect powers are occasionally transferred through conscious affinity between insect and human, those powers are used for dubious purposes. In the film *Creepers* a young girl who likes insects discovers that she can communicate with them telepathically. Like robots, they obey her wishes without hesitation. When a psychopathic killer begins terrorizing her school, she sends hordes of insects not only to stop him, but to punish him in vigilante-like anger and self-righteousness. The film portrays affinity twisted by our cultural beliefs.

The Return of the Flea Circus. Another example of affinity skewed by the culture's condemnation of insects is seen in an artist who became obsessed with fleas and left the art world to create (or reinvent) a flea circus. Featured in a prominent alternative magazine, this artist recently brought her fleas to "perform" at the illustrious San Francisco Exploratorium, a museum of science, art, and human perception.

The artist claims that her act has allowed her to investigate the connection between humans and animals. But one wonders, given the nature of her show, whether the only information she seeks are facts that will allow her to control the performance of the fleas.

As did the flea circus master of three hundred years ago (a man who was considered the epitome of cleverness), this artist also glues costumes to the bodies of every flea in her sideshow. Once attired, she gets the insects to perform feats of strength, such as pulling a toy train thousands of times their own weight by subjecting them to light, something she learned that they greatly dislike. Because the ropes that tie them to the train are glued to their bodies, the captive fleas pull the vehicle as part of their desperate effort to get out of the light.

In the flea orchestra the fleas are not only glued into chairs, they have tiny instruments glued to their legs. As they struggle in vain to escape, the movement of their legs gives the impression that they are playing the instruments. In another act, two fleas with formal attire glued to their bodies are then glued together—back to back. Their panic and frantic attempts to escape make them whirl in a circle like two dancers. Our dislike for these creatures and our vague anxiety about their disease-carrying potential—although the flea that likes to feed on people rarely carries the plague bacteria—suppress what might otherwise invoke a response of outrage to such acts of cruelty to animals.

The artist explains her own lack of feeling for the fleas (she doesn't bother to feed them and says she doesn't care if they die in the show or afterward) as the attitude of a scientist who just makes experiments and observes the results. Equally disturbing is the lack of awareness of the interviewer, who presents the show as artistic and tells her readers that the artist has "brilliantly melded science, ecology, aesthetics, and humor into her current body of work" and that "there's something refreshingly honest in her act—something primal, raw, and unedited."[12] But if there is humor in such a show, it is the dark ridicule associated with shadow material. And what is primal is not abuse of other life-forms, but our genuine connection to all other species, an affinity characterized in native societies by wonder and respect.

Our affinity for insects or any species we have decided is our enemy will always be distorted in some measure until we have done the personal work leading to our metamorphosis into compassionate human beings. Investigating so-called pest species in a heartful way can facilitate the transformation. As Sam Keen reminds us, "In the image of the enemy, we will find the mirror in which we may see our own face most clearly."[13]

Reinventing Ourselves

Real change begins with imagining ourselves in a new way. Contemporary storyteller Brenda Peterson leads groups of inner-city teenagers in role-playing other species. She has discovered that these street-wise, alienated youth identify with other species in a profound way. Their ease and eagerness in assuming the viewpoint of other species suggests to Peterson that nature is not out there, but inside.

Peterson believes it is our own spiritual relationship to other species that must evolve and that imagination acts as a mutually nurturing umbilical cord between our bodies and the planet. When children claim another species as their ally—not just as their imaginary friend, but as the creature that has sought contact with them—it changes their outer world.

The Council of All Beings. A form of group work, based on the writings of leading deep ecologists and educators Joanna Macy, Arne Naess, Pat Fleming, and John Seed, weaves people and the natural world together and is called "The Council of All Beings."[14] Council workshops have had phenomenal success all over the world, awakening in participants the commitment and courage to act in behalf of the Earth community. Participants come together and learn to act not on behalf of themselves or their human ideas, but on behalf of the Earth and its inhabitants.

Several years ago, when I created a curriculum at a local elementary school to teach children how to be in good relationship to other species, I adapted "The Council of All Beings" so I could use it in the classroom. The children already knew that many species were endangered. They also knew about habitat loss and the general plight of the Earth community. But I wanted them to use their imagination and feel what it must be like to be a disliked, a mistreated, or an endangered species. I also wanted them to understand that each creature has different eyes and, consequently, its own view of reality.

Educator Peter Kelly calls this "biological empathy."[15] It differs from anthropomorphism whereby people perceive other species as though they were human. When people insist on the humanness of another species, as they often do with companion animals, they can deny its "otherness" and invalidate its unique worldview and wisdom. I encouraged the children to find elements of behavior and ways of communicating that were similar to how humans respond and then to discover how the creature differed from us and how it adapted to its life circumstances.

The children assumed the viewpoint of invertebrates with the enthusiasm normally reserved for furred and feathered creatures. In fact, the shift to seeing the world through another creature's eyes happened easily for most of them. The uprooting process that encourages our children to detach themselves from the living world and rely only on concepts and language had not yet progressed very far.

Insect Masks Facilitate Identification. In the second week, as preparation for videotaping a "Council of All Insects," I introduced six different insects, intending to assign each child to one group. Once assigned, they would make a mask of their insect and learn enough about it to speak on its behalf.

To introduce the fly, I showed a slide of a gold fly necklace worn proudly by Egyptian officers in ancient Egypt, explaining to the children that officers would present gold flies to their men as rewards for heroic deeds in battle. I also told them about the North American Blackfoot tribe's Fly Society for warriors and ended by sharing a story about a friendship between a man and a housefly, a true tale included in Chapter 5.

As the first class period ended, I asked the children to be aware of their interactions with insects during the week and to note unusual behavior. They were particularly intrigued with the possibility that an insect with whom they had a special connection might make its presence known to them in the course of the day or in a dream.

One animated young boy named Brian was filled with images of the latest horror films featuring insects. A natural comic, he disrupted the class with his antics. One of a few who resisted the idea of identifying with an insect, he preferred entertaining his classmates with imitations of insects as demons running amok. I wondered if he would even stay in the group, as my parent helpers were ready to send him to the office. But the next week Brian came back and pulled me aside to tell me a dream he had the night before. In the dream, he had been threatened by Jason, the psychopathic killer in a series of popular horror films. Suddenly, a human-size fly appeared with a sword and went after Jason, defending the terrified boy.

I congratulated Brian on remembering the dream and having the fly as his defender. He beamed with satisfaction and asked to be assigned to the fly group so he could make a fly mask. He had claimed the fly as his ally—or perhaps the fly had claimed him.

Affinity for a particular creature is always a gift, linked inexplicably to our essential nature. Perhaps the species that mirrors some aspect of our wilderness identity waits, with insect diligence, at the entrance to our conscious awareness. Maybe we have only to become aware of our projections and seed our imaginations with new ideas and stories to create a pathway that permits their entry and helps us return to our native place.

Let's begin now, disarmed by the new realization of our deep and abiding connection with insects—and a good measure of good will and faith—to investigate some common species. We can start, like Brian, with flies.

Part II

Redeeming Pest Species

4

My Lord Who Hums

> Everything in Oneness has a purpose. There are no freaks, misfits or accidents. There are only things that humans do not understand.
> —Regal Black Swan

Caught among our fears of decay, disease, and death, flies are easy to condemn and execute as pests. They personify all the annoyances that prevent us, or so we think, from being really content. Their numbers and a few species' proximity to humans make them easy scapegoats for the uncertainties of life that plague us.

I suspect our lives would change if we understood and could mediate our responses to flies. By redrawing our circle of community to include these winged wonders with kaleidoscope eyes, we automatically expand our comfort zone—one less enemy, fewer battles. Including flies in our circle of concern might even mean that we have made peace with some of the fears that keep us pushing life away. It might also mean, as fourth-grader Brian discovered, that we've opened a door inside ourselves and found an ancient ally.

Since we don't love flies and few of us know anything about them, our opinions emerge from the cultural consensus on flies. Although entomologists are considered the authorities on insects, even these experts are not automatically immune from the power of the collective view. In a recent book, an entomologist who describes herself as an outspoken insect advocate urges her readers to kill houseflies without hesitation—and to consider the act a public service.

A few fly species are already extinct, and many are threatened. It isn't something that concerns people. In fact, some people think that the world would be better off without flies—not a thoughtful or educated opinion, but a common one. It is supported by the official policy that divides all creatures into two categories: beneficial species and pest species. Flies are pests. This designation is a convenient device that retains our belief in flies as the enemy. And what best serves the stereotype of flies as annoying, biting, disease-carrying pests is to eliminate the distinctions that complicate the issue and sweep the varied behaviors of flies into a homogeneous and manageable heap.

The Art of Classification

Entomologists have identified over 90,000 species of flies. How many more exist is a matter of speculation. Classified as invertebrate animals in the phylum Arthropoda, flies, like other insects, have skeletons on the outside of their bodies called exoskeletons. Six legs and the arrangement of their bodies place them in the class Insecta and distinguish them from spiders, crabs, and lobsters. What puts them in the order Diptera is two pairs of wings. Within this order are families that further distinguish and separate flies, and within families there are different genuses, and finally there are the actual species, defined as those that successfully interbreed.

We nonspecialists classify insects differently. Although primitive flies like mosquitoes and black flies share the order Diptera, we usually lump them together because both annoy us and can transfer harmful parasites through their bite. The long-legged and harmless crane fly has the misfortune of looking like a large mosquito and invokes a fear pervasive enough to warrant a diagnostic label—"tipulophobia." That fear binds the crane fly to the mosquito more surely than any classification of body parts or behavior.

Young children, like artists, often sort insects visually, by color and pattern. Older children differentiate by size, number of legs, and method of locomotion. Indigenous cultures had their own systems. The Oglala Lakota grouped together all creatures who could fly. Dragonflies, flies, moths, and butterflies were put in with birds, because each was associated with the great power of the wind.

Beekeeper Sue Hubbell reports that the pragmatic Algonquin tribe divided insects into those that flew and wanted to bite us and those that didn't—a reasonable division. The mosquito, black fly, and biting midge, for instance, were called *sawgimay*, which translates roughly into "small person who flies and bites so fiercely."[1] In this chapter, we'll deal with the ones that don't bite, specifically flies associated with dung and dead matter, and save the *sawgimay* for later.

The Importance of Flies. I knew the picture was skewed against flies when I first started to study them. So I was surprised to discover that the majority of entomologists agree that no group, except bees and wasps, is more important to humanity.

Many flies are primary pollinators when frequenting flowers for nectar, like hover flies who mimic bees and pollinate wild and commercially grown flowers. Flies are also essential to the Earth as master recyclers, rapidly disposing of decaying animal and vegetable matter. An inborn attraction to a variety of decomposition products alerts members of these species to available food. Houseflies, blowflies, scavenger flies, brine flies, coffin flies, flesh flies, and thousands of others recycle decaying animal matter.

Most flies are also an important food source for animals like reptiles, amphibians, bats, and fish. Still others destroy and feed on populations of insects that would otherwise multiply rapidly and upset the balance of countless plant and insect relationships.

The Gift of Maggots

Like many insects, flies hatch from eggs and spend some time as immature larvae called maggots. Maggots and adult flies look and act so differently that in ancient times people did not connect the two as being the same creature. Fly maggots appeared to develop from dead material and were thought to be worms. Today, we know maggots for the opalescent flying creatures that they become, but rarely do we extend appreciation to them in either stage of life.

Throughout the ages, indigenous people valued maggots as food. The Dogrib Indians of the Athapascan tribe of eastern Canada, for instance, considered fly maggots of several species a delicacy. And Yukpa women and children harvested the water-bound larvae of the net-spinning caddis fly (Neuroptera family). Gathering them from beds where the river current was the strongest, they wrapped the maggots in leaves and roasted them over a hot fire.

Housefly maggots consist of fifteen-and-a-half percent fat and sixty-three percent protein. Studies show that even dried housefly pupae (maggots in their final stage of metamorphosis before turning into flies) provide protein of sufficient quality to support the normal growth of chickens. Ronald Taylor, an advocate for using insects to help alleviate world hunger, believes housefly maggots could be a valuable food resource for domestic animals. Mass-reared adult flies could even help solve the world's organic waste problem by converting waste materials into high-quality nutrient supplements.

What a maggot eats depends on what kind of a fly it will become. Some hover fly maggots have an appetite for aphids, eating thousands in a two- or three-week period. Others eat bulbs, roots, and fibers, and some, like the rat-tailed maggot, prefer sewage. The majority of maggots feed on something dead and high in protein like garbage, dung, dead animals, or, less commonly, untreated wounds and the dead surface skin of live animals, including humans. Although we tend to abhor such an appetite, maggots are valuable because they break down and recycle carrion so effectively. In fact, without their services, we humans could not smell the roses for the stench of decaying matter.

Our View of Maggots. In the West, we object to the way maggots look, move, and eat. We don't normally like land creatures without legs—especially ones that ingest foul substances. The author of a recent book on taking a "new look" at hated creatures shares his reaction to maggots in his compost pile: "There I saw one of the most disgusting sights imaginable—the entire heap...was a swarming mass of maggots!"[2]

His chapter on flies, laced with expressions of his repulsion, presents information in line with his belief in flies as "multiplying monsters."[3] Even when he learns that the maggots in his compost pile are creating soil-enriching material, he is unwilling or unable to let them be.

One of his sources is a natural history of flies written by a distinguished entomologist, who writes: "Breeding in dung, carrion, sewage and even living flesh, flies are a subject of disgust…a topic like drains, not to be discussed in polite society."[4]

The information this expert conveys is couched in emotional terms that color the facts. Disgust is, after all, a learned response. Every culture teaches its members what is disgusting and what is not. Children develop their disgust reaction by observing the facial expressions and reactions of their parents and teachers. And what is disgusting to members of one culture may not be disgusting to members of another. What allows us to eat shrimp and escargot, for example, and refuse maggots and caterpillars is the bias of our particular culture.

The tendency to describe another species and its behavior in terms of the emotions invoked in the observer is common and an easy way to transmit prejudice. The adjectives this entomologist chooses to describe different fly species ("aggressively advanced," "little nuisances," "pesky," "maddening" and "infuriating") slant the information he presents and does little to advance our understanding of flies. If his readers share the prejudice, they won't even notice the dark cast. Small wonder that the writer trying to take a new look at flies manages to kill all the maggots in his compost pile. Not a very new look at flies.

A New Look at Maggots

A new look might include the role of blowfly maggots in forensic science. In a homicide, these creatures are used to pinpoint the time of death, because blowflies, important as pollinators of several crop plants, seek dead meat when they lay eggs. Drawn to a fresh corpse, blowflies will invariably lay eggs on it. Maggots hatch in a few days and when forensic entomologists examine a body, they collect them. Counting back through the stages of larval development and adjusting for geographical variables that affect growth rate, these specialists can determine when the eggs were laid and thus the approximate time of death. The evidence provided by these maggots has been instrumental in hundreds of convictions.

Instruments of Healing. A new look at flies would also include the old knowledge that maggots are instruments of healing. In fact their healing abilities have been traced back to the Mayan culture. With the advent of modern medicine, practitioners in the United States discarded and then forgot maggot therapy until rediscovering it on the battlefield during the Civil War. With more casualties than the medical staff could handle quickly, wounded men often waited several days before receiving help. Female flies, attracted to the soldier's decaying flesh, laid their eggs in their open wounds. The eggs hatched in several hours and the maggots began feeding on the decaying and dead tissue—

painlessly to the wound bearer and without damaging healthy tissue. Not only did these tiny creatures eat the bacteria that caused gangrene, they excreted substances later found to accelerate healing and prevent the need for amputation in most cases. The same healing ability was brought to the forefront again during World War I when similar circumstances brought maggots and men together.

Thirty years ago, doctors used maggots in the treatment of a serious bone infection called osteomyelitis. The healing was due to the antibacterial substance that maggots exude—the most important of which is urea. In the forties, when scientists learned to synthesize these substances and manufacture them chemically, introducing penicillin to the masses, the average person all but forgot the gift of the maggots.

Today, as many bacteria prove resistant to antibiotics, maggots are returning to service. A few doctors in the United States are using maggots to clear away dead tissue on sick patients who can't tolerate surgery but suffer severe bedsores, burns, bone infections, and tumors. It is a role largely unknown and unappreciated by a public who continues their loud proclamations of disgust when they entertain the idea of maggots.

The Fly as Skilled Navigator

As adults, flies also have many laudable abilities. In times past, flies were admired in diverse cultures for their remarkable flying skills. Current studies reveal that the tremendous amounts of energy needed for flight are generated from elastic-like springs that turn the fly's wings into efficient motors that capture and reuse kinetic energy.

Recent observation and measurement have also substantiated the impressive flying speeds and navigational abilities of most fly species. Many species can travel at speeds exceeding fifty miles per hour. The male deer botfly, considered by some the fastest creature on wings, is thought to be able to develop speeds of several hundred miles per hour. Other experts discount the claim, giving top honors to the tachinid fly. An unpublished research project using slow-motion cinematography clocked a male tachinid fly pursuing a female at a speed that reached 87 miles per hour.

These inherent flying skills, noticed even when unmeasured, earned flies the respect of warriors in many ancient cultures. The flies that Egyptian officers gave their men as a reward for valor and tenacity in battle were fashioned from ivory, bronze, and gold. To an ancient Egyptian then, being associated with the fly was an honor, as the link between bravery and the fly was firmly embedded in the culture.

In the Fly Society of the North American Blackfoot tribe, members recognized and admired the fly's ability to harass the enemy without being captured or killed. Certain warriors were summoned by a dream or vision to join the Fly Society, where they sought to emulate the insect and gain its power or medicine. An underlying belief in their firm connections with other species let ancient and indigenous people observe flies and other insects without the prejudice operating today toward a range of species we consider pests or junk animals. Identification with another species, invoked and enacted during ceremonial rituals, was an acceptable way to acknowledge, celebrate, and embody another species' unique strengths and talents.

My Lord Who Hums

The hum associated with the flight of all flies varies with the speed of their wing beat. The housefly, for instance, beats its wings 345 times per second, humming in the key of F in the middle octave.

The sound of a fly humming was not always a source of irritation. Scholars report that in ancient times the word "zebub" referred to a fly's humming or buzzing. And since "Baal" meant Lord, some experts maintain that the name for the once revered Philistine god Baalzebub—"Conductor of Souls," healer, and oracular deity—actually translates as "My Lord Who Hums or Murmurs."

The modern meaning of Baalzebub or Lord of the Flies comes from the Judeo-Christian tradition in which this deity, paired with putrefaction and destruction, was derisively called "God of the Dunghill." The medieval folklore propagated by church authorities succeeded in its zealous undermining of the influence of earlier religions and deities like Baalzebub. The resulting religious doctrine left most insects, masters of destructuring life during its down cycle, not only excluded, but despised. While some creatures like the bee and goldfinch were declared good and symbols of the light and spirit of humanity, others, like the fly, spider, and the praying mantis (because of its appetite for live prey and the female's habit of eating the male during copulation), were pronounced evil.

Flies were considered demons or devils, living embodiments of evil, sin, and pestilence. Actual flies suffered from this twisted association of their natural connection with death and decay as stories from the dominant religious traditions transformed them into despicable creatures who deserve to be killed on sight. Savagery was added to the condemnation in William Golding's classic novel *Lord of the Flies*, and a Marvel comic book makes Baalzebub the devil in a hell where flies torment sinners.

Projecting the shadow aspects of Christian ideology into certain creatures provided those who embraced the religion with easy targets for their hostility. Those who bought the mechanistic view of the natural world and the militaristic orientation of medicine also had a target. When flies were implicated in the spread of disease, their damnation was secured. And once accomplished, the contributions of the fly and the complexities inherent in our relationship with them were buried.

Fly as Monster

As flies are no longer a sacred presence to humans, it is perfectly acceptable to kill them on sight, even preferred. Our stores stock fly swatters in decorator colors, and our stories and films help us maintain an adversarial stance and teach it to each new generation.

Monster-making continues to be a popular religious and secular activity. By focusing on a few species of flies and the bacteria that live on and about them, and by calling on our hostile imagination to pull our emotional strings with images of flies feeding on feces and then coming to call, we are all too ready to believe that flies are monsters.

The demonizing of any species depends on the perception and belief in our inherent separateness from other animals and the tangent belief in their evil or amoral intent. "The Fly," a short story written in 1957 by George Langelaan, was made into a movie the following year. It joins human body with fly body in an attempt to horrify us. And it does. We have long ago lost our understanding of transfiguration as an expression of our fundamental unity with nature. Whereas once the complete or partial transformation of human to animal was a perception of the reemergence of life energy in another form, today it is a subject of repugnance and fear.

Mesmerized and delighted with these kinds of horror films, the public welcomed the *Return of the Fly* in 1959 and *The Curse of the Fly* in 1965. In 1986 filmmakers remade the original movie, taking full advantage of studio technology to create even more spectacular special effects. Perhaps our consistent response and fascination with this subject is a distortion of a deeper, more authentic understanding of our interconnectedness with other species.

In the original short story, the scientist-victim meddles with the powers of nature in his investigation of the transmission of atoms. After disintegrating the family cat in an earlier experiment, he decides to confirm his theory by trying the experiment on himself. He is unaware that a fly has flown into the chamber with him and the results, the reader is informed, are a hideous travesty of nature—his human features replaced by feline and fly features. The fly is changed too and now has the scientist's head.

The scientist's wife reacts with horror as "the monster, the thing that had been…[her] husband, covered its head, got up and groped its way to the door." High drama to be sure, but it is a typical omission that no attempt is made in the story or film to describe the fly's reaction (or the cat's). We rarely enter the world of an insect or other species imaginatively and respond with empathy out of that identification. We might be surprised if we did. Imagine the fly's horror when it discovers it has lost its head and has a human head—inadequate to its fly tasks—as a replacement.

If we look at the story symbolically, we move past being entertained by our horror and catch a glimpse of the flies in our psyches who move unbidden and unwelcome into our inner chamber. Their presence is an unexpected part of the equation and fouls our attempts to manipulate the forces of transformation.

The Official View of Flies

The link between flies and filth prevents people in industrial societies from accepting and understanding the presence of flies in the world. We loathe the decay and garbage we help create, and along with this, the fly who recycles it.

Our children also learn to hate flies early in life. A popular Marvel comic book character, "The Fly," is an unsavory man who is transformed into a fly and becomes little more than the voice of our projections. In one issue, the fly man hijacks a barge carrying garbage and begins to eat it with relish until some part of his still-human consciousness overrides his fly instinct. Realizing the "depravity of his act,"[5] he is overcome by self-disgust.

Educational attempts to balance the prejudice toward flies and promote an appreciation for them are also vulnerable to contamination from our biases. A children's book about flies introduces them this way: "The best thing that can be said about most flies is that they are a nuisance."[6]

With their imagination thus seeded and the cultural view clearly spelled out, children have their opinion formed for them. Most children's books also list the kinds of flies that bite and annoy people and the diseases they cause or carry. After reading them, it is unlikely that anyone will have a good feeling toward flies. The statements are presented as facts, and most of us would not think of questioning this simple and straightforward indictment of flies. But the issues surrounding our beliefs about sickness and health and who causes what illness and under what conditions are neither simple nor clearly defined.

The Complexities of Human Sickness and Health. This topic extends far beyond the scope of one book; however, a reasonable objective might be to inject enough healthy doubt into the official version of flies to start us asking questions and opening to the tangle of fears and facts that paralyze thought and impede appropriate action. One of the things that stops people from extending concern and protection to flies is the prevalent fear of disease.

We all learned from our parents and in school that flies spread diseases. We carry the image of flies crawling on revolting substances and then landing on our food, transferring bacteria in the process. When specialists discovered and published the fact that each housefly actually carries over two million microorganisms, the average person did not have a context to process the information. The number could only feed the hostile imagination. And it did. The bacteria count upheld the view of flies as death-wielding, disease-carrying creatures who deserve to die.

In his book *Furtive Fauna*, biology professor Roger M. Knutson presents another view. He reassures his readers that none of the bacteria on a fly is likely to cause disease in people. In fact, he says, we all—flies and humans—have pretty much the same sort of organisms living on us anyway, and the flies just follow our example and move them around from one person to the next with little physical consequence. Good mental and physical hygiene and community sanitation measures are usually enough to provide reasonable protection.

The Bacteria in Our Midst. If we liked bacteria, their high counts wouldn't disturb us. We generally think of them, however, as germs or pathogens—shrewd and malevolent specks of life that ride winged carriers to human hosts where they make trouble for us. But a microscopic view tells us that all surfaces, including our bodies and our food, are already teeming with life invisible to the naked human eye.

If we examine our skin under a microscope, we will see a startling sight—every square centimeter is populated with approximately 100,000 microorganisms. It is a fact that does not engender comfort, but it is something we need to accept. Scientists who study these populations tell us that we can't avoid microbes or destroy them without killing ourselves. Understanding that we are surrounded by bacteria can put the popular bacteria number counts in their appropriate context. Knowing more about bacteria also helps.

The Adaptability of Microbes. Bacteria or microbes are the dominant lifeforms on the planet—our ancestors and contemporary associates. To detail their influence would require several volumes, but in short, all the Earth's life systems depend on them.

Microbes continually adapt to changing environmental conditions by transferring genes to other bacteria different from themselves. The receiving bacterium uses the visiting cell's genetic material to perform functions that its own genes can't handle. Gene exchanges happen quickly and reversibly. Since all the world's bacteria have access to a single gene pool, each strain of bacteria has access to the chemical prowess of the entire bacterial kingdom and can combine its body with other organisms and form temporary or permanent alliances.

This genetic fluidity renders the concept of bacteria species meaningless, but it is still convenient. Outdated too is the traditional medical paradigm regarding cause and effect and the agents of disease. In fact, if modern medicine is to meet the health challenges of today and create effective strategies for dealing with the new diseases springing up in our midst, it will need a radically different conceptual base, and one that would allow us to explore illness and its relationship to the patterns in our lives. Such a new conceptual base will also investigate illness as the result of an upset or deterioration in the balance of an entire ecosystem (or the balance of an immune system) and will not single out certain species as enemies.

No organism exists by itself. Instead it co-evolves with all the elements in its natural surroundings and seeks a state of equilibrium or homeostasis. It is generally accepted that the bacteria and viruses identified in new infectious diseases have been around a long time but were never identified with human diseases. German bacteriologist Guenther Enderlein maintained that certain bacteria are able to take on multiple forms during a single life cycle. When a person is healthy, these microbes live in symbiotic relationship with other cells and play a helpful role in the body's immune system. But with any severe deterioration of the body's internal environment, they may change into disease-forming agents.

When new diseases do occur, we can be sure that some equilibrium has been disturbed. So a thorough treatise on diseases in which flies are implicated as carriers will invariably implicate the burgeoning human population and its disruption of existing balances between species.

Disease as Evolutionary Process. The difficulties inherent in any attempt to prevent disease and promote health are complex and compounded by factors like increased international travel, the influx of armed forces and immigrants from endemic areas, increased numbers of household pets, the popularity of exotic regional foods, and the use of antibiotics and immuno-suppressive drugs. New diseases are on the rise, and old ones are making a comeback as new strains of organisms emerge, responding to the chemicals we have used to battle them. The complexities are overwhelming.

Dr. Marc Lappé, a proponent of the emerging view of disease as an evolutionary process, maintains that every new disease or illness is linked with environmental and ecological factors. The primary factors implicated in new bacterial diseases, for example, are upsets in the natural balance of existing ecosystems. Whenever we clear a forest, build a road, or change a waterway, we abruptly open a habitat. Some species indigenous to the original balance die out, while others enter or leave the disturbed area and find new opportunities to multiply before natural checks are in place again. For example, in Brazil the first occurrence of a flulike illness that affected eleven thousand people happened after the trans-Amazon highway was completed. Scientists spent nineteen years investigating this new disease called "oropouche"—caused by a forest virus and carried by a midge fly—before linking it to human activity.

Ferreting Out the Truth

Most books on flies state without qualification or elaboration that houseflies spread disease because of their appetites for dead things and their habit of softening food by spitting up a substance that liquefies it. It is the rare individual who questions what they read about other species. We expect specialists to be immune to common prejudice, but no one is immune to the influence of repressed material without consciously working to retrieve their projections. The official explanations that justify our aggressive responses to flies collapse when examined, opening into a tangled web of human complicity that invites pause and still further investigation.

Flies and Cholera. In a recent juvenile book on insects, for example, the entomologist author states without elaboration that houseflies are implicated in cholera, a sometimes lethal diarrheal disease. It sounds like a closed case. Yet a closer look at cholera indicates that the bacteria associated with its symptoms thrive in contaminated water. Outbreaks occur during periods of transition, particularly during the growth of new cities when good sanitation is lacking. In recent outbreaks of cholera, researchers have even identified the algal blooms of plankton as carriers—their abnormal growth probably a result of our overuse of fertilizers in farming. Still other outbreaks of cholera have been linked to contaminated processed food from small factories without adequate sanitation, and the risk of cholera always increases whenever dense populations share a common water supply without sanitary precautions.

Deadly bacteria strains that destroy their hosts can only afford to be deadly because something lets them be easily transmitted from host to host—so they don't have to worry about keeping their host alive. If they can't be transmitted easily, they generally evolve toward a relatively mild state of coexistence with their hosts to avoid dying when their host dies.

It might ease our minds about flies to know that deadly cholera bacteria need water, not flies, for transmission. It is water that allows cholera bacteria to stay deadly. Overcrowded living conditions lead to water contamination, which in turn permits an easy transmission of bacteria from one person to another. In fact a direct correlation exists between the deadliness of a diarrheal bacterium and its ability to be transmitted by water. What this suggests is that water purification would transform deadly pathogens into milder ones—and records indicate that has happened.

Unfortunately many clean-water programs have had their funds diverted because field studies did not show a drop in the overall number of cases of cholera, and the drop in the frequency of lethal cases did not get the attention it warranted. Microbiologist Paul Ewald, who has made the link between epidemiology and evolutionary biology, argues that we could eliminate lethal cases of cholera if we supported clean-water programs. Since clean water can prevent transmission of all cholera bacteria, it would prompt lethal strains to transform into milder strains that wouldn't kill their hosts.

Killing flies, however, is often easier and less expensive than addressing overpopulation and sanitation problems or even finding an effective cholera vaccine. Governments rally people to join together against a common enemy. The Philippine government recently offered a bounty on flies and cockroaches, which are blamed for spreading cholera in Manila. The outbreak was traced to an abandoned reservoir used as a public toilet by slum dwellers.

Adding further complexity to the subject of cholera is the fact that people can interact with most strains of cholera bacteria and not get sick. New studies suggest that only a single strain of the cholera bacteria becomes deadly and that happens when it is infected by a virus that switches on a poison-making gene.

No one knows why the virus switches on this poison-making gene, but when it does the bacteria can result in a lethal diarrhea in people. And again, if that victim defecates in a river, or in an unsanitized area, they help the bacteria and the virus spread and retain its deadly nature.

The situation is complex and there is much that is not understood. To simply implicate the fly in the books we write for our children supports enemy-making and does a disservice to both children and flies. Propaganda always argues for a simplistic approach. Putting common conclusions under a microscope reveals not a simple cause and effect but a network of events that renders this type of linear thinking inaccurate and dangerous.

Other Diseases Linked to Flies. The same author of the juvenile book that linked flies to cholera also glibly links flies to two other water-borne diarrheal diseases—typhoid and dysentery—and to chronic and acute diarrhea and intestinal upsets. A brief investigation of the first two diseases indicates that, as in the case of cholera, the presence of flies is not central to these afflictions nor is the absence of flies linked to effectively preventing their spread.[7]

In regard to the cause and transmission of diarrheal diseases, most physicians agree that chronic diarrhea has many causes—like emotional stress and addiction to coffee and other forms of caffeine—and is transmitted in many ways. Acute diarrhea with fever and blood or mucus in the stool indicates infection from bacteria or parasites, especially in areas with poor sanitation. With good sanitation, however, the odds of becoming infected decrease dramatically, and the chain of events is riddled with many possibilities for transmission other than flies—like dirty hands, for example.[8]

Yet, how many people would have questioned the entomologist's statement? Most would assume that the author's indictment was a closed case because it was congruent with what they already believed, and they would agree that flies should be killed as a public service.

Even if we accept the fact that heavily populated areas without adequate sanitation will result in an increase of flies that breed and feed on sewage, in order to present a balanced view of the situation we would need to mention the facts about these kinds of waterborne diseases as well as implicate humans as likely carriers. We would also need to warn the reader about the health risks to humans and other species and the likelihood of creating secondary insect infestations when we allow insecticides to be employed to wipe out fly populations. Given the dangers of pesticide use, improving sanitation or developing vaccines may be the only reasonable courses of action.

I suspect that an investigation of any disease related to the fly would result in the same kind of complexity and web of interrelated events and conditions found in our brief look at cholera. Acknowledging that likelihood would prevent us from making simplistic cause-and-effect conclusions. In the case of flies, the bottom line is that with normal attention to mental and physical hygiene and the benefits of modern sanitation practices, it would appear that Knutson is right: We have little to worry about from houseflies and other species with an appetite for dead and dying matter. In fact we need them desperately. Fly populations increase in proportion to our own, so when we overcrowd areas and cannot dispose of our waste products efficiently, they arrive to help. It's their job.

Invitation to Sickness

From another angle, even if flies carried a new strain of bacteria to us, it doesn't necessarily mean that we will get sick. Susceptibility also plays a big role. Biologist Lynn Margulis's recent work discounts the idea that microbes by themselves cause disease. Too simplistic. She maintains that whether a person becomes sick is dependent on many personal, cultural, and environmental conditions, not just the intrinsic characteristics of a given microbe. "We are infected by potentially dangerous microbes all the time without getting sick, and some 'infections,' such as naturally occurring skin bacteria, are actually beneficial in maintaining good health and proper body metabolism."[9]

It is generally accepted that any weakness or deterioration of the immune system creates a susceptibility, an invitation to sickness. War, famine, and revolution weaken us. Prolonged anger and depression weaken us. It is also well substantiated that nearly all of the ectoparasite-borne diseases like the plague and several louse-borne fevers are not present in insects or in us until there are destructive situations produced by human activity.

Disease agents surround us, not only as viruses, bacteria, and parasites, but also in the multitude of potential irritants such as carcinogenic chemicals, allergens, and toxic plants. A relatively well person can often interact with these agents and not get sick. A growing number of physicians agree that internal factors determine the nature of our relationship with such agents. This means diseases are not single entities but complex relationships between many internal and external factors and are greatly influenced by natural fluctuations in our immune system. It also means that we can relax about our relationship with flies, work on inner and outer harmony in our lives, rein in our imaginations, and permit flies to perform their jobs without unnecessary intervention.

Making Peace. Knutson concludes that getting rid of all flies—viewed as desirable by some—would produce more serious problems than any trouble they might cause. The world needs flies. People need flies. A feeling for them, more natural than the average person might suspect, will quicken and grow after we confront the fears that keep us from extending our concern to them.

By keeping our sights on flies as disease carriers we delude ourselves and may actually impede our progress. We need to understand how propaganda works in service of the collective shadow—simplifying issues, using emotionally charged language, and glossing over details—so we can extricate ourselves from its damaging leadership. Making peace with the bacteria around and on us is as life-serving as it is realistic. It not only relieves the fly of its role as heavy, but it also promotes a healthy acceptance of life that correctly orients us in relationship to the Earth community. It may also help steer science away from its role as controller and master of life and anchor its disciplines to a conceptual base built on interdependence and balance.

If we let our fears about bacteria run wild, the knowledge of a strange and foreign kingdom in our midst could paralyze us with fear. More information, if turned over to the storytellers, might better direct our imagination. Stories, if they did not ease our concerns, could at least promote acceptance of a world in which our planetary elders, omnipotent and life-creating, live like tiny gods in our midst.

The Dark Side of Existence

We have typically upheld with great energy and conviction our behavior toward flies and other species that we have dismissed or judged as hostile, invasive, or life-threatening. The intensity of our judgments indicates that our attitude has roots that extend still further than the issues we have touched on so far. In the final analysis, our responses to flies may be but one thread in a larger pattern that is at odds with the essence of existence itself.

Perhaps one reason we persist, beyond what is rational, in our activities to eliminate parts of our world is that we lack a life-sustaining, supportive context in which to move through our own human passages of life toward the inevitable death of our bodies. We live with impermanence on every level, on a foundation of mystery, and deny it resoundingly, preferring not to be reminded of death and decay. To our youth-oriented culture, aging and death negate life. To scientists and others under the spell of a model of life that excludes consciousness, dying is a process that requires heroic intervention.

Cultivating a feeling for flies may be a simple step that ultimately entails a radical revisioning of our lives. It may require coming to different terms with the impermanent, yet enduring, natural world and its cycles of death and decay, as well as our own changing bodies. When we read about flies softening their food with digestive juices tainted with their previous meal, which might indeed be feces or decomposing flesh, we become nauseated. It confirms our belief in their despicable nature and satisfies the projection. We also don't want to know that honey is partially dried bee vomit or that butterflies seek and sip urine and sweat as well as nectar and fruit juices—facts likely to spoil any positive projections we have accorded these insects.

Life secretes substances beyond our naming. Out of touch with our own secretions—the saliva that accompanies our food, the perspiration that regulates our body temperature, the slime that allows us to procreate, or the protective layer of mucus that keeps our stomach from digesting itself and which traps the dirt and germs that rush into our lungs with each breath—we focus on the fly and hate in this creature the processes and qualities we deny or hate in ourselves.

Healing begins when we dare to become conscious of our violent and irrational ways and the mythologizing that makes flies or any other species our enemy. Healing begins as we begin to cultivate compassion for ourselves and the others who share this planet.

Let the word "hero" be used, Sam Keen suggests—not for those who seek to triumph over the forces of life, but for those willing to take the solitary journey into the depths of his or her self, to "reown the shadow,...[and] discover the power and authority of wholeness."[10] Reserve, too, the word "courage" for the man or woman who leaves the false security of the cultural worldview and lives in the "creative anxiety of...reflective consciousness."[11] From that place of concern and "don't know," we might find the right questions to ask and the willingness to live into the answers.

By adhering to the idea, radical to some, that all life-forms are a valuable and necessary part of the Earth, the change in our culture would be deep. Its rumbling would be felt in all corners of our lives, toppling the inner and outer structures built on false security and misplaced superiority. People would be open to new experiences, have new questions, and encourage scientists to investigate the nature of susceptibility and its role in the treatment and prevention of diseases for which other species act as carriers. Giving up inflated ideas about controlling and dominating the Earth, we might learn to participate in its community and accept that we might not always get our way. And finding our natural place, we could celebrate our interdependence, the miraculous fact of our lives, and the equally miraculous existence of flies.

5

A Feeling for Flies

> [Pray] that we may apprehend and rejoice in that everlasting truth in which the highest angel and the fly and the soul are equal.
>
> — Meister Eckhart

Clearing the channel of the fears that prevent us from accepting the presence of flies in our lives is a good first step. But there is more to the journey back to our native place if we are to celebrate flies and respond compassionately to them. That takes feeling. If we have a feeling for flies, not only will we be concerned with their welfare and be mindful of their right to pursue their life with dignity, we will also understand that kinship with any species enhances our identity and strengthens our ties to life.

Turning again to other cultures, let's explore the beliefs that promoted harmonious relations between humans and flies. In diverse traditions, for example, flies, like butterflies and bees, were souls in search of rebirth. Many myths tell of souls taking the form of flies and seeking to enter a woman's body to be reborn as a human. The pre-Christian Celtic hero Cu Chulaim, for one, was conceived when his mother-to-be swallowed him in the form of a fly.

The idea of the soul's passage at death into another body—human or animal—is called metempsychosis. A common belief in cultures that viewed death as simply a change of form was that passage into another body was not only possible, but also likely. Since physical form was impermanent and the soul was permanent and indestructible, it seemed reasonable that the soul would seek another home. A tangent belief was that all creatures possess this eternal aspect, which is identical in essence to the soul that animates human beings and which survives the death of the physical body. In fact, shamans, well versed in these matters of soul, often practiced temporary metempsychosis when they entered or sent a portion of their soul into an animal body to visit the world of spirit.

Ancient Egyptians, who also believed that a soul could enter any form it pleased and remain there as long as it liked, viewed flies as the *ba* or souls of departed ones in search of rebirth. They created and wore jewelry with fly amulets symbolizing the human spirit and modeled after metallic flies living in the region. They also placed fly amulets on mummies as symbols of the *ba* returning to the body following revivification. Today in some rural villages, local traditions even forbid the killing of the green or blue blowflies, because the villagers believe them to be inhabited by the souls of persons who once lived in the area.

In a similar vein, the South American Auraucanian Indians believe that departed tribesmen, chiefs in particular, take one form of horseflies. They think of them as spirits from beyond, accepting the ones that appear at their frequent ceremonies and celebrations as a sign that their dead kinsmen are sharing in the feast.

In all these cultures, and in religions like Hinduism and Buddhism that teach the doctrine of reincarnation, death is not an extinction of life, but a transition from one form to another. The soul only temporarily inhabits a physical human or animal body, for each is involved in this cyclic process of death and rebirth.

In the West we also associate flies with death. It is one of the reasons we don't like them. Death is not a popular subject, and, despite elaborate descriptions of life after death from our dominant religious traditions, we run from death with great fear. Our dislike for flies has unmistakable ties to their association with decaying, dying, and dead things, essential preoccupations in the great round of living and dying—and subjects we have banished from polite conversation. Perhaps flies are just unwelcome reminders that physical life is terminal and there are bigger forces in the universe that act on and in us.

Flies as Kin

Since native people believed that flies were kin, their general openness to the wisdom of these species influenced their mythology. The Ahanti tribe of Africa, for instance, paid homage to a fly god, and the Kalmuks had taboos against killing local flies, regarding them as soul creatures with supernatural powers.

In the Navajo traditions, Big Fly was a revered mentor mediating between the Navajo people and their gods as the voice of the holy spirit. Big Fly's powers superseded those of the deities on whose behalf it acted, and it taught the people how to make proper offerings to the Gods. Navajo legends tell of Big Fly counseling members of the tribe by sitting on or behind the ear of a person who needed instruction, forecasting future events or whispering answers to questions.

Scholars think Big Fly's physical counterpart is the tachinid fly, a species with the habit of lighting on a person's shoulder or chest just in front of the shoulder. When a fly landed on someone from the Navajo tribe, the stories of Big Fly would mediate their reaction. Perhaps they would take a moment to be still, listening intently for the fly's counsel.

Stories of Flies. Since flies were relatives in tribal cultures, it was natural to tell stories about them as we might about a special cousin or aunt or uncle. In many indigenous stories, flies were heroes that applied their flying or biting skills in service to the tribe. Other tales explained the appearance and behavior of local species. For instance, a long and involved myth of the California Mission Indians explains why the fly rubs its "hands" together in a supplicating gesture. The insect is begging forgiveness for speaking harsh words that resulted in people dying.

The South American Luiseño Indians offer another image. When their first ancestor, Wigot, a beloved guardian and teacher, was dying, the people sent Coyote to search for fire. After much searching, he returned without it. Meanwhile, Blue Fly twirled a stick between his hands, creating fire for Wigot's mourning ceremony. Blue Fly twirled the stick so long and so fast, however, that he couldn't stop and today still makes that movement with "his hands."

Each explanation of this particular behavior of flies arises from a close observation of flies, an acceptance of their presence, and a belief in an animated world. Attempts to find correspondences between people and other creatures and to understand and explain their unique viewpoint was always made within a context of interdependence. It is easy to imagine adults entertaining their children with these stories of Fly, and seeding their imaginations with images of kinship.

Finding correspondences between people and other species is a time-honored tradition in all cultures prior to the industrialized age.

A haiku poem by Issa, the Japanese farmer poet, says this about the fly's movement with its front legs:

> The flies in the temple
> imitate the hands
> of the people with prayer beads.[1]

Images like this one also stir the imagination and create good feelings. As we've already noted, our imaginations are seeded quite differently. Typically we imbue these small creatures with a malicious intent and then respond to that projection. If they are flying around us, we interpret their proximity as aggressive, although we might just as easily interpret it as playful. We don't know how to respond when they violate our boundaries with their quick darting movements, or when their numbers prevent us from tracking their movements effectively.

We like to be in control and prefer our companion species to have obvious points of kinship. The fly offers us little in terms of physical similarities, and its size prevents us from looking it in the eye. Yet, even with these inherent restrictions, if we replace the perception of malevolent purpose by an imaginative, non-aggressive one, we would respond differently to them.

Insects in Fairy Tales. If we also understood the symbolic role of flies in our psyches, and in myths and fairy tales, we might be on friendlier terms with their physical counterparts. In fairy tales, for example, people and animals represent energies within the totality of the human psyche. The dramas portray the universal growth process—which can be viewed as an inner journey toward our true nature. And the outcome of the tale depends on the psychological maturity of the hero or heroine. In these wholly psychological tales, insects are often called in to sort things out and perform tasks that the hero or heroine is unable to perform. As allies, they befriend the hero or heroine and play an essential role in helping him or her overcome challenges. In a Vietnamese story called "The Gentleman of the Flies," the hero allows the flies that invade his house to stay and, being a kind person, even takes care to feed them. Later, when he wants to win the hand of a princess, the flies help him accomplish the impossible tasks set by the king.

In dream and myth, the house typically symbolizes the known and conscious part of the self. So what is to be done when flies "invade" our house/psyche? Although the full interpretation of these tales lies outside the objective of this telling, it might be enough for all but the psychologically savvy reader to say that the story promises that many unforeseen benefits come to one who can befriend the small invaders. These friends of flies are the ones who benefit from the insect's gratitude—usually when they need uncommon help.

A Buddhist View of Flies

In Buddhist teachings, flies are sentient beings. They may even have been our mothers, fathers, or grandparents in previous lives. A 1927 collection of Japanese stories by Lafcadio Hearn includes the story of Tama, a devout maidservant who becomes a fly after her unexpected death so she can prompt her former employers to have the Buddhist services performed on her behalf.

A fly is also the form taken by Maitreya, the Buddha of loving kindness. A Buddhist tale about the Master Asanga illustrates the ideal Buddhist response to flies:

> Asanga went into retreat to develop pure compassion, meditating on Maitreya for twelve years, without results. Disappointed, he finished the retreat and on his return he met a dog whose lower body was covered by maggots. Asanga wanted to remove the maggots, but he knew that they too needed flesh to live. He cut a slice from his own leg to feed them. Then he considered how to remove them without injuring them and decided he must lick them off gently with his tongue. So he leaned over to do so, but when his bent head touched only the ground, he looked up and saw that the dog and maggots had disappeared. In their place was Maitreya Buddha. Maitreya told the awestruck Asanga that he had been with him from the very beginning of his retreat, but could not be seen until that moment of pure compassion.[2]

A Fly in Her Tea. A less extreme lesson in compassion occurred to deep ecologist, educator, author, and workshop leader Joanna Macy when a fly fell into her tea. The memorable event occurred during her first summer with Tibetan Buddhists. She was in the midst of a meeting with local Buddhists, intent on pushing through plans for a craft cooperative. A fly fell into her tea, which was a minor occurrence to Macy. After spending a year in India, she thought herself to be undisturbed by insects. She must have shown some reaction, though, because Choegyal Rinpoche, an eighteen-year-old tulku, leaned forward in sympathy and asked her what was wrong. She assured him it was nothing—a fly had just fallen in her tea.

When he still seemed concerned, Macy assumed he was just being empathic, so she reassured him again that it was not a problem and set her cup aside. But he continued to focus great concern on her cup and finally, while she watched, put his finger into her teacup and with great care lifted the fly out and left the room with it.

The conversation resumed, and Macy's attention returned to the meeting. When Choegyal Rinpoche reentered the room, he was smiling. He quietly told Macy that the fly would be all right. He had placed the tea-soaked creature on a leafy bush near the door and stayed until the fly began to fan its wings. He assured Macy that they could confidently expect the fly to take flight soon.

That is what Macy remembered of that afternoon. Not the elaborate plans and final agreements, but Choegyal's telling that the fly would live. She felt laughter in her heart and delight at the notion of concerning herself with the well-being of a fly. Choegyal's great compassion, revealed in the obvious pleasure in his face, conveyed clearly that she was missing a great deal by not extending her concern to all beings.

What may sputter inside ourselves in indignation at the thought of helping a fly may only be self-importance arising out of a narrow band of awareness. Our sense of self expands when we extend our compassion to insects. The world reveals its animated nature and minute perfection as these small beings intersect our path and move into our awareness. It invokes joy and invites celebration.

Compassion is "an appropriate virtue for our new cosmology," says spiritual renaissance leader Matthew Fox, and "an appropriate response to...[the] reality of interdependence."[3] Interdependence is what this generation is looking at, puzzling out its implications. The question is not how to connect with a fly—we are already connected. The question is how to translate that connection into appropriate behavior. Cultivating a genuine feeling for flies can help.

Surrendering to Flies

The Australian aborigines, like other indigenous cultures, understood that everything in life has a purpose. By living intimately with the Earth and its creatures, these people developed a profound understanding of each creature's role—including the swarms of bushflies that visited them regularly. They responded to the presence of these flies with an acceptance and appreciation that we would do well to emulate.

In *Mutant Message Down Under*, a fictionalized account of an actual event, author Marlo Morgan joined an Australian aboriginal tribe on a walkabout and inadvertently brought with her the baggage of her Western beliefs. At dawn she saw hordes of bushflies traveling in black masses. She gagged and choked as the flies descended, covering her body and crawling into her ears, nose, eyes, and throat. The tribal people around her had a sense of where and when the flies would appear, so they stopped and stood passively and soon thousands of flies covered their bodies.

Finally, after several encounters with the bushflies, the aboriginal leader Regal Black Swan explained to Morgan:

> Everything in Oneness has a purpose. There are no freaks, misfits or accidents. There are only things that humans do not understand. You believe the bush flies to be bad, to be Hell and so for you they are, but it is only because you are ignorant of understanding and wisdom. In truth, they are necessary and beneficial creatures. They crawl down our ears and clean out the wax and sand that we get from sleeping each night. Do you see we have perfect hearing? Yes, they climb up our nose and clean it out too.... It is going to get much hotter in the days to come, and you will suffer if you do not have a clean nose...The flies crawl and cling to our body and take off everything that is eliminated.... See how soft and smooth our skin is and look at yours.... You need the flies to clean your skin, and someday we will come to the place where the flies have laid the larvae, and again we will be provided with a meal.[4]

Regal Black Swan looked at Morgan intently and ended his explanation by reminding her that humans cannot exist if every unpleasant thing is eliminated instead of understood. "When the flies come, we surrender. Perhaps you are ready to do the same."

Taking his words to heart, when she next heard the bushflies approaching she followed the tribe's example. In her imagination, she went to an expensive health spa and pictured trained technicians cleaning her entire body. When the flies left, she returned from her mental trip, understanding that surrender was the appropriate response.

The Descent. Surrender means defeat in our culture. We are encouraged not to give in to anything, taught instead to persevere and overcome obstacles whenever they present themselves. We all understand Morgan's initial response to the bushflies—holding herself in stiff resistance to something unpleasant. We think it will help us endure, but it doesn't. Stressed out and fatigued by the fight, we find ourselves in exactly the same place for the next encounter.

What fights with us during these times is larger than our desires, opinions, and measurement tools. These are the initiatory forces of creation that push us to grow beyond our ideas of who we are and what life is about. They push us toward the soul, toward what is most genuine and unique in each of us. Rainer Maria Rilke reminds us in his poem "The Man Watching" that "what is extraordinary and eternal does not want to be bent by us ...."

> What we choose to fight is so tiny!
> What fights with us is so big!
> If only we would let ourselves be dominated

as things do by some immense storm,
we would become strong too, and not need names.[5]

We are sons and daughters of a lopsided society, too often stuck in the heroic stance when surrender is the required response. Few of us, however, have learned how to surrender. Even the idea is foreign, for Western culture has not provided us with instructions for the life-enhancing psychological descent into the abyss and the journey out again. The map was thrown out some three hundred years ago, replaced with a blueprint for controlling life and directions for scientists to eliminate anything unpleasant. Yet, it is this journey down into the depths, understood by ancient cultures and our wilderness selves, that is a necessary part of all life's passages. Being open to the inevitable descent transforms experience into wisdom.

The Descent of Inanna. There are many myths of descent reentering the culture today through the efforts of pioneers like poet Robert Bly and storytellers Clarissa Pinkola Estés and Michael Meade. And in all descents there is the death of a false self which has possessed us and a turning toward the true self. The oldest myth of this motif is known as "The Descent of Inanna," the Sumerian queen of Heaven and Earth. Inanna's journey to the Netherworld is a story about surrender. As she passes through the seven gates of the underworld, ruled by Ereshkigal, Queen of the Great Below, Inanna is stripped and finally killed. Interestingly, her restoration occurs because Enki, the god of water and wisdom, sends two flylike creatures into the Netherworld. Once there, they commiserate with Ereshkigal as she moans over the dead. Ereshkigal is so grateful for the empathy, for the compassionate response to her grief, that she hands over Inanna's corpse, and it is taken above and restored to life.

Therapist and author Sylvia Brinton Perera considers the descent, ensuing death, and rebirth a mirror of the rhythmic nature of the seasons and a model for our own psychological-spiritual journeys. It is "the essence of the experience of the human soul faced with the transpersonal."[6] Inanna's surrender is not based upon passivity but upon an openness to being acted upon. This active willingness to receive permits us to go down into the depths and allow the death of our known identity. Something familiar is lost, but a life more transparent to our deeper values is offered to us. We are given an opportunity to retrieve values long repressed, reconcile the old with the new, and allow a more expansive identity to emerge.

Psychologically, when we encounter the extraordinary and eternal, an event or experience outside our control and understanding, it can feel negative if we are stuck in the perspective of the hero or heroine and blinded by our projections. If we can't forgo the heroic stance or if we stay caught in a defensive fear mode that makes us run from change and the unknown, we may experience only the stripping aspect of the descent, without its cleansing and promise of renewal.

The forces that fight with us during life's initiations require respect and surrender. They demand that we recognize their validity and accord them a power equal to the powers of growth and expansion glorified in the dominant view of Western society. For without openness to this side of life, to the decay-sensing, death-wielding forces of transformation—the flies' domain—we will be worked on invasively, pitilessly against our personal will, because our will, what we want, is too small.

Leaving Heaven and Hell. Stephen Levine, a pioneer in the investigation of conscious living and conscious dying, says that most people live their lives in an incessant alternation between heaven and hell. Getting what they want, they are in heaven. Losing it, or never getting it at all, they drop to what their mind tells them is hell. But hell, says Levine, is only the stiff resistance to what is. It is the fight against the bushflies' presence, and heaven is their absence.

Trapped in this dichotomy, the mind fluctuates between thinking itself fortunate or unfortunate, and our responses arise from its pronouncement about the situation. And that judgment, congruent with our projections, bars us from experiencing the world as it exists apart from our mental baggage. It also blocks the informative and transformative energies that seek to deepen us. By trying to eliminate what we have perceived as unpleasant and achieve what we believe to be heavenly, we have cut ourselves off from the wellspring of life and declared war against beneficial forces in the natural world and in our psyches.

Perhaps one way around this deadening dichotomy of heaven or hell lies in changing what we desire. If our desire is big enough, it could work for us, instead of having small wants that bind us to a narrow track. If we want to understand Oneness, the fundamental unity of Creation, if we want to understand the worldview of every living thing, if we want to preserve the vast biodiversity of the earth, we may only need to suspend our opinions and stay open to nature and its creatures as they present themselves. We may only need to surrender.

Losing Irreplaceable Wonders

The statistics of declining populations of species have reached alarming proportions. Something is drastically wrong with our approach to life or we wouldn't have allowed, and even participated in, the vast destruction of our planet. Although entomologists have identified thousands of fly species, we are starting to realize that having a dead insect in a collection does not mean we know much that is important about that creature and the part it plays in its ecosystem.

Entomologists in California recently tried to get official endangered or threatened status for twenty invertebrates. On the list were seven arachnids, two crickets, three katydids, five grasshoppers, one moth, and two butterflies. They were all rejected, all deemed "insufficiently endangered" to make the list.

"Wise-use" groups like the National Wilderness Institute don't want to protect what it calls "ugly little life-forms." In fact, they want the nation to abandon the Endangered Species Act whose purpose is to protect all species, and let the public's preference for certain creatures guide the official decision-making process, a regressive move that would harm us all.

Experts tell us that when a species declines it is a warning and the first indication that something is drastically wrong. Not to act at that point is a crime of omission that will return to haunt us.

The Flower-loving Fly. Several years ago, an agricultural entomologist proposed putting the inch-long, golden brown, Delhi Sands flower-loving fly on the endangered list. Despite grumbling from some developers and others unwilling to entertain the fly's importance and the larger need to preserve all species, the fly won official status. In fact, as of this writing, it is the only fly on the government's endangered species list.

The Delhi Sands fly has eyes that change colors with reflected light and lives only a few weeks in its San Bernadino, California, habitat, playing a key ecological role in its short life. With ninety-seven percent of its habitat destroyed by agriculture, dairy farms, roads, recreational vehicles, housing tracts, and commercial and industrial sites, this strong flier may not survive on its remaining five hundred acres, despite its protected status. Its cousin, the El Segundo flower-loving fly, vanished in the early 1960s.

Even if the Delhi Sands fly can survive in such a small area, its struggle is not over. Protecting a fly outrages a lot of people. A state lawmaker has written a bill to have it taken off the endangered species list. The National Association of Home Builders has sued to do the same. Even though the species now claims only a small area of land off a busy interstate highway, many industrial developers consider it an infringement of their right to develop any piece of land they want.

No Protection for Pests. Any insect classified as a fly will have a hard time getting protection and keeping it, regardless of its role in its habitat. Arguments against protecting it will draw power from our unexamined consensus on flies. Propaganda will reduce the issue to people versus flies or jobs versus flies.

The Endangered Species Act excludes insects "determined by the Secretary to constitute a pest whose protection under the provisions of this Act would present an overwhelming and overriding risk to man." But the risks are primarily economic, not life-threatening. We don't want to risk losing money or risk being inconvenienced or made uncomfortable. With sad irony, the real overwhelming and overriding risk to people is not flies or other insects labeled pests, but the years of waging war against them with toxic chemicals that damage entire ecosystems. Rachel Carson gave us a wakeup call. It wasn't enough. Pesticide use is at an all-time high. Perhaps we're too willing to be lulled into complacency, believing that doctors and scientists will eventually solve our health and environmental problems.

Taking back the energy given individually and en masse to our projections of flies as enemy, we might use it to find our voices, call out "the king is naked," and reject the official version that insists only our desires are important. Our persistence in dividing up the "Oneness" of the world into good species and bad species and then trying to eliminate those we have judged to be bad ruptures the web of relationship on which our lives depend. To wage war against any species is to wage war against ourselves. Likewise, to refuse aid to a species threatened by extinction is to abandon an aspect of ourselves and diminish all life on Earth.

If we allow science, big business, and politics to stay harnessed to economic values, they will never protect biodiversity, but will continue on a course that endangers us and our children and their children. These economically driven institutions can't help but feed us to the chemical giants that run agriculture, and feed other species to the developers who think all expansion is growth.

To respond appropriately to the crisis of endangered species, we must reexamine and discard current scientific practices and models of the natural world that have not served life and refocus our efforts. We must reeducate society and not allow a handful of policymakers to determine who lives and who dies. For starters, we would make far fewer mistakes if we assumed every insect was a valuable part of its habitat and essential in maintaining its ecosystem's equilibrium. That assumption alone would alter the status quo while we worked on cultivating a feeling connection to these master recyclers and aerial wonders.

Communicating with Flies

The evidence of our multifaceted interdependence with other species, outside what traditional scientific investigation has revealed, has led to another phenomenon generating interest and gaining acceptance in the popular culture: interspecies communication.

Interspecies communication, a rich area of research, has been dismissed by institutionalized science because it does not fit within conventional paradigms. Although modern physics and many other branches of science have moved beyond the mechanistic theory of life, traditional biology is still dominated by it, reducing other species to complex machines.

The phenomenon of interspecies communication, however, has not gone away. In fact, we hear more accounts than ever before of such communication and the uncanny behavior of animals. Biologist Rupert Sheldrake believes it is an area where scientific research can be carried out by amateurs—people who do something simply because they love it.

Many people who love animals have been drawn to this field already, and a growing number of interspecies communication workshops are offered around the world. Pioneer interspecies communicator Penelope Smith has instructed hundreds of people in the art of communication with other species. Smith "communicates" with insects like flies and mosquitoes in the same manner that she does with furred and feathered species—telepathically. Tested by skeptics and ignored by scientists, Smith has demonstrated repeatedly, under many circumstances, that when she engages other species in communication, something occurs between them to affect their behavior.

According to Smith, communicating sincerely with insects yields harmonious results. When she or the people who have learned and practiced her techniques approach insects as intelligent beings with whom cooperation is possible, they achieve the results they want. With a perspective that closely resembles the Malaysian Chewong people's orientation to other species, Smith believes that the most important element in any communication attempt is attitude. Successful communication requires an attitude of nonresistance and a willingness to listen to and really see another's viewpoint and work out a mutually acceptable solution.

When Smith first had an opportunity to communicate with a fly, she was outside relaxing. A fly repeatedly landed on her hand, finally catching her attention. She decided to try and communicate with it. When it landed again, she brought her hand close to her face and told the fly that if it wanted to keep touching her that she would like to do the same. Smith felt the fly's surprise, but it stayed long enough for Smith to gently touch its back with her finger. Then the fly flew away but returned quickly, and they repeated the touching activity five times. Smith reports feeling the fly's trust and its wonder about the encounter.

When Smith communicates with any species, including those of the insect kingdom, in order to resolve a problem, she initially listens to their reasons for doing what they do and agrees to help them meet their needs. Even the listening process, however, is a foreign one for those not familiar with quieting the mind and stopping its whirl of thoughts. It can be cultivated, Smith insists, and in her popular workshops she trains people in this art of communicating with other species.

Universal Mind

Smith's ability may be evidence of deep ecology's nonlocal, extended self that encompasses all other creatures. Physician and author Larry Dossey believes that some kind of a universal mind connects all living things:

> It makes good biological sense that a nonlocal, psychological communion might have evolved between humans and animals as an asset to survival, then the stories of...talented flies may be more than amusing parlor tales. They may be indicators that nature in its wisdom would, in fact, have designed a mind that envelops all creatures great and small.[7]

In indigenous cultures, it was the shaman who had a special rapport with other species and advised the tribe on how to be in right relationship to them.

The shaman's hard-won knowledge arose in part from initiatory dream experiences with other species in which he or she was killed and often eaten in order to assimilate the powers of these nonhuman allies. Even in daily matters, through disciplined awareness and the ability to shift consciousness and enter other levels of awareness, the shaman knew things about other species that made life easier.

Without contemporary shamans available to guide us today, we are left with the challenge of developing our own shaman-like abilities. We don't have to wait until our scientists discover the physical basis for Smith's communication or explain the energy fields that native shamans have detected and mastered. A personal experience can do more to uproot old habits of perceptions than anything we might hear or read.

Befriending Flies. Some precedents for befriending a fly exist. Colman of Galway, friend and correspondent to Saint Columba, reports that this hermit-saint lived in a clay hut with room for only himself, a cock, a fly, and a mouse. He fed and spoke to them kindly, and they became his friends. They each had a job. The fly, it was said, kept the saint's place in his books. And when his friends died, the saint wrote of his grief.

Closer to home is Charles A. Lindbergh's alleged conversations with a fly during his historic flight across the Atlantic in 1927. In the movie version *The Spirit of St. Louis*, Lindbergh discovers that a fly is in the cockpit with him as he begins his forty-hour flight. Once he determines that the fly will not add more weight to the airplane, he relaxes and talks to it periodically. The fly is even credited with helping Lindbergh when he dozes off and the plane begins to lose altitude. Flying from the altitude gauge to Lindbergh's cheek, the fly walks around on his cheek until Lindbergh wakes up and stops the plane's downward movement. Finally, as Lindbergh flies over Greenland, the last land before a stretch of water eighteen hundred miles long, he reminds the fly that this is its last chance to leave and still have land beneath its wings, and the fly flies out the window.

Fly as Teacher. Following native wisdom, Sara Willow, a writer and the founder of a web site called "AnimalSpirit," encourages others to take the time to learn from the creatures they encounter in the course of their daily life. From the fly, Willow learned how to hold herself still, because the fly would only stay on her extended finger when she was practicing stillness. Later when confronted by yellow jackets, she used that ability to be still and discovered that then she could let the wasps buzz around her without panicking in fear.

For most of us, learning from insects or communicating with them is a possibility that only direct experience can verify. At some point we must be willing to set aside our self-consciousness and fear of looking and acting foolish, and try it. Psychiatrist and author Gale Cooper did just that one summer day while sitting in her dining room. A fly buzzed past her, making a futile attempt to penetrate the window. Without thinking and without changing her position, she offered to take the fly outside and extended her index finger. The fly circled, then landed on her finger. Cooper told the fly that it would have to trust her because to get to the door she would have to walk through three rooms. The fly stayed on her finger, and when she opened the door, it flew out. Cooper knew inexplicably that the fly had understood her, and that knowledge brought tears to her eyes—she had touched the great mystery of our connection to other species.

Few published stories counteract the common view of flies as pests, much less suggest the possibility of communication or even friendship. Yet, when any individual stops acting in accordance with a wrong perception, the possibility for something new, something positive, emerges. Novelist J. R. Ackerly writes that when his mother "was losing her faculties"[8] she formed a friendship with a fly:

> One of her last friends...was a fly, which I never saw but which she talked about a good deal and also talked to. It inhabited the bathroom; she made a little joke of it but was serious enough to take in crumbs of bread every morning to feed it, scattering them along the wooden rim of the bath as she lay in it.[9]

Without a context that would allow for our having a relationship with a fly, we are likely to condemn or dismiss those who claim to have entered this terrain. Ackerly assumed his mother was simply losing her faculties, but perhaps she was only losing her judgments and opinions and returning to a natural state of mind, cleared of projections and open to the possibilities of the moment. And maybe we don't have to be afraid of being accused of losing our wits to begin cultivating a friendship with a fly.

Freddie the Fly. The late J. Allen Boone provides us with a model for befriending a fly (or any other creature) in *Kinship With All Life*. In this classic on interspecies communication, he relates how he was called into relationship with a housefly he named Freddie.

Boone's experiments in interspecies communication began when he was asked to take care of a highly trained dog named Strongheart. Boone realized after only a few days that the dog, making use of his unique canine sensory apparatus, demonstrated a wisdom and intent that surpassed the generally accepted view of dogs. Boone also learned that if he approached the animal, and then other species, with a genuine willingness to learn, he found an intelligent being on the other side of the relationship.

One day, after many successful adventures in communication with a variety of species, Boone noticed that a fly appeared to be following him from room to room. He wondered if communication was possible with this life-form. As Boone had done with all of the species with which he wanted to communicate, he wrote down the fly's admirable characteristics. With list in hand, he sent the fly appreciative thoughts, mentally asking it to land on his finger. The fly responded, flying to Boone's finger and walking up and down its length with a vigorous step.

Over the next couple of weeks, Boone and Freddie played games and generally explored their relationship. Sometimes Boone would mark his finger with different colored inks, sectioning it. He would then ask Freddie to fly to a particular color, and Freddie, demonstrating his ability to understand what Boone asked, would do it without hesitation.

When word spread about Freddie, visitors came to meet him. One visitor was an actor who at Boone's suggestion asked Freddie to land on his finger. But each time the man directed his request to Freddie, the fly flew instead from Boone's finger to the ceiling. Freddie's actions puzzled Boone. Freddie was usually as curious about visitors as they were about him. Finally, Boone questioned the man closely, and the actor admitted sheepishly that he had always hated and killed flies. Boone decided that Freddie was aware of the man's real feelings about flies and demonstrated that awareness by his refusal to interact with him.

Before Freddie, Boone didn't like anything about flies, and with echo-like precision his thoughts dictated his experience. He expected flies to be unfriendly, and they were. He expected them to annoy him, and they did. He expected to be bitten, and he was. When he changed his attitude about flies, he reports that he was never bothered again, even in fly-infested jungles.

There is a reality and a wisdom to any creature outside of our fears, judgments, and opinions about them. Indigenous people knew that every creature, including the fly, had the power to impart a revelation and reveal an aspect of the whole. Freddie the fly taught Boone that kinship and communion become possible when people extend courtesy and appreciation to another species, addressing this other being as a "thou" instead of as an "it."

By following Boone's example and adopting a genuinely friendly view of flies, we will discover that cultural pressure will no longer be able to shame us away from responding compassionately to them or any other insect. Our wilderness selves will once again initiate the celebratory response toward them that has been sorely missing in our habitual reactions. And finally, when the gnat asks us what insect we rejoice in where we come from, we might say with enthusiasm, "The fly!"

According flies appreciation and respect is a radical approach to coexistence that requires some initial effort and perseverance on our part. But when flies are acknowledged individually with a simple gesture of courtesy, or as a species for their superior recycling and pollinating abilities, we have the foundation for an environmental ethic based on a feeling for flies. We also open the door to the possibility of rejoicing in flies. And knowing them by their proper names, we might call on their empathic presence to witness our inevitable descents and help secure our release and return. We might also seek Big Fly's counsel as we navigate the complexities of these times and, once again, hear My Lord Who Hums in the aerial visitor who enters our awareness.

6

Divine Genius

> The roach/its shining back and hair thin feet
> creaks the tiles night's music/which means we are safe we are never alone.
>
> —Linda Hogan

An insect that has lived on the Earth 320 to 400 million years is still among us, smaller now, but otherwise unchanged. Fossil imprints testify to its long reign. A twelve-inch version lived during the Carboniferous era 280 million years ago, sharing the Earth with equally ancient mayflies and immense dragonflies with wingspans of two feet. Six-inch fossils of this insect have also been uncovered in places to the exclusion of most other life-forms.

The elder in our midst is the cockroach, the oldest of living insects and one that scientists believe has changed the least of any living form. Today this creature faces its greatest test of adaptability. It must share the world with a hostile and burgeoning human population, or try to.

A popular subject for contemporary science writers and entomologists engaged in changing the public's view of insects, the cockroach has earned the begrudging admiration of the scientific world because of its remarkable survival abilities. Yet, it is known more often for the loathing it elicits in the general public. When species were ranked according to their popularity in a 1980 study by Yale University researchers, no one was surprised that participants put the cockroach last, with the mosquito one notch above it.

In the West, most people are repulsed by the appearance of cockroaches and the way they move as they dash for cover and safety when alarmed. A love of warmth, darkness, and our leftover food has placed a few species in a proximity to us that some find unbearable. The thought of their unseen activities, carried on in darkness, disgust or scare many. We imagine them living sordid lives behind our walls, and companies who make millions of dollars selling products to kill them tell us that thousands wait hidden in our midst for an opportunity to spread filth, transmit disease, and take over our living spaces. These messages and a reservoir of unexamined feelings prompt people in the United States alone to spend over 240 million dollars each year on cockroach-control products.

The cockroach's association with filth—our filth—overrides the fact that these benign creatures do not bite or sting or have any capacity to directly harm humans. Biologist Ronald Rood, champion of species maligned by others, maintains that cockroaches are harmless to human beings and seldom carry any harmful bacteria that would affect us—a statement most entomologists would agree with. Yet, our response to their presence is so strong that of the four thousand known species, the four or five species that live almost exclusively with humans account for approximately twenty-five percent of the total insecticide use in the United States.

Cockroaches and Dirt

People with cockroaches in their homes often call them by other names such as water bugs, croton bugs, Bombay canaries, or palmetto bugs to avoid being accused of keeping a dirty home. The belief that the presence of cockroaches in our homes points to slovenly habits and a less than stellar character runs through our literature and movies. In the 1987 Japanese cult movie *Twilight of the Cockroaches*, for instance, cockroaches overrun a Tokyo apartment, home to a slovenly man, and in a 1996 film, *Joe's Apartment*, resident cockroaches befriend a young man simply because he does not clean up after himself. A variation on the theme is also played out in George Romero's *Creepshow*, a horror film of the eighties featuring thousands of cockroaches that drive a cleanliness fanatic insane.

In Daniel Evan Weiss's novel *The Roaches Have No King*, cockroaches cohabit with a sloppy lawyer, another link to careless living, and Donald Harington's novel *The Cockroaches of Stay More* revolves around a weak-willed, alcoholic writer, thus tying the presence of cockroaches to personality weaknesses and addiction.

This persistent connection between the presence of cockroaches and a homeowner's character and habits may explain the results of a recent survey in which people named cockroaches as the most embarrassing insect. Their presence is a cause for shame and accusations. For people living unexamined lives, ridding their homes of these insects is a way to try to escape judgment.

Ironically, no matter how clean we keep our homes, in warm regions cockroaches are likely to be our roommates. They rarely miss an opportunity to eat and can exist easily on a wide variety of substances including the glue from book bindings. Like their relatives the grasshoppers and mantises, cockroaches have mouths with working parts evolved from legs and equipped with taste organs. They can chew almost anything, hard or soft; have gizzards to grind up their food; and they can even synthesize their own vitamin C. Because they can sense just a few molecules of food, they rarely miss even a crumb of food, and in dire circumstances they have lived without eating for three months and without water for one month.

The Myth of the Dirty Cockroach. Since some cockroach species live amidst our garbage and in our drains, most people assume they are dirty. In fact, the name cockroach has become almost universally associated with uncleanliness and disease. But it's a reputation they don't deserve—scientists have discovered that the cockroach is a clean insect in its personal habits. By performing contortions similar to a cat grooming itself, cockroaches pass every accessible part of their feet and their antennae through their mouthparts. They also wash themselves vigorously after being touched by human beings, a fact incorporated in Mary James's novel *Shoebag,* where a young cockroach wakes up one morning and discovers that he has turned into a little boy. As a bacteria-laden human being, he is then shunned by his insect family and friends.

Instead of imagining that cockroaches are up to no good behind our walls, we would be more accurate to see them cleaning themselves for hours. Since these activities occur in the darkness of hiding places, we are unaware of the hours they spend on these ablutions each day. We are also unaware because pesticide companies don't want us to make comparisons between cockroaches and creatures we like—like cats. It's not good for business.

Phobic Reactions to Cockroaches

Reactions to cockroaches frequently border on hysteria. In a recent book, science writer Natalie Angier shares her terror of cockroaches as a young girl, feelings that eventually developed into admiration as she studied them. But what about those who do not have a penchant for entomology—those whose terror, if not fed by the popular culture, is allowed to exist unexamined?

Fear of insects or entophobia is a common psychological disorder in our society. Irrational fears often immobilize people or, if the fear is of insects, prompt them to use insecticides unnecessarily. A phobic reaction has an intensity that marks it as a disorder. When the object of the terror can be avoided, the phobia's crippling potential is also avoided and the person can exist reasonably well without examining its roots. In Sue Hubbell's *Broadsides from the Other Orders: A Book of Bugs*, a college professor confesses to a lifelong terror of cockroaches.

> I see one in the kitchen and I am terrified, paralyzed, unable to speak or move. It is so small and I am so big. That is part of the horror. It is not the least afraid of me...Kill one? That would be impossible. It is psychically, too big to kill...Besides, even if I were able to kill it, another would come...And that is too frightening even to consider.[1]

The intensity of his feelings puts him in the phobic category, where he is far from alone. Phobias and related anxiety disorders are the most common psychological problems in the West. Typically phobias are treated with a behavior-modification technique that desensitizes the individual by pairing relaxation exercises with images of the object of the phobia. Yet, depth psychologists tell us that the intensity of our response to a person or situation is a sure indication that something else is at work. It is also generally accepted by these specialists that what we refuse to look at in ourselves, the projections we cast out into the world, dominates us, directing our energies from behind the scenes. The source of this influence is no wizard behind a velvet curtain, but the now-familiar shadow. We must only draw back the curtain and bring awareness to projected material to eventually strip it of its power and free ourselves to respond appropriately.

Them-Against-Us Mentality. Most of us have responses to cockroaches that aren't phobic but have an intensity beyond what we understand. Unexamined, these feelings are the reason we don't object to comments like the one below from an exterminator-turned-urban-entomologist whose specialty is cockroaches. In an article in the garden section of the local newspaper entitled "Good Riddance to Roaches," this man says,

> Trying to get rid of cockroaches has become a preoccupation for many Americans.... The problem is that most people don't have the right game plan. Spraying roaches or stomping them gives us a feeling of satisfaction, but it's impossible to wipe out the thousands of roaches living behind the walls by employing those methods.[2]

Spraying or stomping on cockroaches may make some of us feel powerful and in control. With a them-against-us mentality, every cockroach killed moves the person closer to some imagined safety zone. But the satisfaction that comes from killing our perceived adversary is always short-lived. There is never enough safety for the individual who has made other creatures the enemy. And projections like these keep us fugitives on the planet, operating from our suspicions and killing what we don't like or understand.

Learning to Hate Cockroaches

Too often negative feelings about cockroaches appear reasonable. After all, they look and act very differently than we do. We think them too small, sporting too many legs, and moving too fast. Worse, they are maddeningly indifferent to our wants. Not only do they move differently than we do, but they also move into places they shouldn't be—like on us as we sleep (or so we fear) and in our homes. Blatantly ignoring property lines, they also ignore our psychological boundaries. Violations are invasions. When they opt to live with us, their presence serves as a judgment on our housekeeping, their numbers an assault to the eyes and imagination, and their indifference simply insolent. It appears natural and acceptable then to ridicule and kill them in return for these crimes.

The fact that not all cultures feel and act this way toward cockroaches is evidence that our responses are learned, not innate. East Indian and Polynesian people created jewelry and ornaments devoted to the cockroach, and Jamaican stories frequently include cockroaches cast in a positive role. The Nandi tribe of Africa even has a cockroach totem, and in parts of Russia and France people think of this insect as a protecting spirit, its presence in the house lucky. In fact, in these communities, if the cockroach leaves, its departure is taken as a sign of bad luck.

Learning Our Aversion. Since an aversion to cockroaches is not natural, it must be learned. Studies show that even children who have never had contact with cockroaches learn to hate them when the adults in their lives hate them. Researchers have also demonstrated that up to the age of about four, children have absolutely no aversion to cockroaches. In one experiment children less than four readily drank liquid from glasses with plastic cockroaches floating in them. Children over four, however, would not bring the glass to their mouths, having already learned that cockroaches are filthy and should not be touched or put into their mouths.

The fact that cockroach aversion is learned means we can unlearn it. It also means we can stop transmitting our prejudices to our children by showing genuine curiosity (and admiration) about cockroaches, and by rewriting our children's books. As we noted in Chapter 1, children's literature is written without awareness of the cultural bias. The negative slant in so many introductions to insects takes many forms. One common form is starting the text with the general consensus about the insect, like "Cockroaches are really yucky,"[3] and then moving to a liberal scientific stance, "but like them or not, cockroaches are truly amazing creatures."[4]

Another way to transmit a bias is to use images of people assaulting cockroaches when describing their body parts or abilities. In one book, cockroach locomotion is introduced this way: "*Try to kill a cockroach* as it scurries across your kitchen floor. It isn't easy"[5] (emphasis mine). The same author, obviously immersed in the culture's bias, describes a recent import to the United States, the Asian cockroach, as "the most repulsive species" and "disgustingly different" from the German cockroach.[6]

What disgusts the author is the ability of this species to fly. In fact, they can fly to a measured distance of 120 feet, a feat that could be a wonder to behold. And unlike most cockroaches, it seems that the Asian roaches are attracted to light and fly toward television sets and toward people wearing white. Our repulsed author writes: "Outside, they fly up like grasshoppers and cluster on the ground so compactly that a person can *step on* twenty-five or thirty at once"[7] (emphasis mine).

Instead of a descriptive phrase, an image of assault is used to illustrate how closely these cockroaches commingle on the ground, revealing our fear that the insects have joined together portending some misdeed. If it was another animal involved, kittens perhaps or a gathering of songbirds, the inappropriateness of the statement might be more evident.

By reining in our suspicions of their ill intent, the questions evoked by this clustering scene might follow this line: Is there an order and purpose to this insect assembly, a common mission that unites them? Do they move as one thought, synchronized as a flock of birds or a swarm of bees? What do they sense as they congregate, individual insect boundaries abandoned? These questions reflect the spirit of scientific inquiry more than images of humans killing non-humans indiscriminately.

More information could also ease our fears about flying species. For instance, cockroaches are strong fliers, but poor navigators. In Hawaii, the males of some species often fly on warm nights and frequently collide with people and other objects. If we knew about their lack of navigational skills, we wouldn't interpret their bungling attempts at flight and frequent collisions with people as acts of aggression, as many Hawaiians do, even calling them the "B-52 cockroach."[8]

Once we see the ways in which we perpetuate the cultural bias as well as transmit it to our children, we free ourselves to respond differently. Awareness is a powerful tool for transformation. And once aware, we will be less apt to take our feelings as fact and will recognize when fear, prejudice, and hatred is disseminated under the cloak of reason and science.

Cockroach Communities

As our awareness increases, we also realize that those with a vested interest in having us view cockroaches as the enemy want us to believe that these insects are life-forms that respond robotically to the environment and live without purpose. A cockroach "combat" manual says,

> Cockroaches are somewhat gregarious and may hide together in clusters. They share no outward affection with each other....Cockroaches are basically survival machines that hide by day and forage for food by night.[9]

It helps to know that others who don't make their living on substances that kill cockroaches report that cockroaches appear to live in organized communities. Although few insects beyond social insects like ants, bees, and termites actively care for their young, many cockroach species display some signs of having a family life. On occasion, active youngsters even return to the nest site to share a drop of food still remaining on the mouth parts of an adult.

When at rest, cockroaches lie close to each other, antennae touching and moving slightly as though sampling the air for danger, or to reassure each other, or perhaps even to communicate. Scientists say that cockroaches have a strong need to be touched on all sides, a predilection they call thigmotaxix. It's a phenomenon that probably supports their love of community. But it doesn't explain why cockroaches groom each other. Their togetherness is an interesting oddity in the insect world where most insects either ignore or fight one another.

Another mystery is the fact that cockroaches occasionally travel in large groups, moving by the thousands from one building to another some distance away. In 1894, a group of German cockroaches, most of them females carrying egg sacks, caused a sensation when they crawled en masse out the door of a Washington, D.C., restaurant, and moved into a building across the street.

Since there had been no recent insecticide treatments at the restaurant or vigorous house cleaning activities, no one knows what precipitated the move.

Learning Experiments. Contrary to the idea of cockroaches as machines, researchers often credit cockroaches with being "smart enough" to solve a maze. As early as 1912, C. H. Turner used electric shock to train cockroaches in a laboratory experiment. The same questionable technique was used by the special effects technician for the cockroach invasion in *Creepshow.*

Positive incentives for learning, which have been shown to be more effective and more durable in humans and other animals, are also more effective for cockroaches. Ronald Rood reports that a classmate of his taught roaches to distinguish right from left in a simple maze—without electric shock. They passed each test with flying colors.

Cockroaches and Disease

Many research efforts have attempted to link cockroaches to specific diseases as though eager to find a justification for the culture's hatred. Yet, all evidence is circumstantial. *Never* have cockroaches been linked directly in the transmission of a human infectious disease. Most scientists agree that their disease-carrying reputation is quite undeserved.

Recently, however, cockroaches have been linked with human allergies. A major newspaper reports with a familiar venomous tone that now "there is another reason to hate cockroaches" and that scientists have cited these "universally hated" creatures as a major cause of winter allergies and asthma. The news item was based on recent research studies that indicate that when chemicals in the crumbling outer shell of a dead cockroach are inhaled, they can have the same effect on allergic people as ragweed pollen or cat dander.

The correlation is not as cut and dried, however, as one might think. Allergies are complex afflictions, and their source is not easily identified. In fact, the root cause of allergies remains a mystery. What is known is that they are definitely on the rise, in fact two to ten times more common than forty years ago.

A recent book on environmental medicine attributes the increase to three reasons. The first is the great and rapidly increasing load of toxic chemicals in the air, food, and water, which, in turn, builds up in the body. The increase is a direct result of having gone in a hundred years from using 150 chemicals to over 100,000. When these toxins reach a certain threshold, determined by the person's genetic makeup and general state of health, an allergy occurs.

A second reason for the rise of allergies is the vastly increased indoor air pollution levels, a result of materials used in the construction of modern hermetically sealed buildings. And third is our modern farming production methods that produce food laced with chemicals and deficient in essential nutrients. An increasing number of physicians agree that a lack of vitamins, minerals, and fatty acids make us susceptible to a host of ailments. Consuming non-organic meat that contains hormones, antibiotics, and food additives upsets our body's chemistry further. Run-off chemicals that seep into the ground water and the chemicals we use to treat our water also provide us with a steady diet of pesticide residues and heavy metals. Add mental stress to the total load of neotoxins in the body, and the result is a highly suppressed immune system which increases overall susceptibility to allergies.

Because the influence of the individual and collective shadow will affect even the most ethical reporter and the most prestigious newspaper, we need to bring a discriminating awareness to what we read and believe, especially when the conclusions pit us against a plant or an animal in the Earth community. A 1997 report by Brazilian allergy researcher Gilberto Pradex states that the exoskeletons of cockroaches comprise only two or three percent of substances that could cause allergies. The percentage is hardly a "major cause" of allergies as the cited news item asserts. Pradex is currently working on a vaccine based on substances found in cockroaches, a line of research linked to the premises of homeopathy.

In Indonesia, cockroaches have also been used to cure or relieve asthma and other respiratory ailments. Russians transform a common large species into a homeopathic medicine to cure dropsy (a fluid buildup in tissue), and in parts of Europe, powdered cockroaches have been used for pleurisy and pericarditis (an inflammation of the membrane surrounding the heart).

Cockroaches are also part of the ingredient list in other folk remedies. Some American Indian tribes used cockroaches to avert or cure whooping cough. Jamaican Indians drink a mixture containing cockroach ashes as a cure for certain diseases, and they mix cockroach bodies with sugar and then apply the mixture to ulcers and cancers as a healing aid.

The Illusion of Objectivity

A legitimate tie exists between cockroaches and allergies, as evidenced by their use in homeopathic and folk remedies, but when we turn a microscopic eye to the cockroach to find a justification for our dislike, we magnify and blur the context through which we might understand and treat allergies. Studies in quantum mechanics have demonstrated that absolute objectivity in research is an illusion. Biologist Rupert Sheldrake says that "what is being looked for, and the way it is looked for, affects what is found. Moreover, the expectations of the experimenter affect what is observed…"[10]

Hostility influences what we observe and how we interpret our findings. Without awareness of its nature and influence, we will continue to support the chemical companies that encourage us to battle these insects and continue to accept the findings of their research projects. But hating and killing cockroaches will not stop an allergic reaction to a world flooded with petrochemical fumes and pesticide residuals. Nor will it permit us to escape the consequences of enemy-making and the distortions of the propaganda that uphold it.

The Shadow of Scientific Research. Cockroaches are frequently used as laboratory specimens by scientists studying behavior, anatomy, and physiology. Neurobiologists claim the cockroaches' tactile sensitivity, combined with a nervous system built of exceptionally large cells, makes this insect an ideal experimental organism for studying how nerve cells work (perhaps these same exceptionally large nerve cells indicate that cockroaches have an uncommon sensitivity to life).

Although empathy with cockroaches or any other animal is considered a contaminant in traditional scientific research, hostility is not recognized as a contaminating factor. One researcher reports that a cockroach's head will live and respond for at least twelve hours after being severed from its body. He concludes the article with a statement tinged with smugness lauding the cockroach as an ideal experimental animal, because few animal rights activists will disrupt a laboratory to defend cockroaches—no matter what we subject them to.

The shadow has also seeped into state-of-the-art research like that being conducted at Tsukuba University, a leading science center in central Japan where a micro-robotic team and biologists have created what reporters call the "Robo-roach." Researchers have surgically implanted micro-robotic backpacks into cockroaches to control their movements. They fit them with this equipment after removing their wings. Then they replace their antennae with pulse-emitting electrodes. When a remote researcher sends signals to the backpacks, it stimulates the electrodes and the pain forces the cockroach to turn left or right or go forward or backward. The altered cockroach can survive for several months, although over time they become less sensitive to the electronic pulses. The scientist in charge is optimistic that even that can be overcome.

In his concluding comments the mask of objective scientist slips and he confesses, "They are not very nice insects. They are a little bit smelly and there's something about the way they move their antennae. But they look nicer when you put a little circuit on their backs and remove their wings."[11]

Celebrating Our Hostility

Outside scientific circles, the shadow is even more apparent, because hostility toward cockroaches is the norm. Companies interested in selling cockroach-killing products capitalize on our feelings by staging events, often held at science centers, that focus on killing and ridiculing cockroaches. The Midwest Science and Technology Museum, for example, teamed up with an insecticide company to demonstrate creative cockroach-killing techniques to youngsters under the guise of education.

Distorted Creativity. Contests are another type of cultural event that is used in service of the hostile imagination, and creativity within this context also takes on a twisted expression, as it must. Several years ago, for example, the International Roach Finals in Florida awarded a prize for creativity to a man who displayed dead roaches dressed up as wind-surfers and sunbathers. Not an isolated incident, The Cockroach Hall of Fame, a Plano, Texas, enterprise, sponsors a contest for cockroach "art" where people are encouraged to find and kill cockroaches and then dress them up as human personalities and display them in shoebox scenes. A World Wide Web site also includes an on-line gallery of dead roaches dressed up as humans and placed in various scenes.

Dead and captive roaches are a big draw in Purdue University Department of Entomology's "All-American Trot."[12] Since 1991, students and staff have invited the public to a cockroach race and a tractor pull where three Madagascar hissing cockroaches, an exotic species, are forced to pull miniature tractors across a finish line. Dead roaches dressed up as fans with sunglasses and baseball caps glued to their bodies are posed in a grandstand and invoke great mirth from the thousands of people who come to enjoy the event and bet on the race.

Another popular event involves finding the largest cockroach. The Bizzy Bees Pest Control Company of Dallas, Texas, started cockroach contests in 1986 by offering $1,000 to the person who brought in the largest cockroach. Orlando, Florida, quickly followed suit. By 1989, other U.S. cities also had contests to find the largest cockroach, most of them sponsored by roach-control manufacturers. Today the search is international in scope with South Africa and Bermuda participating.

Sighting an Elder. Contrast these events to a scene shared by environmental psychologist James Swan in his book *Nature as Healer and Teacher*. Swan was driving along a dirt road in the Yukon Territory when he stopped by an Indian salmon-fishing camp set up beside a glacial river.

The Indians were taking fish from their trap as Swan approached. When they had taken out most of the six- to eight-pound fish, someone called out, "Look!" In the bottom of the trap was a giant salmon nearly four feet long. One of the older men looked at it briefly and said, "Let it go, it's an elder."[13] They opened the gate and the giant fish shot out and up the stream.

"You have to respect the elders in all the tribes," the older man told Swan. "The plants, the animals, the mountains, the birds—they all have many tribes. If you don't respect them, you'll pay later."[14]

Swan considered the experience a lesson in seeing truly—understanding that the visitation by the great fish was an intentional meeting of human and nonhuman, engineered by forces our culture no longer acknowledges in this society. According to writer-attorney Vine Deloria, Jr., a member of the Standing Rock Lakota, these sightings are "glimpses into another dimension of consciousness,"[15] a dimension where our kinship with nature and other species has its roots.

Perhaps the bid to find the largest cockroach or the largest one of any species is a severely distorted attempt to glimpse the elder of a species and touch the bedrock of our most fundamental selves. The fact that we no longer understand the impulse and blindly follow the cultural pressure to hate and kill the largest—and not only kill them, but accept a reward for doing it—is a sad commentary on our profound alienation and general lack of respect for elders even in our own species.

Alien or Other?

It is common to call cockroaches (and other insects) aliens and monsters. Focusing on the obvious structural differences between us, we justify the separation between us, calling it natural. But are they aliens and monsters—which implies exclusion, opposition, and hostility—or just Other?

We're adept at making monsters, especially from creatures that look so different from ourselves. I wonder if this monster-making propensity has its roots in something more life-serving than creating separation and fostering alienation. The word *monster* comes from the Latin term for a portent of the gods. If cockroaches are monsters in this sense, it means their presence signals that something momentous is underway. Joseph Campbell said it is always the disgusting and rejected creature that calls us to some undertaking, some spiritual passage which, when finished, amounts to a dying of what is familiar and outgrown, followed by a rebirth. The creature, whether frog, snake, or insect, represents the depths of the unconscious where all the rejected, unrecognized, unknown, and undeveloped elements of life reside. The appearance then of a creature we despise like the cockroach may be a call that informs us it is time to pass through a threshold, time to separate from what we know in order to move toward what we don't know. And all threshold crossings, all initiations—when we turn, or are turned, toward our soul or true nature—produce anxiety. Depending on the importance and duration of the change, an element of dread and the fear of the unknown also marks these times.

Boundary Inhabitants. Ecopsychologist Paul Shepard believed that the creatures outside our orderly classification system signify chaos and threaten our efforts to create order and maintain control. The fact that cockroaches inhabit boundary places, like drains, the cracks under our refrigerators, and the spaces between our walls might then feel threatening and thwart our desire for clear distinction.

Philosopher David Spangler says boundaries are the places in people, systems, and organizations where change is possible, the places where new things emerge. Chaos lives there, but so does life. In chaos theory, there is a domain called "the edge of chaos." It is the boundary place between stability and chaos, and the place where a system or organization loses enough of its coherency and stability to be open to change, transformation, and reorganization without collapsing into chaos. Spangler calls it the essence of life, "the domain where the known and the unknown, the familiar and the unpredictable come together in a cocreative way."[16]

So maybe cockroaches are messengers of this spirit of the boundary, this push from inside to change—the call from our deepest nature that fills us with anxiety. By triggering in us an awareness of our shifting inner ground and introducing disorder and anxiety into our thoughts, they act as initiators that disrupt the quiet stability of the status quo. Maybe their presence nudges us to approach this edge of chaos, so that we may touch a primal force of self-organization and unfolding—a force of newness and emergence.

Awakening. Seeing a cockroach nudged (or perhaps catapulted) a forty-three-year-old woman named Bryon Katie into an enlightened way of seeing and being. It happened in February 1986. It was early morning, and Katie was lying on the floor in a locked attic room of a halfway house for women recovering from eating disorders. She had spent the last two weeks largely incoherent and consumed by terror and rage. As she tells the story, "A cockroach crawled over my foot—I was awake."[17] She remembers first that she just felt the cockroach moving over her foot. Then she saw her foot but didn't know it was hers, and in the next instant she saw the cockroach and thought she was seeing herself. Suddenly joy and wonder replaced all the fear in her. The chaos of her old life dropped completely away, and she found herself in a heightened state of awareness that many now call an enlightened state. She calls it being in "the Truth"—aware of the love and unity behind all apparent separateness. Today, Katie teaches others how to stay in "Truth" and release their pain. When she reflects on her turning point, on that moment when a cockroach crawled on her, she calls her awakening a resurrection. She awoke from pain that had become unbearable, and began to live the reality of joy.

If we consider that cockroaches are "Other," not alien, we could view Katie's experience of awakening as a moment of grace in which the potent medicine of this species allowed her to finally see that all otherness emanates from an underlying and essential Oneness. And maybe we could move toward our own resurrection or awakening and invite grace into our lives by courting this quality of otherness in whatever form we find it.

We typically repress what is different or "other" within us in order to fit in. Once repressed, we project it onto others and then view them suspiciously. Perhaps we have just lived too long afraid of the cockroaches' shadowing presence beneath the teetering scaffold of beliefs we have constructed to elevate us above the natural world. What unknown or despised aspect of ourselves do they (or any other insect's appearance) invoke? And how long can we afford not to invite them into our awareness and teach us what is required? How many times will we refuse the call to grow, and when will we understand that we must entertain all kinds of otherness to discover our essential selves?

The Power of Empathy. Entertaining the quality of otherness can mean empathizing with one that appears very different from ourselves. According to psychologist Daniel Goleman, an ability to empathize with another is a characteristic of "emotional intelligence." This ability to participate in another's worldview presupposes connection and, contrary to popular opinion, requires participation and a blurring of the boundaries that separate and distinguish one from another—not structural similarity.

When learning to empathize with an insect, it also helps to remember that empathy is neither anthropomorphism nor behaviorism. The goal is not to make cockroaches into people like journalist Don Marquis's poetry-writing cockroach Archy or the new Internet Rodney roach. Neither is the goal to turn cockroaches into machines. They would resist both transformations. The real objective is to understand them or learn from them by participating in their world through the use of our imaginative faculties and to communicate with them at some level, which assumes there is a being on the other side of the relationship.

The Intelligence of All Brains

In traditional science, empathy is frowned upon and thought to distort reason because its influence often interferes with the expedient course of action. Empathy, however, is a critical element to understanding ourselves and others. Scientist-initiates, who are shamed and mocked when they empathize with other species and express concern for their lives, typically must dissociate from their feelings or leave the field.

Fortunately not all researchers succumb to the pressure of trying to appear totally objective, and a few even manage to escape the spell of the culture's shadow—like biologist William Jordan. When Jordan was a graduate student, he was responsible for the department's cockroach cultures. His close observation of the activities of a variety of species triggered questions about their intelligence and similarities to humans. What he observed didn't line up with the official version of cockroaches as survival machines. He saw them as intelligent beings adapting to their life circumstances with flexibility and innovation.

He noticed that many common behaviors in cockroaches were also behaviors in people. He reasoned that because humans have evolved from animals, it was only logical that our human mind is an extension of animal urges, that is, that reason was driven by emotion, not the other way around. Like all animals, humans spend the majority of their lives pursuing food, territory, social position, and mates. Even qualities like courage arise from primitive parts like the urge to resist domination by another.

So Jordan decided that all brains are intelligent and generate behavior remarkably similar in dealing with the basic tasks of life. In humans, our conscious mind merely overlays our instincts or innate intelligence and watches them operate, making connections among memories, mechanical comprehensions, emotions, urges, and the like, which we then experience as rational thought.

The late Willis Harman probably would have agreed with Jordan. He said that goal-seeking and purpose in other species (dismissed in traditional biology as our tendency to project human traits onto nature) appears to be a universal process through which Spirit contemplates and directs its own evolution.

A number of current studies also support Jordan's ideas, including research that reveals the primitive roots of human sexual behavior. Sex-attractant pheromones or chemical signals, which have been proven to play an important role in tens of thousands of species (including cockroaches[18]), also operate in us.

Although human beings have many expressions to describe the chemistry between two people, research cited by entomologist Gilbert Waldbauer in *Insects for All Seasons* indicates that pheromones are probably the basis of the physical attraction. In one study women were shown to be physiologically affected by a pheromone that can be extracted with alcohol from the hair tufts in the armpits of men. The study of human pheromones, although still in its infancy, has many convinced already that humans produce chemical aphrodisiacs that initiate and sustain sexual behavior.

For Jordan, the long hours he spent observing the university's cockroach communities made a profound impression on him. He decided that since evolution's basic goal is survival, the final goal of that process would be eternal existence. The ancient cockroach, around in recognizable form for eons, is approaching that eternal state. Add the fact that the basic purpose of the brain is to aid survival and, by that criterion, the cockroach is without doubt "divine genius."[19]

The Masters of Living

Fascination might temper our responses to the few species of "divine genius" that elect to share our homes if we taught our children what entomologists have actually discovered about the cockroaches of the world. The cockroach is a virtual master at living, having evolved behaviors and survival strategies specific to all climates except the arctic regions. Some species native to Asia even spend part of their lives in water. But most live in the equatorial belt. In fact, in dense rain forest areas where there are no honeybees, cockroaches help pollinate plants.

Cockroaches are also important in their ecosystems because they remove and recycle dead plant and animal matter from the forest floor and are food for reptiles, amphibians, fish, birds, other insects, and many mammals like monkeys and bats.

Built-in Warning System. Scientists credit the cockroach's evolutionary success to two appendages—the cerci—located at the posterior end of the insect's abdomen. The cerci alert the cockroach to breezes and other events too far back to be sensed by their antennae. This built-in dual alarm system also alerts them to danger, ahead or behind, so they can flee to safety.

To complement this warning system, sensors on their feet and antennae also pick up vibrations on the ground and in the air so they can feel when something is coming their way. In fact, cockroaches are known to predict earthquakes, by a significant increase in their activity, long before people feel any movement.

Cockroaches can also hear sounds over a broader range of frequencies than humans can, and big compound eyes with a crystal-reflecting layer make them sensitive to the smallest shard of light and keep them safe under a cover of darkness. Biologists who study cockroach cultures agree that it is nearly impossible to sneak up on a cockroach, regardless of the hour. They are always stroking the air with their delicate and refined antennae, as though divining the news from molecules in the air.

Generalists of the Insect World. The versatile cockroach has survived, not by finding one tiny, ecological niche, but by being unspecialized, like us. When they were forced into our early biosatellites, they withstood the confinement and the blast into space with apparent ease. They have even survived our radiation tests. When researchers gave an American species 830 rads (a rad is a standard unit of absorbed radiation), they appeared unaffected and survived to die of old age. The average dose that killed them was approximately 3,200 rads (by comparison, half of all humans will die after receiving a dose of 250 rads).

Ironically, as we persist in hurting ourselves with toxic chemicals, the cockroaches' ability to resist pesticides continues to baffle and amaze scientists and the public alike. Peter C. Sheretz, a teacher at Richmond's Virginia Commonwealth University, reports that the cockroach can dine on lethal cancer-causing toxins without any harmful effects and hopes this finding may help us discover a cure for cancer.

Those who study cockroaches agree that almost everything about these insects is a lesson in survival that goes far deeper than their adaptation to our presence on Earth. Instead of feeding cockroaches to the hostile imagination, cursing them for surviving and imagining them as adversaries, we would serve ourselves and them better by cultivating a genuine admiration for these remarkable creatures. "You're a veritable cockroach," we could say to one who has survived despite great odds.

Relating to Giant Madagascar Hissing Cockroaches

One survivor and an embodiment of divine genius that could help us bridge to other insect species is the Madagascar hissing cockroach, a three-inch-long insect with an air-punctuating hiss that startles predators from twelve feet away. The shell of a Madagascar cockroach has the color and look of polished wood, and people with private breeding colonies sell them as novelties. Many end up in colorful plastic houses entertaining humans for a short period of time, although their life span under better circumstances is about two years.

With a tongue-in-cheek consoling tone, the author of the children's book on cockroaches mentioned earlier concludes his book with the statement that cockroaches are here to stay and so is our dislike. But the child reader is assured that it might be worse, because living on Madagascar is a cockroach that makes loud hissing noises: "Aren't you glad it doesn't live with you?"[20]

How many children reading this and innocently absorbing its prejudice will be able to open to the possibility of a relationship with a Madagascar hissing cockroach? They will have little capacity to enter their world empathically. They are likely to accept the common belief that there is no consciousness inside a cockroach body, no being on the other side of the relationship, and thus operate from the assumption that relationship is just not possible. But is that true?

Communicating with Cockroaches. In *Animal People*, psychiatrist Gale Cooper introduces us to people who have relationships of unusual intensity with animals. She includes people who have ignored the superficial differences between humans and other species and have concentrated instead on making connection. One of the memorable people Cooper introduces us to in this delightful book is Geoff Alison.

Blind from infancy, Alison reports that he has always had a sense of connection with insects, feeling their delicate movements with an equally delicate touch. As a child, he was gradually introduced to the animal world in his backyard when his brother placed insects in his hand. All along through an assortment of pets, he was very aware of each as a living entity and empathically felt their misery when they were in an adverse situation. He explains his affinity this way:

> Being blind made me connect to animals through touch, and it put me more at their level. The visual power that human beings have allows them to analyze and act on an animal quickly, and it puts the animal at quite a disadvantage. I am in much less of a power level.[21]

When he turned eighteen, a friend gave him four giant Madagascar hissing cockroaches. In time, he got to know them as individuals with distinct temperaments. Given identical situations, they responded differently. Some were born curious, others cautious. Another one was always fearful and still another demonstrated a remarkable memory.

According to Alison, when someone first meets a cockroach, the insect's response level is less than when they become accustomed to the person. Aware of their need to feel surrounded and have a firm surface beneath them, Alison begins by picking them up in a way that doesn't force them off the surface they're already on. Then he tries to move his hand to conform with how they are moving. It has been his experience that regardless of how they were feeling when they were picked up, there comes a point where they start exploring. After exploring, they relax completely.

Intuitively, Alison sensed that for these creatures every step, every move of their antennae, is a stimulus that kindles a kind of amazement or wonder in them. His intuition arises from his willingness to participate in their worldview.

Clearly in touch with his affinity for insects and having structural dissimilarities minimized by his blindness, Alison was available for close and constant interaction with his cockroach community. He learned that they formed friendships outside the mating bond and passed the responsibilities of leadership from elder male to younger male in an orderly communication that lasted several days. For example, in the days before the current leader died, a young strong male stayed with him. They curled up together, walked together, and it was usually that male which took over as the leader when the current leader died, standing guard over the rest from an elevated position. To Alison their behaviors were not exceptional when we are aware of the intelligent awareness that permeates and directs all species.

The Death Pose. The death pose of cockroaches—on their back when death comes naturally—has long puzzled scientists. In David Gordon's *The Compleat Cockroach*, Cecil Adam, the creator of a newspaper column "Straight Dope," recants some popular ideas on the phenomenon. When asked why dead cockroaches are found on their backs, he suggests that cockroaches may suffer heart attacks when climbing a wall and the aerodynamics of their bodies are such that when they fall, dying, they land on their backs. He also suggests that when they die by ingesting an insecticide, the poison causes them to twitch and flip themselves onto their back where they can only "flail helplessly until the end comes."[22] A third idea proposed by Adam was that after they died, a passing breeze blew over their dead bodies.

An entomologist commenting on Adam's theories said that all three explanations were possible, although he said that cockroaches can't have heart attacks. He also added the not-so-helpful information that cockroaches don't always die on their backs. When they are poisoned, they may only be able to crawl back to a covered place before dying in a legs-down pose.

Contrast these hypotheses with Alison's discovery that cockroaches die in stages, progressing into a gradual euphoria or death state reminiscent of a yogic state in which all involuntary body processes are controlled. In the first stage, the cockroach stops eating and moving. It appears completely relaxed yet totally unrelated to the world. As the gradual disengagement with the physical world proceeds, the cockroach voluntarily rolls over on its back, legs limp (maybe preferring to have the weight of its body off its legs). It may lie that way for several days before it actually dies.

Communication with Cockroaches

Adversity often brings opportunities for relationships that otherwise might not occur. Alison believes that the potential for a psychic bond between any two beings always exists, but unusual circumstances often bring it to the foreground, as it did when a cockroach in his colony became disabled. She was an old female, and she lost her feet to a fungus infection. Alison gave her special care. When he brought food for the others, for instance, he gave her the best of everything and put it next to her. He noticed that the other cockroaches were careful of her, giving her a place to sleep undisturbed by their activities. In time she learned to trust that her needs would be met. Alison could feel her calmness and a certain kind of "shared mutuality."[23]

Alison also felt she rose above her essential instinctual responses to meet him. She would telepathically communicate to Alison about specific things like when he brought her food she liked, but also about less concrete things like she enjoyed his company and was aware that he was responding to her situation.

When the time came for her to disengage from the world, she signaled Alison. She was already on her back when he entered the room. He picked her up gently and she conveyed through mental pictures that the time of death had arrived. She was starting to vibrate or express through the whole of her being a pitch rising in volume. So, before that compelling process drew her in further, she wanted to say good-bye to him. She conveyed that it had been good being together. It was her last communication. She became limp and died several days later at two years of age.

Although Alison's relationship to insects exists outside our cultural models, his is a story that could open the possibility of relationship. It is a story for the children of our time.

Connections. Stories of positive connection between humans and cockroaches are few. But occasionally one emerges, although it is usually published as a novelty. For instance, in 1938 a prisoner in an Amarillo, Texas, jail told of how he had trained a cockroach to come to his solitary confinement cell when he whistled. In 1995, the *Weekly World News* featured another inmate incarcerated in a southern jail who had a pet cockroach to whom he fed bits of cheese and whom he walked around his cell using a tiny thread as a leash. It only became news when a guard killed it and the prisoner sued the prison.

Brenda Marshall, editor of *Light*, the journal of the College of Psychic Studies in England, and her husband enjoy a close communion with other species and believe that not only is the animal kingdom open and welcoming of contact, but that compassion toward them and an ability to coexist in harmony is a prerequisite for knowing ourselves. "That animals recognize empathy is well attested," she says.[24]

When Marshall and her husband were in Brazil for a few weeks, they ran exceptionally hot water in the bathtub one morning and a large cockroach popped out of the drain. After that, they were more careful about keeping the water tepid. One night not long after, they were sitting in the living room, which was separated from the bathroom by a long corridor, when through the doorway the cockroach appeared. It walked straight across the room to Marshall's husband, climbed up the chair, and stayed there. He fed the insect and discovered it liked tiny pieces of hard-boiled eggs. During the rest of their visit the insect would often join them on quiet evenings.

Mother Hathor, an English woman and one of the original members of the Foundation Faith, a self-styled religion with headquarters in New York, became conscious of telepathic communication with animals when she lived on a coconut plantation in Xtul, Mexico, with religious friends. Her first communications were with the six dogs they cared for. In time she also communicated with other creatures living there, including cockroaches, which she found to be "friendly creatures" who always accommodated her requests for them to leave. When she later lived in Chicago, Illinois, a cockroach became a regular visitor to her bathroom. The insect would appear and then wait—as though waiting for her greeting. Once she acknowledged him or her, it would move away again.

Taking Heart from Cockroaches. More stories exist of people gaining self-insight, assurance, strength, and courage by watching cockroaches. One insect enthusiast undergoing chemotherapy for breast cancer even chose the cockroach in her accompanying visualization therapy as the animal to seek out and transform the cancer cells.

Eddy Rubin, a young man whose parents placed him in a mental hospital when he was fifteen, writes in *The Sun* magazine that he spent the first couple of evenings in a padded cell. Stripped of all his possessions including his glasses, although he was legally blind without them, he could hear taunts from other patients. After some time, he noticed a movement out of the corner of his eye, a brown blur. He got closer and closer until he could see the cockroach. Although he'd never liked cockroaches before, he was grateful for the company. He followed the cockroach around the floor, watching it run up and down the mattress, clean itself, and search for food. Suddenly it disappeared into a small crack in the wall. It had escaped. Rubin knew inexplicably that it had modeled an important lesson. If the cockroach could discover a way out of apparent imprisonment, so could he. And he did.

Elisavietta Ritchie writes of a lonely year in Malaysia during which time her marriage of twenty-three years failed. Through a succession of long evenings as she sat up late at night writing, a cockroach appeared and stayed near her. The insect seemed to observe her, or perhaps it was merely keeping her company. She "admired his ability to decamp when threatened; run; survive in a crack; be patient; hide from the searchlight of sun, and then later, resume your station."[25] And she felt a sort of gratitude for its presence.

A Lesson from a Cockroach

In one of my classes on "Thinking Like a Bug," after a general orientation and discussion about how to enter a positive and respectful relationship with insects, I scheduled a young arthropod enthusiast to come and bring a variety of creeping creatures. He brought a giant Madagascar hissing cockroach, among others. As he talked about the cockroach, my class clown Brian asked with one eye toward me, "If I step on it, will its shell crack?" The class laughed, giving Brian the response he wanted. I interrupted the speaker to admonish Brian for the way he asked his question about the strength of the shell. Fully aware of what he had done, he just grinned sheepishly and assured me he was just kidding around.

After the presentation, I decided to take the cockroach, called Cedar because of his coloring, around the room while the guest speaker let some of the children hold his tarantula. Most of the children wanted to pick Cedar up, but I suggested that we could best demonstrate our respect for him by letting him choose to be held or not. I asked them to hold their hand palm up next to mine. Cedar, at ease in the palm of my hand, might choose to walk over to their hand, or not. They liked that idea and lined up for a turn.

Brian was first in line. I told him since he couched his question about the strength of Cedar's shell in an aggressive manner, Cedar might not go on his hand. Although I had shared the story of Freddie the fly with them, I could tell that Brian didn't really believe that the cockroach might have been aware of his attitude—just as Freddie the fly seemed to know that one of his visitors didn't like flies and routinely killed them.

Brian held his hand, palm out, next to mine. Cedar crawled to the edge of my open palm until his antennae briefly touched Brian's hand. Then Cedar withdrew and turned around, walking away. Brian was disappointed and suddenly not so sure that I hadn't been right. He stayed and watched as child after child came up and put his or her hand next to mine. The cockroach moved slowly, without hesitation, onto each child's hand, exploring.

After a short while, Brian was back in line, asking for another chance. I agreed and again he put his hand next to mine. Again, Cedar moved to the edge of my hand, antennae moving. When he reached Brian's hand, he stopped and turned around, moving away. Brian was visibly upset now. I suggested that he make his peace with the insect by sending him his sincere apology and desire for connection.

More children lined up for a chance to have the cockroach on their hand. Each time he obliged. Then one little girl approached, exclaiming: "Ooh! He's ugly, he's ugly!" Nothing I said got through to her or stopped her from reciting this shrill chant of distaste. But despite her words, she held her hand next to mine. Cedar refused to go on it—wouldn't even move toward it. She left without absorbing the fact that the cockroach acted as though he knew she was insulting him, for the insect crawled without hesitation onto the hand of the next child in line. It was a powerful lesson for all of us.

When the class period was almost over, I told them we had to stop. Brian begged for one last chance to have Cedar walk on his hand. He said he had been really working on his thoughts and was ready. He held his hand next to mine. Cedar again moved to the edge of my hand and then crawled onto

Brian's hand. Judging from the look of pleasure and triumph on the boy's face, you would have thought that he had just received a great gift—and he had. This discerning insect had honored his efforts and let him know that trust was restored.

Experiences like this one in the classroom raise more questions than they answer. Perhaps they merely serve to orient us correctly. A feeling for cockroaches is both the path into the mystery of our connection with other species and the lantern that lights the way. There is much to know about these creatures and the resiliency they embody—secrets accessible through intuition and empathy, delivered like pearls when least expected.

7

The War on Sawgimay

> There is nothing in life too terrible or too sad that will not be your friend when you find the right name to call it by, and calling it by its own name, hastening it will come up right to your side.
> —Koba in *A Far Off Place* by Laurens van der Post

"People who have not been in Narnia sometimes think that a thing cannot be good and terrible at the same time," wrote C. S. Lewis in *The Chronicles of Narnia*. Lewis could well have been talking about those creatures whose bite is painful or whose bloodsucking habits can transmit harmful parasites. It's easy to see only the terrible aspect of the Algonquin tribe's *sawgimay* or "small person who flies and bites so fiercely" and their cousins, the wingless biting and blood-dependent insects. To balance the scales and temper our responses, we need to know the good, and if not the good, the way to coexist with this aspect of creation and respond compassionately to them and perhaps ultimately to rejoice in them.

Life as a Battlefield. For starters, let's tackle the dilemma of sharing the world with these creatures by looking at how we translate the experience of being bitten or stung and our feelings about sharing our blood with them. Most of us view being bitten as an assault, an act of war. Our interpretation comes from viewing life as a battlefield between forces of good and evil. In fact, this perspective is the context for all wars between human beings, as well for all wars against other species. It is also the underpinning of the attempt by industrial societies to eradicate, albeit unsuccessfully, the mosquito and other bloodsucking and biting insects. When this view is dominant, enemy-making is supported in all its variations.

If we operate out of this context, we perceive the act of being bitten or stung as unprovoked aggression. Since this perception is congruent with our belief that insects are our adversaries, we don't question it. The pain or itching then invokes anger, and a desire for revenge. Projecting malice into a creature lets us feel justified in countering its attempt to bite us with an equal or greater violence.

The now familiar, heroic and aggressive stance so characteristic of our culture simplifies the situation. So does our readiness to divide all other species into good ones and bad, friend or foe. Once we've determined just who the bad species are, we can pursue them with a reactive vengeance. And we have.

Vengeance is Mine. Vengeance is the word that Jim Nollman, author of *Spiritual Ecology*, uses to describe his behavior when yellow jackets, attracted to the smell of fir sawdust, flew close to him when he was cutting a board with a chainsaw. That week his daughter had been stung three times by wasps, so when he saw them, he swung the saw at them.

After four or five blows, six yellow jackets were on the ground. Two were dismembered, two more were whole, but unmoving, their skins broken and vital fluids staining the soil where they lay. The last two were still alive, but stunned and quivering on the ground. More yellow jackets appeared and again he took the saw to them. Fourteen more lay on the ground broken and dead or dying. Soon more wasps appeared, but suddenly it seemed to Nollman that their buzz was different. They ignored him and flew to their fallen comrades, gripping their bodies and lifting them in flight. It looked as though they were coming to rescue their comrades.

Nollman imagined them taking their injured comrades back to the nest and administering aid to them, and this image prompted him to stop the assault. Only then did he admit that he has always known that his daughters get stung because they are frightened by the wasps and so flail at them with their arms—a "classic case of interspecies miscommunication."[1] And a typical scenario with which most of us can identify.

The Myth of Objectivity

A second context for translating the experience of being bitten or stung is closely related to the view of the world as battlefield. Its roots arise, however, from a paradigm that has separated us from nature and then fashioned a lifeless world of creatures, wired to respond to particular stimuli. It includes the myth of objectivity and the belief that everything can be explained in terms of linear cause and effect and understood by our rational faculties. This context strips intentionality and intelligence, however defined, from insects and reduces them to mindless and soulless life-forms programmed to pursue goals regardless of circumstances, other forces, or human behavior. So it is just unfortunate to be bitten or stung. The act does not invoke a particular response or invite the "victim" into deeper reflective waters. To kill or curse them is fine, but basically irrelevant.

This context also denies the possibility of a consciousness-directed universe. It denies an inspirited world, all noncausal symbolic connections, and all paradoxical and seemingly illogical relationships and observations that lend depth, meaning, and mystery to our lives. If we operate from this framework of beliefs, we must ignore a lot of evidence that contradicts it, including recent discoveries in quantum physics, an increasing number of experiences in interspecies communication, and our own intuition.

Nollman's interpretation of the wasps coming to the aid of their companions as an act of compassion implies some kind of insect consciousness and thus carries the stain of anthropomorphizing. When he later called on the expertise of an entomologist, he was told authoritatively, scolded even, that wasps do not care about one another. The expert patiently explained to Nollman that it is only chemical signals (pheromones) that determine their behavior, because their brains are tiny, and they are little more than robots. But Nollman knows what he saw and how he felt. He decided that if we insist that nonhuman animals cannot possess the attribute of consciousness because of the size of their brain, despite evidence from clinical death experiences that consciousness can exist apart from the physical brain, then we may be deluding ourselves.

Being Bitten or Stung as Punishment

A third context has benign roots and arises out of a genuine desire to live authentically. The growing minority who hold this view cultivate harmlessness and try to live lightly on the Earth. They may also seek, to one degree or another, communication and kinship with other species. So, being stung or bitten feels like a betrayal—or a judgment. Perhaps the insect perpetrator has judged the human victim and found him or her lacking. The pain is punitive then, noting and punishing some unknown transgression. "But I always liked insects," they will exclaim, confused and upset by a painful or irritating encounter.

Many native tribes also viewed the sting or bite as a judgment—usually a transgression noted or a warning. But they didn't believe the bite was meant to cause the person to collapse in remorse or self-blame. Rather it was a call for the recipient to reflect on the situation and make a correction. For instance, they might ask themselves if there had been a violation of right relationship and if so, what needed to be done to rectify the situation? This kind of reflection would move the recipient of a bite from confused victim to someone willing to take responsibility for upholding his or her end of the relationship.

Another way to look at being bitten or stung is to view it as part of a larger process. We simply can't assume that if we just act right, we will not be bitten—that is too simplistic. Our conscious intention and behavior is only part of the equation and not always the final determinant of an encounter with another species. To think otherwise is to indulge in wishful thinking.

The Trickster. Good intentions and affirmations will not save us, for example, from the disruptive energies of the Trickster, a mythological character and element in our psyches that routinely upsets the plans of the personality. Recognizing the energies of the Trickster is important, for we all experience its influence at those times when our best intentions are thwarted, and we are thrown into the anxiety that accompanies unwanted events. The Navajo people acknowledged these forces and personified them in their Insect God, a Trickster named Beotcidi, who was in charge of stinging insects. Irreverent, erotic, and outrageous, Beotcidi brought into people's lives a chaotic or disruptive element. As a universal figure or archetypal pattern common to all individuals, Beotcidi hovers on the border between the conscious and the unconscious. As a messenger of the potential wholeness of the psyche, he personifies the energies of the shadow and the unconscious. This idea of creeping creatures as initiators who trick the complacent human being into growing is a theme we will return to when we look at the spider and scorpion.

Given the limitations of the three contexts through which we typically interpret being bitten, it's fairly obvious that we need a new one. The first two traditional views of other creatures as malevolent or robotic simply prevent goodwill and impede understanding and cooperation between our species and others. Even the third context, which is a big step away from overt abuse of other species, still leaves us feeling victimized, with our willingness to coexist peacefully negated by the bite. A new context must emerge from our understanding of the interdependence of all species, a recognition and acceptance that we are food for some other species, and that some pain and discomfort is a necessary part of living. It must also come out of an evolving maturity and psychological sophistication that helps us acknowledge the elusive forces that move into our lives and push us to grow.

Other Views of the Sawgimay

Buddha suggested that if we consider all living beings as our mothers from previous lives, we will be more likely to generate love and compassion toward them. So if a hungry mosquito comes for us, we are advised to consider the kin relationship and act accordingly.

In accord with this perspective is an anecdote about bed bugs (another blood-loving, biting species) from a letter written to the editor of the *Pakistan Times* in 1985. The irate letter writer had recently ridden an express train during which he was bitten by bed bugs that had infested the train. He writes,

> On being tortured by the bug's bite, I started killing the insects. A shrewd passenger said to me, "Why are you killing these bugs?" I retorted: "They have sucked my blood." The man said, "Well, human blood is an object of love for these bugs and they are thus, our kids." I was constrained to deduce that this very idea is, perhaps, preventing the Railway authorities from taking any remedial measure to eliminate the bug menace."[2]

We are apt to agree with the exasperated letter writer. Thinking of mosquitoes as our mothers from a previous life and bed bugs as our offspring are not acceptable points of view for most Western individuals. We don't trust the compassionate view, thinking perhaps that insects will take advantage of us if we give them any latitude at all.

Native Views of the Sawgimay. When we look at how indigenous people understood and dealt with being bitten, predictably, we uncover a more compassionate view, a product of the nonhostile imagination and a willingness to acknowledge other species' worldviews as well as the shortcomings of human beings. In many native stories, for instance, mosquitoes originally helped humans, but when the people forgot their responsibility to help other species and use the resources of the natural world in a balanced way, mosquitoes were given the ability to bite us to remind human beings of the imbalance.

In tribes like the Kwakiutl of British Columbia, biting creatures (including the bee, wasp, midge, mosquito, and gnat) play a role in creation, and the Kwakiutl carved wooden masks to invoke the spirit of each insect. Wearing the masks in ceremonies revealed the person's ancestral links while indicating kinship, rank, and privilege. The primary function of the masks, however, was to connect the Kwakiutl with the powerful supernatural insects and other animals that once ruled the earth. The wasp masks, for example, represent a clan whose members are linked by heredity to an ancient wasp being. Knowledge of this kinship and contact with this being helped participants identify with local wasp species.

To be stung by your ancestral kin was viewed as a transmission of power, a warning, or a call to action. The tribe's council of elders usually determined what being bitten or stung meant in any particular circumstance and whether a response was necessary. So pain didn't prompt a counterassault, nor was it treated as a random act or as a judgment. It served to alert the person and initiate a thoughtful reflective process.

Stories from other tribal societies also teach respect for biting creatures by reminding members of how the biting talents of local species were enlisted in times past to help the tribe. In the creation story of the California Miwok tribe, Fly assists the Miwok in defeating a cannibal giant by biting the sleeping giant all over his body until he found his vulnerable spot. Then he flew to the people and told them where the giant was vulnerable, so they could set a trap for him. A story like this created good feelings. When someone from this tribe is bitten by a fly, the experience is tempered by the knowledge that the creature once saved the entire tribe.

An Unholy Joy. Contrast this attitude with naturalist Steward Edward White's view of biting flies. In his 1903 book *The Forest*, he calls killing black flies a "heartlifting…unholy joy…[that] leaves the spirit ecstatic…[3]

> The satisfaction of murdering the beast that has had the nerve to light on you just as you are reeling in almost counterbalances the pain…."[4]

In this statement, White upholds the dominant viewpoint of the last three hundred years. The vindictiveness of this stance and the tinge of self-righteousness identifies it as a reaction emerging from the belief in the world as a battlefield, and we can see the same reaction today in a variety of situations. For example, in an award-winning television commercial advertising Tabasco sauce that aired during the 1998 Super Bowl, a man who has just smothered his sandwich with the condiment and is relishing the intensity of the heat in his mouth watches without concern as a mosquito lands on him and takes a blood meal. When the mosquito is finished, it takes to the air. A few seconds later it explodes while the man smiles wryly and applies more sauce. And in an article published in *Discover* magazine on why mosquitoes suck blood, the author concludes the piece by saying if a mosquito "picks on us" we can always get even by using a trick she learned from a researcher. It involves stretching our skin taut around a feeding mosquito so that it traps the mosquito's feeding instrument or proboscis. When it can't withdraw, it is forced to suck until it explodes.

These kinds of reactions are typical responses from individuals raised in a society that has embraced the myth of superiority and dominance and cast its repressed hostilities over the natural world. They are also responses that have dire physiological consequences.

Carolyn Myss, a new voice in the field of energy medicine, says that the energy of vengeance, of "getting even," is one of the most toxic emotional poisons to our bodies and causes afflictions ranging from impotence to genital cancer.

Many clinical studies support Myss's contention that toxic emotions threaten our health. Clinical evidence for a direct physical pathway that allows emotions to impact the immune system is growing. Chronic, intense, and negative emotions are known to cause stress hormones to flood the body. And when our dominant approach to life is antagonistic, that is, when our personal style is hostility, mistrust, cynicism, temper flare-ups, and rage, we have "double the risk of disease—including asthma, arthritis, headaches, peptic ulcers, and heart disease."[5]

The Need for Respect

A context that honors our interdependence with biting insects must acknowledge that relationships are reciprocal and that respect is an essential ingredient. So as we noted previously, when we are bitten by an insect, it behooves us to evaluate our behavior and determine if we've been respectful in our actions. Stumbling into a wasp's nest and being stung, for example, should be taken not as punishment but as a clear message to be more careful. In other situations, when our actions have been circumspect, being bitten or stung might be a signal that something in the larger environment is out of balance.

Native people routinely engaged in this kind of self-reflection, and, when they determined that they had acted in an inappropriate way, they promptly offered an apology, said a prayer, or performed a ceremony to return the relationship to harmony and good will. As we saw with fourth-grader Brian and the cockroach Cedar, an apology and sincere desire for a continued relationship may be necessary to restore trust between two individuals—regardless of species.

Big Biter. To the Eastern Canadian tribe called the Montagnais, the overlord of fish, particularly of salmon and cod, was a fly known as Big Biter. This species of fly would appear whenever fish were being taken from the water, and it would hover over the fishermen to see how their subjects, the fish, were being treated. Occasionally, Big Biter would bite a fisherman to remind him that the fish were in its custody and to warn him against wastefulness.

Indigenous people were willing to be held accountable for their interactions with the natural world. It was necessary for survival and for continued receipt of the benefits of their plant and animal kin. Big Biter's bite was viewed as a communication, and native people tried to understand it and correct their behavior if necessary.

Today, Big Biter is likely to be sprayed or swatted if it comes near fishermen. Our culture doesn't require accountability, unless it's economic accountability. As a result of overfishing, coho salmon was placed on the threatened species list in 1996. Fewer than 6,000 wild salmon are left in an area that had an estimated 50,000 to 125,000 five decades ago.

The legendary steelhead trout is also an endangered species. In thirty years, it has gone down to ten percent of its original numbers because dam building, logging, and gravel mining have damaged the streams where the fish spawn.

If we were to adopt the native practice of holding ourselves accountable, it would help return accord to the web of relationships of which we are a part. For both the Miwok and the Montagnais tribes, biting flies were understood and accepted in a nonadversarial context. These people would, for example, recoil at White's words and actions and consider his behavior and attitude a horrendous act of disrespect, punishable by having the fish avoid his hook. We scoff at those notions. We are too sophisticated to believe that the natural world can hold us accountable. Our science is too advanced. Yet the same science has finally evolved enough to verify that we are interdependent with all other lifeforms, and that insects are the stream keepers, the messengers of the level of biotic integrity. Mistreatment of any part of the community will, in time, adversely affect us.

It doesn't really matter whether the current decline of salmon and other staple species like the sturgeon is related to our overwhelming disrespect toward insect species or the result of polluting the streams and overfishing. The tie to insects is an imaginative truth, a way of acknowledging our interdependence and the insect's role of messenger. Letting them prompt us into self-reflection, we might become our own critics, evaluating and righting our relationships to the finned creatures before our disregard results in further imbalances and the demise of still other species.

Our conditional reverence for life continues to hurt us. As noted previously, the Endangered Species Act is the primary regulatory law that upholds our egocentric view of who is valuable and who is not. It allows companies and individuals to eliminate pieces of the natural world that compete with us for our food and for our blood. It is not surprising then to find that among those condemned are mosquitoes.

The War Against the Mosquito

Mosquitoes are on the receiving end of our hostilities—most of it never questioned. Over one thousand agencies in North America collectively spend more than $150 million each year on mosquito controls, including the use of pesticides that directly or indirectly harm mosquitoes' natural predators and other nontarget species. Thousands of acres of swamps, drained and filled in the name of mosquito control, sentence the creatures that live in the water, like dragonfly larvae, to death along with those that depend on mosquitoes for sustenance.

In the United States the mosquito control effort was not initiated to curb the population of the kinds of mosquitoes that carry malaria or other fly-transmitted parasites. We have poisoned our streams and wetlands because we fear *all* mosquitoes and the possibility of being bitten excessively. In fact, through property taxes and parcel fees, we pay for mosquito control without having to make a conscious decision about it, giving tacit license to regulatory agencies to eradicate the species that inhabit our region.

Ironically, the primary source of large numbers of biting mosquitoes has not been addressed, at least in California, and no doubt elsewhere. Entomologist Richard Garcia points out that agricultural practices have created most of the mosquito breeding sites. For example, the 600,000 acres of rice fields in California are the source of two or three species of biting mosquitoes considered pests. In these fields where hybrid crop plants require intensive irrigation, the water doesn't drain off but collects in small pools. The pools aren't permanent enough to support aquatic predators, but they can and do breed tremendous numbers of mosquitoes.

Vector Control. Each year the war against the mosquito is justified in newspaper and magazine articles whose propaganda-like strategies warn us of the potential danger of a burgeoning population of mosquitoes. After arousing us with sensational headlines and copy, they assure us that vector control agents have everything under control and our best interests at heart. And we believe them.

In my area, news articles report each spring that we are due for a bad season of mosquitoes. Some years they blame the drought and the increase of small stagnant pools of water along dry creeks and rivers that provide breeding grounds. Other times they blame the rainfall. This year it is the El Niño rains that supposedly will result in excessive numbers of biting mosquitoes.

The author of a newspaper article called "Bugmen Slap the Beasts Before They Bite" sets the stage for the ensuing combat by telling readers that recent rains have encouraged the mosquito population to double. He warns that the salt marsh mosquito poses a threat to shoppers at the local mall, as well as to hikers, fishermen, and people coming to the area to watch fireworks. Although he admits that only two mosquitoes were spotted at the mall the previous year, he points out that nearby streams are filled with the larvae of the region's twelve species. And although he is careful to note that none of these twelve species carry malaria, he implies that their existence is the offense. The fact that they might bite us is apparently enough to sanction the war.

Ending on an upbeat note, the reporter reassures us that the Mosquito and Vector Control agency—a group of entomologist heroes whose stated objective is to kill all mosquitoes in a 360-square-mile area—has the situation under control. Although their weapons in this instance are not toxic chemicals, but pellets steeped in a deadly bacterial solution and inch-long mosquito-eating minnows (called mosquitofish), the point is that the pellets and predator fish will be used as weapons to eradicate all twelve species. And when they make public the battle plans, nothing is mentioned about what will happen when we rupture the food chain by eliminating the larvae of twelve mosquito species and introduce a new predator, or what will happen to the species that depend on mosquitoes for food.

Mosquitofish Tales. The mosquitofish, a native of southeast United States, is a minnow that feeds near the surface of the water. Its appetite for mosquito larva makes the use of this fish sound like a great solution to the anticipated problem of too many mosquitoes. Yet its introduction into a new environment has been shown to create new problems. In 1905 authorities shipped the fish to Hawaii as an experiment in mosquito control. In that state it not only ate mosquito larvae, but it ate harmless insect larvae and the young of desirable commercial and game fish. In India an experiment using mosquitofish to control mosquitoes also failed when this predator ate the local fish that also consumed millions of mosquito larvae. And the history of its use shows that where predatory fish or birds are scarce, the mosquitofish overpopulates new habitats.

When our hostile imagination is in control, we don't check the facts. We're too willing to believe that the experts know what they are doing. Militant headlines like this one from my local newspaper, "Swat Teams Rev Up Battle Against Larvae," arouse us; so does copy like "Prepare for the annual onslaught on the human flesh. There will be more of them than ever."[6] Then having agitated us, reporters reassure us—as if we were children—that "Swat teams...are mounting an aerial attack a month earlier than usual."[7] Unaware that this is propaganda, we throw caution aside and vote for all kinds of programs that promise safety from those that might assault us.

Florida's Mosquito Control Program. California's news reports on mosquitoes are not so different from those in other states where mosquitoes thrive. A Florida news report warns the reader that the mosquito eggs that lay dormant during the drought are now hatching with the advent of rain, and the mosquitoes are "almost as big as alligators and ten times as ferocious."[8]

At the Florida Mosquito Control center, scientists are exploring solutions other than pesticides for killing mosquitoes. They have finally learned that insecticides can't be sprayed in sensitive environments like the Everglades where most of the mosquitoes breed, as the chemicals kill too many beneficial insects. What authorities still seem unable to grasp is that mosquitoes are essential for the health and continuance of the Everglades. They are food for the fish of the Everglades who, in turn, feed the alligators, whose activities keep channels through the marshes open. Without mosquitoes the Everglades would grow over.

One biological control strategy gaining acceptance in Florida is importing mosquitoes ten times the size of the local mosquitoes. These giant, iridescent blue mosquitoes from Southeast Asia were expected to eat several species of freshwater mosquitoes found in Florida, and researchers released a test group in an area close to Miami. So far the only drawback is that these big ones have no appetite for the black salt-marsh mosquitoes that inhabit Miami and its region. What they do have an appetite for, and how that will impact other species, is not known.

Another strategy currently in the news is euphemistically called a "diet pill" for mosquitoes. It is a formula that alters the mosquito's digestion so it can't feed. When the concoction is placed in water where mosquitoes breed, the larvae eat it, starving to death within seventy-two hours. And again the role of mosquitoes in the food chain is ignored, dimmed by the clamor of the public for elimination of whatever might cause them discomfort.

Learning about Mosquitoes

We can't make educated decisions about mosquito control if we rely on those making money from the control programs to tell us the facts of either the species or the situation. We need unbiased information about mosquitoes, but it isn't a subject that people want to learn about. In fact, when *Natural History* magazine devoted an issue to the mosquito, one outraged subscriber wrote to the editor and cancelled his subscription, proclaiming hotly that the issue was an absolute waste of effort and of no interest to anyone.

By learning something about mosquitoes, however, we leave the black-and-white opinions of the culture and restore a desperately needed complexity to the adversarial relationship between ourselves and these species. If we start with the basic facts about mosquitoes, we enter unknown territory right away. Scientists don't know where mosquitoes originated. Their best guess is that they evolved in the tropics 100 to 200 million years ago.

Entomologists have identified almost 3,500 mosquito species, adding about 18 new species each year. Today, mosquitoes occupy every continent and seventy-five percent live in the subtropics and tropics, where a single square mile can support more than 150 species. Away from the tropics, the number of species drops dramatically. More than 240 species live in North America, and in the Arctic, which has less than a dozen species, the greatest concentration lives in its thousands of square miles of tundra pools.

A little-known fact about mosquitoes is that, like other flies, they are pollinators. In the Arctic, bog orchids are pollinated exclusively by mosquitoes. Mosquito researcher Lewis Nielsen has collected mosquitoes whose bodies were dusted with pollen grains from more than thirty species of flowering plants. Nielsen, one of the contributors to the controversial *Natural History* issue on mosquitoes, believes mosquitoes are more important as pollinators of wildflowers than anyone has previously realized.

If you consider that each plant species supports ten to thirty animal species, eliminating the insect that pollinates it not only destroys the plant, it destroys all the other insects who depend on the plant and hurts the predators of those insects. The web is intricate.

Wrigglers and Adults as Food. Many mosquito species are also important to the normal life of the pond or stream, where they begin their life as "wrigglers," swimming in the still water where the female mosquito lays her eggs. As wrigglers, mosquitoes are food for water tigers and other predatory insects and they also feed salamanders, ducks, and geese.

The mosquito wriggler becomes a pupa in a week to ten days. Looking like a tadpole with a jointed tail, the pupa tumbles through the water for a few days before turning into an adult. Biologist Ronald Rood likens the metamorphosis from tumbler to adult to transforming a jeep into a jet plane in two days while driving the jeep at sixty miles per hour. In short, it's an amazing feat.

Adult mosquitoes feed many organisms including minnows, dragonflies, game fish, warblers, swallows, and bats. Many mosquitoes also trap bacteria in stagnant water, and some catch and kill other mosquitoes. In Malaysia, certain blood-drinking moths depend on mosquitoes. These moths follow mosquitoes, lapping up whatever blood they spill, and one unusual moth actually pierces the skin of mammals to obtain blood.

Although we typically associate mosquitoes with their bloodsucking habits, plant sugar is their main food for flight and survival. In fact, many mosquito species exist only on fruit juices, nectar, and tree sap from broken branches. The larvae of an Ethiopian mosquito preys on the larvae of other mosquitoes, including the ones that can carry malaria, before changing into an iridescent adult that drinks only plant juices. The tree hole mosquito feeds on birds when it needs blood and does not bite humans. And a tropical species feeds on liquid that a certain kind of ant regurgitates, while other mosquitoes sip nectar, flying from flower to flower like bees.

Wraparound, compound eyes let mosquitoes see in almost all directions at once, and tiny ears found at the base of their antennae provide them with good hearing. They are guided to us even on the darkest of nights by a sophisticated sensory system that lets them detect the carbon dioxide that we and other hosts exhale. We know they are near when we hear their hum which is caused by their wings whirling at an amazing six hundred beats per second. In fact, males looking for females are said to cluster around a tuning fork of this frequency.

Mosquitoes also find us from the chemicals our skin exudes, and those chemicals are affected by our diet. One substance that attracts them is octenol, which was first detected in ox breath and was later found to be produced by grass fermenting in the stomach of these mammals. One theory says that an individual who eats a lot of greens produces excess octenol which in turn attracts mosquitoes to them. If it is true, these people might counter their attractiveness by eating garlic, a strategy that has worked historically for the natives of Tanzania. Recent studies also confirm that garlic, which is exuded in sweat, repels malaria-carrying mosquitoes in particular, although other species are not necessarily repelled by it.

Mosquitoes are also particular about where they bite a host. A Dutch research team discovered that some mosquito species like feet and ankles and others like our heads and shoulders. The ones that liked feet liked smelly feet and found their feeding site by tracking foot aromas. Those smells are in turn linked to odor-producing bacteria. And the bacteria that attract some species of mosquitoes are the same ones that give strong cheese its flavor.

Blood to Raise a Family

In some environments where hosts are scarce, a few mosquito species have evolved the capacity to generate eggs from protein reserves carried over from the larval stage. But most of them need blood. The hum of the female of those species that do take a blood meal sends us running for cover, activating our fears of being the object of what we perceive as blood lust. But if we were to imagine ourselves a female mosquito for a moment, we might understand that the need for blood stems from an instinctual push to gather the resources to start a family and perpetuate the species.

Without blood, egg production may go from one hundred eggs to fewer than ten, and in some cases to one—hence the persistence of the female mosquito. If we adopt the Chewong people's concept of med mesign, or different eyes, where the intent and behavior of any species arises from its worldview, it would encourage us to bring compassion and understanding to our encounters with mosquitoes. In malaria country, where mosquitoes may carry the malarial parasite, we could then choose, without undue anger or vindictiveness, to use netting and not allow a mosquito to feed on us. And in our own neighborhoods we could share our blood with an occasional mosquito or two without feeling victimized or angry, or, in the midst of many, use a topical treatment or extra clothing to dissuade them.

Owning Our Complicity. Although it is easy to feel like a victim when mosquitoes in our vicinity are looking for blood, the view of mosquito as tormentor and relentless pursuer of innocent humans ignores our complicity in creating the current state of affairs. Consider the fact that the handful of species that has developed a preference for human blood has only done so in the last couple of hundred years as their natural hosts, other warmblooded animals, have diminished. With the destruction of wildlife and the corresponding increase in the human population we have become, of necessity, a primary host for some species. The mosquito has fewer and fewer options as we continue to wipe out much of their ancestral forest habitat and the animals that they once depended on for blood.

A creative solution for both humans and mosquitoes will require a vastly different attitude on our part. As we stop the chemical onslaught—a necessity even before we have alternative plans—we must address the implications of our interdependence. In the 1950s, for example, when the World Health Organization (WHO) used the pesticide dieldrin to eliminate mosquitoes in northern Borneo, they didn't realize or didn't think it important at the time that it also killed local wasps and other insect predators. A week or two after spraying, however, the roofs of the villager's homes began caving in, eaten by caterpillars whose population had previously been checked by the wasps. Also, hundreds of lizards died from eating the poisoned mosquitoes, and then, when the village cats ate the poisoned lizards, they died too. Without the cats to keep the rat population in check, the numbers of rats mushroomed, and they ran around the villages carrying typhus-infested fleas. How many more disasters do we need to document before we really believe all things are connected and refrain from uninformed killing strategies?

Malaria and the Mosquito

If we evaluate our progress in finding the good in mosquitoes, we now know that mosquitoes pollinate wildflowers and serve as an important link in the food chain. Taking it a step further, we can agree that female mosquitoes are not vicious or malevolent, but merely bite us when they need blood to start a family. And perhaps we can also admit that we have unwittingly become host to some species because of our direct and indirect destruction of other species and their habitat. Taken all together, these facts may help to shift us into an uneasy acceptance of mosquitoes, but more is needed if we are to turn this relationship around.

As we circle the issues that snarl our connections with mosquitoes and block our feeling connection to them, we come to the big topic of insect-transmitted disease. Although the issues involved would take several books to explore thoroughly and are not readily dissected or disposed of in a linear progression, we must enter the territory and briefly look at the history of our war against mosquitoes and the current state of affairs. We also need to know a little about the malaria parasite and its connection to the mosquito.

Our schemes to eradicate the malarial parasite and its carrier, one species of mosquito, have failed. In fact, overall, despite advances in biotechnology, the health of tropical people has gotten worse in the past twenty years. In *Malaria Capers*, Robert Desowitz, a leading researcher of tropical diseases, reports that in many places malaria is more prevalent, less treatable, and less controllable than it was twenty or thirty years ago because of intensive spraying with DDT and overuse and improper use of antimalarial drugs. Worse, before our chemical intervention many adults in tropical regions were clinically immune to malaria. Today they have lost that hard-earned protection and are frequent victims of malaria's intermittent fever. The threat to young children from acute, life-threatening malaria is also greater because of the recent appearance of drug-resistant strains.

Health officials in every country where malaria occurs are resorting to anti-malarial drugs like quinine (derived from the bark of the South American cinchona tree) and a synthetic product called chloroquine that can keep malaria from being fatal.

In 1972, the WHO formally declared the war to eradicate the mosquito a failure, and many involved feel a sense of defeat. Some are resorting to blaming and making excuses for the fact that malaria has returned in full force. A small minority realize that defeat can be instructive. Political historian Gordon Harrison suggests that such widespread and pronounced failure is trying to tell us something—either "we have proved ourselves incompetent warriors [or]...we have misconstrued the problem."[9]

The Bid for Land and Supremacy

The facts indicate we have misconstrued the problem. To begin with, if we look at the history of the first all-out war against another species—the mosquito—we encounter not a humanitarian crusade as some propagandists would have us believe, but a chain of events motivated by greed and riddled with widespread corruption.

The war against the mosquito begins at the peak of colonialism. Historians of this period report that its roots are not grounded in an altruistic concern for the well-being of tribal people in the tropics but rise out of racism and a common desire among colonists to have domination over land and people. The colonizing mind saw every land it conquered as empty (of people with value) so it didn't feel a need to respect anyone's rights—human or mosquito.

Equatorial Africa resisted settlement by the white colonists, as did tropical Asia and Central and South America. Heat and diseases like malaria took their toll on the foreigners. The number of colonist victims was so high that many believed that the hostile and "savage" possessors of the tropics, that is, the native people of those lands, had put a spell or curse on them.

This belief that they had somehow been bewitched was supported by their observation that malaria, so often deadly to Caucasians, scarcely seemed to affect the native people. In 1870, an Italian doctor speculated that these "inferior races"[10] shared the immunity to malaria of the lower animals. The apparent immunity of indigenous people then was used as evidence of a racial hierarchy in which Caucasians believed themselves to be first and thus master over all others.

Although the self-deceptions and theories about the cause of malaria continued until the malarial parasite and its mosquito carrier were identified, the threat to colonialism from malaria was the primary motivation for scientists at that time to focus on the disease and then try to eliminate the mosquito. Concern centered on keeping tropic-based administrative, military, and economic rulers alive and healthy. Concern for the native population was limited to maintaining a healthy workforce for commercial enterprises.

Battling the Mosquito and Malaria. Because the military needed protection from malaria, World War II renewed the motivation to find a malaria vaccine while continuing efforts to eradicate the mosquito. When DDT was developed in the forties, it looked like a gift from the gods.[11] We now know that it was at best a Faustian bargain.

The failure of DDT to eliminate the mosquito shifted the focus from prevention to treatment. People once again turned to chloroquine—although today, due to its overuse, it is no longer effective against many new resistant strains of malaria. A "new" antimalarial drug called qinghaosu is actually two thousand years old and is made from sweet wormwood, which has fever-reducing properties. It has also been found to cure certain deadly strains of malaria in humans more rapidly and with less toxicity than chloroquine or quinine. It has even been effective against chloroquine-resistant strains. Now chemically analyzed, it could be synthesized for large-scale production, which may occur shortly, now that global warming has expanded the malarial parasite's territory and brought the likelihood of contracting malaria closer to those in a position to fund this kind of research.

The Positive Side of Malaria

As terrible as malaria and its fevers could be, it was good for Alfred Russell Wallace, a nineteenth-century naturalist. Wallace initially contracted malaria in the Amazon. Later when he was living in Malaysia and suffering from a severe attack of its intermittent fever, he conceived of the idea of natural selection. It was as though the thought had been suddenly inserted inside his feverish brain. Unaware that Charles Darwin had already devoted decades to an unpublished work on evolution by natural selection, he wrote up the idea. Wallace's version was more spiritual than Darwin's version and reflected his own explorations and belief in a higher power, but he acquiesced to Darwin and later they presented Darwin's version of the theory together.

Malaria-caused fevers were also good for some cases of mental illness. It was known even in ancient Greece that sometimes the insane recovered their sanity after an attack of malaria. But it wasn't until after World War I that this information was used to create a program to systematically treat mental illness.

In Germany in 1905, Fritz Schaudinn discovered that the microorganism in the blood, brain, and tissues of people with syphilis couldn't survive temperatures several degrees above normal body temperature. Since the malaria parasite was a temperature-elevating agent, it was reasoned that the fever it caused could stop the downward progression of syphilis, and afterward, quinine could cure the malaria. Patients wouldn't necessarily get better, but after having malaria they would not get worse, and those with minor neurological disorders could return to a normal life. The reasoning proved to be correct. The success of Viennese psychiatrist Wagner von Jauregg in using malaria to treat syphilis earned him a Nobel Prize, and other nations, including Britain and the United States, set up special treatment units where tens of thousands of syphilitics were saved from a sure and agonizing death.

Susceptibility and Malaria

Some people are immune to some strains of malaria because of an abnormality of hemoglobin (the "working" constituent of red blood cells).[12] What the colonists interpreted as an immunity to malaria based on racial inferiority was actually strain-specific, naturally-acquired immunities, all of which involve genetic modifications that make the red cells less appetizing to the parasite.

Besides naturally acquired immunities to particular malaria strains, susceptibility is the single most critical factor in whether a person gets sick or not. Susceptibility factors also determine why one mosquito succumbs to the malarial parasite and another does not.

Few people appreciate the fact that a mosquito carrying the malarial parasite may suffer as much as the person or creature she bites. When a mosquito takes in a blood meal containing the malarial parasite, the microscopic invader penetrates the insect's stomach. Once in the mosquito's stomach, the parasite multiplies, forming cysts on the stomach wall. At this point the mosquito gets ill from malaria and sometimes dies. If the mosquito survives, the cysts break and release hundreds of new parasites. Some of these make their way to the salivary glands where they are transmitted when the mosquito gets her next blood meal. Although our war on mosquitoes is an attempt to eradicate *all* mosquitoes, only a few mosquitoes that have had a blood meal from a sick person carry disease parasites in their saliva and transport them to a well person.

As we saw with houseflies and disease, traditional Western medicine has fostered dangerous delusions about cause-and-effect relationships between diseases and agents of disease. We are taught, for example, that the malarial parasite is an agent of disease, and therefore, it is the cause of malaria. But, when the malarial parasite is injected by a mosquito into the blood of a susceptible person, the key word in the equation is not "mosquito" or "malarial parasite," it is "susceptible." Not everyone exposed to the parasite gets sick with malaria. In some individuals, the illness fails to develop, either because it is checked by the immune system, or for reasons yet unknown. Natural fluctuations in our state of health or balance may determine whether or not an agent of disease finds fertile ground in which to develop.

New Battle Strategies

Although we claim to have renounced the war against the mosquito and settled for control, our war mentality keeps us ever on the lookout for weapons. A battle is now being waged using a bacterium discovered in the Negev desert in 1976. The WHO is spraying this microorganism over vast areas of Africa, Asia, and South America, expecting it to kill malaria-bearing mosquitoes. The bacteria are known to kill black flies as well. Although lauded as a humanitarian milestone, the verdict is still out as to whether the mosquito or the parasite will mutate around the attack or whether the bacteria will take advantage of its new territory and feast on other species as well.

Altering the mosquito genetically is still another line of research. One group is trying to alter the mosquito so that it cannot inject the malarial parasite into a person. Another group is trying to alter it so it can resist the parasite. How the malarial parasite or mosquito will adapt to such manipulations has not been explored, and those concerns get buried or dismissed in the overwhelming enthusiasm that greets the announcement of this kind of research.

Yet another line of research involves administering a "benevolent vaccine" to people that will not prevent them from getting malaria, but will help prevent them from transmitting it back to a mosquito taking a blood meal. It works this way: When a mosquito takes a blood meal from a person with malaria who has been vaccinated, the insect will not only take in the infected blood but she will take in the antibodies that will protect her from becoming infected. This in turn, it is reasoned, will prevent small malaria outbreaks from reaching epidemic proportions.

Although this research is not motivated out of a concern for the well-being of mosquitoes, one has to wonder if a solution would have been forthcoming had our objective been to keep the mosquito from harm, as well as protect people from malaria's deadliest strains. Even more radical would be an all-inclusive reverence that would honor the balance among mosquito, parasite, and human beings, working on behalf of all three to maintain a balance where all thrive, and where no one lives at the expense of another.

Ending the War. The relationship between mosquitoes and certain parasites is a fairly stable one. Successful parasitism depends on each side keeping the other in check. Ecologically, the man-mosquito-parasite relationship is a stable and balanced system we have tried to upset, and can't.

Evolutionary medicine proponent Marc Lappé, introduced in the fly chapter, draws on the ideas of anthropologists, evolutionary biologists, and population geneticists when he argues that part of our inability to correctly formulate the problem of malaria and find appropriate strategies for treatment has to do with the pervasive belief that we are somehow apart from the forces of evolution. Lappé reminds us that "Disease is a part of life. Diseases do not arise fully armed to strike unsuspecting hosts. Patterns of disease and illness have millennial roots."[13] Modern medicine has not understood the complex interplay of disease, human and microbial coevolution, and modern therapeutics. We can no longer define progress and civilization as conquests over nature's diseases. Progress can come only when we understand and cooperate with the forces of nature, realizing that both we and mosquitoes are molded by the same forces.

This new perspective on disease results in radically different treatment approaches. For example, sometimes a "cure" may mean dropping control measures altogether and focusing instead on strengthening innate resistance. In large populations, the best prevention is to prevent deadly forms from being transmitted so they will evolve into a milder strain, and we can help the process by making sure—using netting, isolation, or any other practical method—that mosquitoes don't get to people who have malaria.

Protectors of Habitat

In *Fruits of Darkness*, Joan Halifax proposes that certain creatures and plants are protectors of place, allies to their habitats. In this view, the mosquito and other biting insects are animal guardians of the planet's rich resources.

We have been destroying the protectors of wild regions for a long time, justifying our actions and insisting that we are only protecting ourselves from harm. Intimidated and perhaps outraged by their power to hurt us, we have made such species the enemy, trying to eliminate whoever or whatever requires us to change our ways. And whenever we have been successful, we have lost the integrity of the ecologies these protectors guarded on behalf of all species.

Mosquitoes and other bloodsucking insects have been viewed by some as heroes of the ecology because for decades they made tropical rain forests almost uninhabitable for human beings, delaying the great destruction of these tropical forests that are home to an abundance of wildlife.

The tsetse fly is another insect protector who needs blood to lay eggs and sometimes carries a parasite that results in African sleeping sickness, to which cattle, horses, and donkeys are particularly susceptible. This insect has stopped the settlement of Botswana's lush Okavango Delta, a paradise for wildlife, but the central swamp is the tsetse fly's kingdom. Attempts to eradicate these flies are fueled by the desire to open the central wilderness area to the cattle industry—a questionable objective. Making the Delta safe for cattle and humans would allow towns to get the water they need to thrive, but it would destroy wildlife habitat in the process.

As of this writing, the tsetse fly has been wiped out in certain areas, like on the island of Zanzibar. Its demise is due to a new technique of using doses of gamma radiation to sterilize insects, which are then released en masse into the wild. I suspect the same tactics will be used in the Delta if they haven't already, and with the tsetse fly gone the wildlife it protected will follow.

The Myth of Progress

Most of us don't consider biting insects as guardians of habitats. We have traditionally considered them an affront, impeding what has been called progress. It's a viewpoint we teach each generation. The author of a juvenile book, for instance, presents the modern view of these protectors of place:

> Mosquito bites alone are reason enough for constant war on these pesky insects. Great swarms of them have delayed agricultural or industrial developments until the swamps and bogs where they were breeding were drained. Areas that would otherwise make ideal vacation spots have been abandoned or left underdeveloped just because of the mosquitoes' itching bites.... [14]

Behind this narrow line of thinking are the assumptions that human-centered plans have supremacy above all else and that to be bitten is an outrage that justifies war. Anything that interferes with human plans is reduced to an irritant, a pest, and an obstacle to eliminate.

We assume that we have a right of access to the entire planet, that no place by right of its own nature is off-limits to humans, including native and nonindustrial societies. If an area resists human settlement and our technologies, it becomes even more desirable, because it challenges our sense of superiority and dominance. Attempts to conquer and subdue wild places are framed in heroic terms. When environmentalists suggest that humans respect the integrity of inhospitable lands, they provoke outrage. The suggestion questions some deeply held beliefs about our status on Earth and who has rights of access to the planet.

The Rules of Coexistence. Those who know and love the land tend to live in harmony with its cycles, respecting and accepting the protectors of place. In rural areas, beekeeper Sue Hubbell reports that people typically think differently about black flies and other species that city dwellers categorize as pests. There is an easy acceptance of these biting flies by those on intimate terms with the countryside, for they have learned how to dress appropriately and plan their outdoor activities around the flies' presence or absence. It means that human-centered desires are sometimes delayed or not met at all. Behind this kind of behavior is a give-and-take attitude that acknowledges the right of every species to live on Earth, and the right of black flies to occupy and even dominate the woods during certain seasons.

So how are we to view protectors of place, those species that are also intimately involved in our relationship to disease-causing bacteria? Can we allow them their guardianship of our remaining wild places without resentment and forego our belief that we have the right to inhabit and change every area on the Earth?

As far as coexistence is concerned, in areas where the chances of getting malaria or other mosquito-transmitted diseases are great, it would behoove us to take protective measures to keep from being bitten and to keep mosquitoes away from people who have already contracted the disease. We could also step up our efforts to reintroduce wildlife back into our environments so mosquitoes have more choices. And in the West, where the chances of mosquitoes carrying the malarial parasite are still slim, we might accept that we are food for some species and find practical strategies to balance their need against our ease. We might, for instance, decide to avoid areas at times when black flies or mosquitoes are abundant or use a topical solution to dissuade them when we are unable to leave the area—both effective strategies. For the more adventuresome individual, communicating with mosquitoes might also be an option.

Communicating with Mosquitoes

Most communications with mosquitoes are triggered by a desire to stop them from feeding on us. In Jim Nollman's book *Dolphin Dreaming*, he introduces us to Nicholas, an old man, half Native American and half something else. Long after mosquitoes would drive his family and friends indoors, he could stay outside and the mosquitoes wouldn't bother him. When questioned, he held out his arm in front of Nollman's face and clenched his fist three times, explaining that this got the blood to flow through the body in a certain controlled way. "The mosquitoes," he said, "...know the language of the blood better than anyone else. I've spent forty years here, so I've had a lot of time to learn that language too. Now I know how to tell the mosquitoes to stay away."[15]

Rolling Thunder has another, more benign method. He teaches that the emotions that produce good feelings also produce a repellent that keeps the mosquitoes away.

> These attitudes [good feelings] make vibrations; and they have a smell to them. That's what keeps the mosquitoes away.... [16]

Those vibrations may be what the New Age community has latched onto as the ideal stance in dealing with mosquitoes. They may also be similar to the vibrations produced by Francisco Duarte, a young Brazilian man we will meet in the chapter on bees, when he communicates with insects. Such accounts certainly warrant further investigation and experimentation. But there is still another alternative—sharing our blood.

Feeding A Mosquito. After my first speaking engagement on the insect-human connection at a local conference, I fielded questions from the audience. A young man asked what I thought he should do about being bitten by mosquitoes. I suggested applying a topical solution to discourage them if there were many and if he couldn't avoid being in the area. But if there was only one, I added, he might consider feeding it. Then I told him what had occurred to me a few days before the conference.

I was home alone working at my desk when I heard the familiar sound of a mosquito around me. I looked up and saw the tiny creature flying in front of me. I knew she needed blood to start a family and, given my focus and intention to speak on behalf of insects, I thought I should offer my arm. So I extended my bare arm and she promptly lit on it. Within a few seconds the incessant itching began, and I blew gently on her until she flew off my arm. Sheepishly, I offered my apologies, explaining how hard it was to be still because it itched so terribly. Then I asked her to try again while I braced my arm and willed myself to stay still, offering my arm once again. This time she landed on my palm and stayed there. I didn't feel anything so I brought my hand up close to my face to see if she was just resting. But I saw her body fill with my blood, her proboscis still embedded in my skin. Still I felt nothing.

It didn't make sense. I knew from my reading that the properties of a mosquito's saliva affect how fast they can tap and take up blood. The speed is correlated with its antigenic activity. In other words, when their intake is slow, there is a compensatory delay in what the host feels—it gives the mosquito time to complete the act before alerting the host and putting itself in danger.

But in this case the mosquito's first attempt to feed on me triggered major itching right away and the second time nothing. It couldn't be related to the properties of her saliva. And the discomfort wasn't just delayed, as I didn't feel any sensation at the moment or later. It was as though she had purposely, mysteriously, fed this way in deference to me. I recognized the gift and gave thanks.

When I finished the story, the young man thanked me. The next morning he pulled me aside, excitement written on his face. He had sat outside after dinner the night before, and a mosquito had come humming around him. Remembering my story, he had extended his arm, offering her the blood meal that she sought. She had descended on his arm and taken her fill without causing him any discomfort then or later. He was filled with the wonder that coming up against the mystery of our connection always brings. The experience brought up new questions about his relationship to this kingdom. Like the Cree warrior Mama-gee-sic in Chapter 3, who was befriended by the Queen mosquito, perhaps this young man had found his affinity and entered the territory of rejoicing in mosquitoes.

When we position ourselves in accord with the other inhabitants on this Earth, when we view the world through another creature's eyes and understand its motivation, we can move in compassionate response to that awareness. Doing so, we open ourselves up to the powers that support this planet and to the unexpected response of a gracious mosquito. Perhaps we also realize that the world, at heart, is more benevolent and friendlier than we have recently supposed. Give thanks.

8

Lords of the Sun

> Maybe if we love them, they will not have the need to be aggressive.
> —(Roc) Ogilvie Crombie

If you study beetles, the idea that God is in the details is sure to take on new meaning. The variety and beauty of the over 350,000 species identified by entomologists (who suspect that there are several million species of beetles altogether) has captured the imagination of many a professional and amateur naturalist, as well as a few artists.

That God must have "an inordinate fondness for beetles"[1] is what biologist J. B. S. Haldane allegedly told a group of English theologians who asked him what one could infer about the Creator by a study of creation. What kind of God (or Goddess) indeed would love beetles so much that he or she would populate the Earth with more of them than any other animal? That question is what Edith Pearlman asks in an essay she wrote for *Orion Nature Quarterly*, pointing out that beetles comprise two-thirds of the total number of insect species and almost eighty percent of the total number of animal species.

While acknowledging the strictly ecological response—which explains the abundance of beetles in terms of the enormous number of suitable niches available for them—Pearlman turns to the imaginative realm to answer the question. She decides that one possible Creator with a fondness for beetles would be a wise and compassionate deity who would love beetles for their flexibility and humility. Another kind of deity concerned with details like color and pattern would love beetles because of their variety and beauty—particularly the brilliant metallic and luminescent ones. Perhaps it is this deity's love of beetles that inspired the proverb: "The beetle is a thing of beauty in its mother's eye."[2]

And finally, Pearlman decides there could be a Creator who is a beetle. This omnipotent one would have created beetles in his or her image and then made suitable homes for them. Placing them on the Earth approximately 250 million years ago, he or she would also have given beetles their first job—that of pollinating flowers like the magnolia and water lily, two of the world's earliest flowers. And perhaps as this almighty beetle looked over creation with antennae finely tuned to detect beetles in rotting tree trunks and stored grain, buried in dung, wood, and under leaves, nibbling on plants, and crunching on insects, he or she would also have observed how other creatures were faring, admitting, however, that it was the beetle for whom the universe was made.

There is precedence for a beetle god in the ancient Egyptian scarab deity in its Lord of the Sun aspect, representing light, truth, and regeneration. Modeled after a dung beetle, the sacred scarab personified the sun, which symbolized consciousness, for the sharp projections on its head extended outward like the sun's rays. In its Lord of the Morning Sun aspect, Khepri, this god was depicted as a regular dung beetle, or with a human body and a beetle head.

The South American Lengua (Chaco) tribe also had a beetle god, a big dung beetle named Aksak, who created the world and produced the first man and woman from particles of soil. Likewise a dung beetle was the Sumatran Toba's creator, bringing a ball of matter from the sky to form the world.

In Cherokee myth, a diving beetle played an important role in the creation of the world by diving down into the primordial seas that covered the planet to bring up matter. And an aquatic beetle assumed a creator role for the early peoples of India and Southeast Asia when it plunged down to the original liquid core and returned to the surface with matter to form the Earth.

An Inordinate Degree of Beetleness. If we follow Pearlman's example and consider a creator who is inordinately fond of beetles and a world with more beetles than any other species, we might learn something about ourselves. From the perspective of deep ecology—the paradigm that says our extended identities encompass the world—it means our true identities have an inordinate degree of beetleness. Or to use James Swan's metaphor, we have more beetles in our "inner zoo" than any other creature. But if we have so much "beetleness" in our makeup, why are our current relationships with beetles based only on how they serve or thwart our agricultural and gardening objectives?

Our relationship to beetles is predictably shallow, shaped as it is by the them-versus-us mentality that characterizes our relationship to all insects. In the popular culture, Marvel comic books portray Beetleman as a villain. And in J. M. Alvey's story "The Green and Gold Bug," an evil magician uses an African beetle to destroy a young man's human qualities until he is driven mad and murders his wife.

Franz Kafka's 1915 *Metamorphosis*, one of the most famous works featuring a scarab beetle (although some scholars insist Kafka was referring to a cockroach), is the story of a young man who wakes up and finds he has changed into a "monstrous vermin." So begins this existentialist classic.

One of the few beetle images with godlike powers is a reworking of the Egyptians' sacred beetle in a British horror story written by Richard Marsh in 1897. Called simply "The Beetle," it is a tale of sorcery about a high priestess who can transform herself into a giant beetle and who oversees a religious ceremony with human sacrifices.

Most of us neither like nor dislike beetles. There isn't a separate category for the fear of beetles, and so any fears we might harbor usually fall under entophobia, the fear of insects in general. If we are not afraid of insects, we may learn to like a few beetles, particularly the ones that are beautiful and have presence, like those of the scarab family or those with a captivating talent, like fireflies. We also like those that are familiar, like ladybugs, unless of course their familiarity stems from eating our crops, like the Japanese beetle or the cotton boll weevil. Then we reduce them to "crop pest," which seals their fate in our fields and our imaginations.

Children, unsullied by our views, are quick to feel a connection to insects and quick to be inspired by the presence of all beetles. Any hesitation about beetles they might have is readily diminished or exacerbated by how the adults in their lives react to them. Recently I took my five-year-old niece on a walk to look for bugs, which is one of her favorite pastimes. As we walked down a country lane in rural Michigan, she noticed a large number of insects on the green thickets that lined the path. I told her they were beetles. Since I always encourage her to name the insects she sees, she told me she was going to call them "Christmas beetles," because they were red and green and beautiful. I didn't tell her that adults call them Japanese beetles, an imported species that has received the full weight of our chemical arsenal and still breeds and eats furiously in our fields.

Correspondences with Beetles

What capacity could we kindle in the young for seeing correspondences between ourselves and beetles if we encouraged them to apply their imaginations to what they observe? In the state of California where I currently reside, a teaching unit on Egyptian mythology is part of middle-school curriculum, although the connections these ancient people made between beetles and matters of life and death don't make much sense to the teachers—much less to the students. The seed ideas underlying ancient philosophies conflict with our current way of looking at the world. The former are based on an animated, interdependent world and the wisdom of looking for the macrocosm in the microcosm. The desire to control life underlies our contemporary mythologies. We value the ability to control almost everything through technologies, power, and management, with economic values guiding the process.

What if we told the children that, just like the beetle who has wings concealed under its glossy shell, we too have a hidden life tucked away within our physical bodies, a hidden spirit with wings? If we helped them make that single comparison, the beliefs of the Egyptians and other cultures might make more sense. Perhaps they could see then that the beetle's astonishing transformation from a wormlike larva (called a grub) to a winged creature has always made it a powerful and appropriate symbol for rebirth and eternal life.

Ancient Egyptians, who took the dung beetle as a symbol of the body's strength, the soul's resurrection, and the omnificent and unknowable Creator, also revered other beetles. A wood-boring beetle, for example, signified transformation and rebirth. Scholars believe its symbolism arose from an actual wood-boring beetle that feeds on the tamarisk tree. Just as this beetle emerges from the trunk of the tamarisk tree through small holes bored by grubs, so Egyptians believed the imprisoned Osiris, lord of the afterlife, emerged from the confines of the tamarisk, freed by his sister/wife, Isis.

A wood-boring beetle may also be the species referred to in one of the most picturesque images of transformation. The image comes out of an old story retold by Henry David Thoreau in the final chapter of *Walden* and is relayed below in Thoreau's words:

> Every one has heard the story which has gone the rounds of New England, of a strong and beautiful bug which came out of the dry leaf of an old table of apple-tree wood, which had stood in a farmer's kitchen for sixty years, first in Connecticut, and afterward in Massachusetts,—from an egg deposited in the living tree many years earlier still, as appeared by counting the annual layers beyond it; which was heard gnawing out for several weeks, hatched perchance by the heat of an urn. Who does not feel his faith in a resurrection and immortality strengthened by hearing of this? Who knows what beautiful and winged life, whose egg has been buried for ages under many concentric layers of woodenness in the dead dry life of society, deposited at first in the alburnum of the green and living tree, which has been gradually converted into the semblance of its well-seasoned tomb,—heard perchance gnawing out now for years by the astonished family of man, as they sat round the festive board,—may unexpectedly come forth from amidst society's most trivial and handselled furniture, to enjoy its perfect summer life at last![3]

Who is not inspired and given hope by the idea of a hidden life? Like countless larvae and dormant eggs that live underground or in trees for months or years, waiting for some beneficent conditions to stimulate their transformation, perhaps our potential is also hidden inside us, waiting—buried under layers of societal and familial beliefs. And perhaps our thoughts and actions can help create the welcoming conditions they need to emerge.

A few years ago I read a news item about a little-known North American beetle that was found in the vault of a New York securities firm. Apparently the insect had arrived years earlier in the chestnut, oak, or other hardwood used to construct the vault. This species has since appeared in the vaults of jewelry stores, banks, and other businesses. Although the official view judges it only as an infestation warranting an insecticide spray, what could we make of its appearance in the midst of the financial and business world? Hope for a new way of doing business, I like to think, for as we will see when we look at the problems inherent in big business agriculture, it is evident that our current way of growing crops and reaping profits cannot be sustained.

Beetles as Messengers

As messengers, insects often convey information simply by their presence. A case in point is the aforementioned beetle that flew to Carl Jung's window, inspiring him to investigate meaningful coincidences which he called "synchronicity."

When beetles are out of balance, that is, when their populations are greater than normal, their number and presence also convey messages telling us that either a natural cycle of transformation and growth is at work, as when masses of beetles emerge in the spring, or that a problem exists.

When large numbers of insects appear in growing fields, we are quick to judge them a problem, thinking that if we eliminate them we have solved the problem. But what if they are merely the bearers of a message that a problem exists? Many agricultural experts have demonstrated that in a balanced environment insects primarily attack weak and dying plants. One of the first and most prominent scientists to take this view was Dr. William A. Albrecht, chairman of the Department of Soils at the University of Missouri. Through extensive experimentation, Albrecht demonstrated that "insects are nature's disposal crew, summoned when they are needed, repelled when they are not."[4]

Albrecht provided the scientific support for organic farming and proved conclusively that declining soil fertility resulted in less nutritious crops that lacked vitality. Calling the use of insecticides (and weed killers) a desperate act, Albrecht insisted that insects and disease "are the symptoms of a failing crop, not the cause."[5]

Up against the clout of the chemical industry, which supported the public's view of insects as adversaries, the opposing view of insects as messengers didn't receive the following it warranted. And scientists had not yet answered how insects could know which plant was sick and which one was well. Today, due primarily to the work of Dr. Philip S. Callahan, senior entomologist with the United States Department of Agricultural (USDA), and others before him who laid the groundwork for his findings, they know.

Communication Between Plants and Insects. Callahan has shown that beetles and other insects are well aware of what goes on around them. They communicate with each other and plants using a variety of antennae tuned to the infrared band of the electromagnetic spectrum. Callahan explains:

> Insects "smell" odors electronically by tuning into the narrow-band infrared radiation emitted both by sex scents and by plants they desire as food ...Sickening plants signal the news of their impending death to waiting bugs by means of the same infrared radiation.[6]

His studies also demonstrate that the sicker the plant, the more powerful the scent it emits, and the easier for the insect to hone in on it. So when large numbers of insects attack crop plants, we would do well to ask if these plants, often modern hybrids that need excessive fertilization and irrigation to grow, have signaled the insects, "telling" them that they are sick.

Another message communicated by large numbers of beetles or other insects ravaging our crops is that the current use of toxic chemicals in agriculture has eliminated their natural population suppressors, creating not only a fast resurgence of their populations, but secondary infestations of other insect species that were once in balance and thus harmless.

Ignoring the messages, farmers and the agricultural conglomerates that control how they farm—what seeds, fertilizer, heavy equipment, and pesticides they buy—have traditionally tried to fix all signs of imbalance by killing the messengers. As a result of this approach, which has replaced natural systems with totally alien systems, they have created more serious problems. And many of these problems, like our chemically soaked environment and the degradation of the soil, are becoming too big to ignore.

Cosmetically Pleasing Produce

One assumption underlying current agricultural practices is that many insects compete with us for our food and so require that we kill them with insecticides and other techniques under the cover of Integrated Pest Management (IPM). It seems straightforward enough, and few of us outside the arena of agriculture question it. Yet, in many instances, the rationale for having to combat insects is merely a cover-up for selling more insecticides or selling a particular hybrid seed.

Growers also use heavy applications of pesticides to satisfy our demand for uniform and cosmetically pleasing fruits and vegetables. As consumers we have been adamant about not wanting to see insects on our produce or see holes made by insects, or other signs that these creatures had first access to our food. To accommodate us, farmers, who know that all fruits and vegetables have whole insects or insect body parts on them, spray just enough pesticides to eliminate the evidence. The price we pay for our unrealistic expectations about the way fruits and vegetables should look is that we buy and consume food with pesticide residues we can't remove.

Perhaps if we were more willing to allow surface imperfections, we could send a different message to those who grow our food. We could demand a generally more wholesome and nutritious fruit or vegetable. As it is, agrichemical companies playing on our willingness to accept anything as long as it looks good plan to maximize yields by restructuring and altering DNA in our produce—as they have already done with the tomato. And genetic manipulations will produce a counterfeit appearance of freshness and desirability long after the product's natural shelf life has expired.

The Green Revolution. The cosmetic issue is only one of many. Behind a variety of new approaches and techniques, we see a now-familiar military and profit-hungry mentality at work driving the agricultural industry. It is the same mind-set underlying the war against the mosquito, and the parallels are apparent. Even the Green Revolution, touted as a movement by developed countries to supply food to Third World countries by introducing them to modern agricultural methods, has been called "colonialism in a green cloak."[7]

Rather than taking a simplistic approach and separating ourselves from advocates of current practices, we'll accept our complicity in creating or allowing these kinds of enterprises to operate as they do. We can agree, perhaps, that it is our beliefs and assumptions that have given the industry license to ignore human values and pursue the bottom line at great expense to people, other species, and the environment. Once we have done our shadow work and accept that we too are the exploiters, the controllers, and the ones who compete at the expense of others, we can choose to act differently and send a forceful directive to those who control our food supply.

The War Against the Cotton Boll Weevil

Space considerations require that we limit our investigation and keep our sights on beetles as we briefly examine the roots of current agricultural practices, evaluate biological control tactics (using insects to fight insects), and summarize recent genetic engineering efforts and issues. These are big topics, but the message from the beetles, which in this chapter represent all insects branded crop pests, is short and to the point: We are not the problem, nor are we your enemies.

In the United States, the hostilities toward plant-eating insects increased a hundredfold when farmers began monoculture, the large-scale planting of one kind of crop. Although monocultures are inherently unbalanced ecosystems, their promise of high yields and, therefore, high profits overrode other considerations, and the government funded farmers who agreed to specialize in one crop. Whereas a balanced ecosystem keeps populations in check by predators, limited food supplies, and even diseases, monocultures destroy these natural population suppressors. And without them monocultures are vulnerable to the slightest disturbance.

The monoculture of cotton was responsible for the United States waging a war against the cotton boll weevil. Those favoring the war (and most feel there wasn't any choice) are quick to point out that boll weevil damage in the United States has been estimated in the tens of millions of dollars annually. Those who question the hostilities point out that monoculture was and is the problem. In fact, monocultures today are not just single crops, they are single strains of a crop, once bred, and now genetically altered, to produce the highest yields—and all vulnerable in exactly the same way to insect damage, disease, and weather conditions.

The Cotton Boll Weevil. The cotton boll weevil belongs to one of the largest families in the beetle or Coleoptera order, with forty thousand weevil species described and countless others unnamed. The adult weevils have a very distinctive look, with a long, curved snout, half the length of their body. On the snout are their antennae and at the tip are their jaws.

Weevils are specialists. Each species likes a specific plant, and most common trees, shrubs, and herbaceous plants are host to one weevil species. For cotton, it is the cotton boll weevil. The damage to a cotton plant by the adult cotton boll weevil is minimal, but their offspring, which hatch from eggs laid in tender young cotton buds called "squares," either destroy the square before the cotton is produced or damage the seed fibers of the cotton boll.

We branded the cotton boll weevil our enemy when it took advantage of the monoculture of cotton, thinking perhaps when it saw the fields upon fields of its favorite plant that it had died and gone to weevil heaven. It couldn't know, and wouldn't care, that the United States was preparing to enter the world of international trade in cotton.

Given a cornucopia of fields planted with its beloved plant and few predators, the weevil multiplied rapidly. After all, its remarkable reproduction abilities, like those of most insects, were designed for a balanced environment with natural population suppressors in place. As the weevil spread, those aiming to make their fortunes in cotton could see their dreams of becoming an international power disappearing. They started using insecticides to contain the weevil but with limited success.

When DDT was invented and its production exceeded the amount the military needed to protect World War II troops in the 1940s, the United States government okayed it for civilian use. Its acceptance was ensured by skillful propagandists who convinced the public that they didn't have to settle for cohabitation with fiendish insects when an insect-free world was now within reach. And to emphasize the demonic character of insects during this time period, a political cartoon even showed a beetle with Adolf Hitler's head.

Allowing DDT to be sprayed extensively in yard and field was inevitable, given the chemical manufacturer's skillful advertising campaign, our belief in DDT's harmlessness to all but insects, and the mindset of war and righteous indignation that arose from the fact that a mere beetle could threaten promising enterprises like cotton. Entomologists, chemical companies, state extension services, and big land grant universities recommended its use. In fact, many urged farmers to spray their fields every week. It didn't matter if the beetles were present or not; it was just good insurance. Frank Graham, Jr. in *The Dragon Hunters*, a book on biological control on which this section relies heavily, says, "The proponents of this approach urged that for the nation to let down its chemical guard would be to expose itself to hunger and disease."[8] But as Graham points out, most of the insecticide use had nothing to do with preventing hunger or disease. It was directed to cotton and a bid by the United States to lead the world in its production.

The power of pesticides and the belief in the eventual subordination of nature and its creatures prevented any serious consideration of changing the practice of monoculture to polyculture or using any other methods like crop rotation to introduce balance back into the growing fields. We thought we didn't have to. Monocultures were the road to great yields and great profits. Most people thought hybrid seeds and the chemical fertilizers and pesticides they depended on could override natural laws and counteract the inherent imbalances of monocultures. Agriconglomerates (transnational corporate concentrations of food-related industries that make fertilizers, pesticides, seeds, and equipment) bet on it. Only insects like the weevil, disease, and unpredictable weather patterns stood in the way. And so the boll weevil's large numbers and devastating effect on cotton crops, clear messages of imbalance, were ignored.

Fortune Disguised as Misfortune. Even when the best efforts of farmers and chemicals didn't stop the beetle from devastating their crops, most, lured by the promise of big money, wouldn't give up the fight. Yet some had to in order to survive—like the farmers in Enterprise, Alabama. After the weevil's activities resulted in a sixty-percent loss in yield on Alabama's cotton crops, Enterprise farmers surrendered. Forced to diversify, they planted peanuts, corn, and potatoes. The results of this change of direction were wholly unexpected. From the new crops, these farmers realized substantially greater profits than they had in their best years of cotton, and their economy began to thrive. Better yet, the citizens of Enterprise knew they were indebted to the boll weevil and erected a monument in its honor with the inscription:

> In profound appreciation of the Boll Weevil and what it has done to the Herald of Prosperity. This Monument was erected by the Citizens of Coffee County, Alabama.[9]

Today cotton is still being grown in the region, and the crops and boll weevils continue to thrive. The rest of the country didn't follow Enterprise's example and instead kept their faith in technology's promises of better chemicals and practices that would allow them to keep their cotton monocultures. Bucking weather, disease, and insects, millions of cotton plants were planted even in areas not suited for growing cotton—simply because they were available. And it seemed to have worked. No one in power was pointing out the hidden costs of these practices.

Resistance to Pesticides Ignored. So driven were we as a nation to become and stay an international power in cotton that policymakers ignored or dismissed data as early as 1954 that indicated that boll weevils were developing a resistance to insecticides. In fact, Frank Graham says that the chief of research on cotton insects at the USDA denied the evidence of weevil resistance because he feared it would damage the insecticide and cotton industries.

As the USDA went through one chemical after another, untargeted species of plants and animals paid a heavy price. And as both insect predators and parasites died, other insects that had never been considered a threat to cotton, and that found ways to resist the onslaught of poisons, emerged as major threats to cotton plants—like the bollworm and the tobacco budworm that still threaten monocultures today.

The agricultural community kept spraying, despite learning that pesticides didn't work, that no matter how many insects were initially killed, some always survived and passed along their immunity to the toxins. And eliminating insect predators and parasites always led to secondary insect infestations. Still they believed that using pesticides would let them survive for the short term until scientists found permanent solutions. Meanwhile monocultures and the heavy equipment, pesticides, fertilizers, and irrigation they need produced great yields and profits.

The cost of this "success," however, has not been widely publicized, although in the area of human health, the links between pesticides and various kinds of cancers, including breast, bladder, and prostate cancer, are being exposed in well-researched books like Sandra Steingraber's *Living Downstream: An Ecologist Looks at Cancer and the Environment*. Steingraber points out that since Rachel Carson's *Silent Spring* was published, pesticide use has doubled in the United States, and women born in the United States between 1947 and 1958 have almost three times more breast cancer than did their great-grandmothers at the same age. Another study in 1988 showed a significant association between agricultural chemical use and cancer mortality in nearly 1,500 rural counties in the United States.

As a woman diagnosed with bladder cancer, Steingraber is concerned that the uncertainty about relatively minor details has been used to throw doubt on the real connections that do exist between human health and the environment. And remember that the effect on human health is only one of the costs of our success in producing great yields. Other authors have documented the declining soil fertility, the loss of plant and animal health and biodiversity, and the loss of small and medium farms.

As laypeople, what we hear about current practices make us vaguely anxious and confused about the issues. Glossy brochures created by chemical companies and biogenetic enterprises still talk about hope, satisfaction, and partnership. They show happy families with good-looking children playing against a backdrop of green fields. Their advertisements justify the chemicals and genetically altered seeds they sell to farmers by referring to a war between the insects competing for our food and the valiant farmers with their allies—those in applied entomology and the chemical companies—trying to thwart them. We tend to believe the propaganda because it confirms our own misgivings about insects. Even after many exposés have revealed widespread corruption in agribusinesses by those seeking profits by downplaying the danger of their products,[10] our doubts about insects are so great, our belief in insects as the enemy so firm, that they overrule our reason and common sense.

Although we admit we're concerned about pesticide safety, the majority of people believe that to have enough food, we have to keep fighting crop-eating insects with chemicals. But we don't. In 1945, when grown in rotation with other crops, corn had few insects that bothered it. In fact, corn lost to insect damage was only 3.8 percent. In the 1990s, after forty years of chemicals, it was 12 percent. And there are many other examples.

Today, even with the advent of other insect-control techniques, 35 percent of all insecticides used on major crops are applied to the fourteen million acres of cotton grown in the United States. What does it say about us that we keep fighting cotton-loving insects with chemicals and risk getting cancer in order to have twenty-five different brands of cotton blue jeans available to purchase at the mall?

And today the subterfuge is still going on. The battles against crop-loving beetles and other insects continue, as do the monocultures and chemical uses that led to and still sustain the imbalances. Keeping our war mentality, the only real change we've made in response to the threat of pesticides is that we've added biological control and genetic engineering tactics to our Integrated Pest Management arsenal.

Biological Control

The primary tactic of biological control is to import and introduce insect predators to check the population of other insects eating or in some way damaging our crop plants. It's a widely praised, environmentally friendly solution, but for the record its use in Western culture is still fueled by a war mentality and is aimed primarily at shoring up monocultures. On the plus side, biological control mimics nature to a certain degree and decreases pesticide use. It has to in order to allow the new predator to do its work. On the minus side, it sometimes backfires when complexities in the ecosystem are unknown or ignored. And when a mistake occurs, the potential risks to both biodiversity and ecological stability are high.

Biological control first caught the attention of the agricultural community when entomologists imported an Australian ladybug to check the population of cottony-cushion scale in California citrus groves. When the beetles brought the scale insect's populations down and kept them in check, it elevated not only the status of biological control in the United States, but the status of ladybugs.

Beetle of Our Lady. Ladybugs are beetles that have a long association with crops. Two thousand species of ladybugs are spread throughout the world. People from different areas call these insects by different names including ladybird, ladycow, ladyfly, and ladycock. The lady prefix was given to these beetles by grape growers of medieval Europe who dedicated them to "Our Lady, the Virgin

Mary," and often called them "Beetle of Our Lady." In Sweden, ladybugs are sometimes called the "Virgin Mary's Golden Hens," and in France, they are spoken of as the "Cows of the Lord."

Since the success of the Australian ladybug in controlling scale populations, USDA officials have introduced over 179 species of ladybug, and there are many success stories. Other beetles have also been used successfully. In Florida, for instance, flea beetles were introduced to curb the South American alligator weed, and a weevil that only eats the Australian melaleuca tree has been released into the Florida Everglades to help check the spread of this tree. Recently a species of jewel beetle has also been imported for release into the Pacific Northwest in the hopes that it will control seed production by Saint-John's-wort, despite its identification as a valuable antidepressant herb.

One of the most successful biological control campaigns brought nonnative dung beetles to Australia to recycle the waste produced by cattle. The imported beetles checked the increase of horn flies, whose population skyrocketed when cattle dung piles used for their egg laying weren't degraded by native beetles and were only minimally broken down by native termites.

The Risks of Biological Control. Despite the enthusiasm for biological control tactics, introducing a nonnative species is always risky to an ecosystem's biodiversity and stability. In Arizona, for example, preliminary observations show that an introduced beetle has significantly displaced a native species, and no one knows what the short- or long-term effects of this displacement will be.

Another disaster reported in the fall of 1997 indicated that a weevil introduced in North America in the late 1960s to control fast-spreading European milk thistles has started preying on native thistle plants, significantly reducing the seed production of its flowerheads.

Still another downside of biological control concerns the ladybug. Lauded as the solution to ridding garden and field of unwanted insects, today it's sold in every nursery in the United States. Our exploitation of its appetite for aphids, however, may endanger it. Ladybug prospectors, out to satisfy our demand for ladybugs and make a profit, search for ladybug hibernation sites, and when they find one they capture millions of the beetles when they are the most vulnerable. After transporting them down the mountain in boxes or sacks, they keep the ladybugs in refrigerators. Those that survive being collected (the mortality rate is 20 to 40 percent) are distributed to orchards and garden shops. Sue Hubbell reports that 100 million ladybugs were distributed in the state of Washington alone.

Ancient Roots of Biological Control. Although ladybugs first demonstrated the potential of biological control to farmers, it's not a new tactic. The association between beetles and crops and the use of natural predators to control insects is a long one. Ancient Chinese texts even mention using ants to control stinkbugs, citrus-damaging insects, and insects that eat stored products.

In many countries, biological control was carried out with human beings playing the role of population suppressor. In Thailand, for instance, farmers and their families and friends ate the beetles that damaged coconut and sugar palms, as well as caterpillars that fed on coffee plants, and bees that damaged woodwork and telephone cables. The strategy also worked for locust swarms. In fact, African natives looked upon great plagues of locusts with delight and gratitude. They believed that the locust had been sent by their gods in order to provide them with food. In many poor regions of South Africa, locust flights (a survival strategy for the species when facing starvation) are viewed as such a blessing that the medicine man sometimes promises to bring them instead of rain.

Biological Control of the Boll Weevil. Eating boll weevils or other beetles was not a viable option in the United States. While most entomologists in league with chemical companies were still favoring the use of insecticides, a few progressive individuals as early as 1930 started looking for boll weevil predators in other parts of the world. Two predators brought over and introduced couldn't adapt to the climate and a third, a parasitic worm that infects the boll weevil, couldn't control its populations.

Recently new strategies have emerged as entomologists have applied their knowledge about insects to help eradicate or control them. For instance, they have learned about insect hormones that are helpful at one stage of an insect's development and disruptive at another, and about pheromones (chemical signals conveying information to others of the same species); both have been used to battle insects. Sterilization and radiation techniques have also been fine-tuned and harnessed for the war effort. All of these tactics were supposed to reduce or eliminate the need for pesticides, but none has worked as well as expected.

If we return to the premise that insects are messengers, aligned with the environment and natural forces, consider how different our approach would be. What would we discover if we were looking for information that would allow us to coexist with insects? What insights into the nature of life would reveal themselves if we were looking to create balance and sustainability and if we were willing to change our methods of growing food just to end the war?

Genetic Engineering

We are still, however, trying to maintain the status quo. In the last ten years, biological control strategies have been augmented by genetic manipulations. Genetic engineering (GE), also called genetic modification (GM), is a process by which genetic material is transferred from one species to another. Artificial breeding across species lines produces a "transgenic" plant or animal species. Genetic engineering of a plant or animal so that it can be a more effective weapon is almost commonplace today. A standard practice is to alter an insect so that it can control other insects more efficiently, withstand new climactic conditions, or "self-destruct" at a particular age. Release of some of these altered creatures has prompted lawsuits to stop further releases until ecological risks have been determined.

Altering Plants to Produce Insecticides. One major genetic alteration to crop plants that affects beetles and other insects involves inserting a gene that allows the plant to produce a natural insecticide, reducing the need for pesticides. A little-touted fact, however, is that many of these genetically altered plants (GM hybrids) now have toxic substance levels that could reclassify them as pesticides instead of vegetables.

GM hybrids are advertised as important weapons in the battle against crop-eating insects. They are also lauded as the best hope for a solution to world hunger, even though most experts agree famines are not caused by lack of food but by lack of access to food. Since most of us find ourselves confused by the issues and wary of the claims of both biotechnology companies and environmental activists, we tend to distance ourselves from the issues.

As of this writing, the EPA has approved seven plants altered with this toxin, despite the fact that the plants' new ability to kill the insects that feed on them will diminish in just a few years as the insects develop a resistance to the substance. It is another short-term solution to produce high yields that will let a few reap large profits—apparently justification enough for its use. And farmers in the United States have already planted millions of acres with this altered seed.

The toxin in these altered plants comes from *Bacillus thuringiensis* (Bt), a soil microbe that produces a natural insecticide best known for its effect on caterpillars. Bt exists in countless natural strains, each fatal to a specific insect. Its natural precision, however, is seen as inefficient and, therefore, a handicap compared to synthetic poisons that kill insects indiscriminately. Another drawback to natural Bt is the fact that sunlight destroys it, so it has a brief field life. Yet its limitations are the very traits that make it sustainable and environmentally friendly.

We tend, however, to value efficiency over sustainability. As a means to power and profits, efficiency has a long history. Only the machinery changes. Instead of DDT, we now have genetic engineering. To make Bt more efficient, one company has overcome one of Bt's natural limits so it now has a greater field life. Nothing is mentioned about what it will do in its longer life. Its limitation has been removed and thus also a degree of sustainability and stability.

Bt in spray form was the first genetically engineered pesticide to win EPA approval for release, and organic farmers and gardeners have used it for almost forty years. As insects adapt and resist Bt in hybrid crops, not only will all Bt varieties be rendered obsolete, but also organic farmers will be left without one of the main tools they believe they need for insect control. Ironically, although many lawsuits have been filed, Bt hybrids will probably be obsolete—due to insect resistance—before the legal battles have been settled.

Another cause for considerable alarm is the fact that potato plants genetically modified to produce Bt (to kill aphids that try to feed on them) may harm ladybugs as well. Research conducted in Scotland found that female ladybugs placed on the transgenic plants to eat the aphids *not killed* by feeding on the juices of the Bt plants soon lay fewer eggs and die earlier.

Add the recent surge of herbicide-resistant plants[11] on the market, which permits greater quantities of chemicals to be applied to our fields, and other health-related concerns, and the result is a hazardous situation for both beetle and human.

The Tyranny of Efficiency

After all these efforts, most of which have had disastrous impact on the health of beetles and other species, no major attempt has been made to sacrifice the efficiency of monoculture and chemicals. In *Kinds of Power,* James Hillman says that "efficiency is a primary mode of denial."[12] It elevates the job being done—in this case agricultural productivity—above all other considerations. The Nazis killed people efficiently, Hillman points out, and Richard Nixon justified Watergate and all the cover-up operations because of the importance of running the country efficiently and moving it toward world peace. And now we're raising food efficiently—supposedly to end world hunger and at a cost we have yet to comprehend.

It is not science and technology themselves that are to blame for our present critical situation but rather the way in which we have allowed them to be used. Hillman observes that we have raised efficiency to an independent principle—a dangerous course of action because efficiency deadens sensitivity and favors short-term thinking. Time is efficiency's enemy. Any action that takes time threatens efficient operations unless it is shown that the time taken to do it will result in even more efficiency. Poisons are efficient. Spraying and dusting methods currently used in our agricultural fields to efficiently kill insects and weeds were adapted from methods developed to kill people efficiently in concentration camps during World War II.

A panel of members of the American Association for the Advancement of Science has made public its concern that if current trends continue, the United States, which exports more crops than any other nation, will no longer be able to export food by the early part of the twenty-first century—it will have exhausted the resources it relies on, including the petroleum from which the fertilizers and chemicals are made. The productivity won't be there without a major revamping of mainstream agriculture.

Any long-term solution can only emerge out of a radically changed identity. If we change our ideas of who we are individually and as citizens of Earth and return to what we already understand about the functioning of the natural world, we may discover sustainable technologies that honor our extended identities. Perhaps embracing our inordinate degree of beetleness can also help lead us to regenerative methods of growing food. A perspective that recognizes the individual, the society, the Earth, and all its inhabitants will automatically gravitate toward holistic, nonexploitative, and ecologically sound long-term practices.

Sustainable Agricultural Technologies

If we look to the beetle for guidance, we can see that its food preferences and ability to multiply indicate the need for checks and balances in the growing environment—specifically a diversity of plants and a variety of natural population suppressors.

"Beetles epitomize diversity...which eloquently extends beyond the physical, encompassing strategies of behavior, defense, reproduction, and adaptation,"[13] say Arthur Evans and Charles Bellamy in their beautiful tribute to beetles. So, again taking our lead from beetles, we could replace narrow-focused genetic manipulation with seed-stock diversity to balance life in the fields, in the soils, and in the environment at large. Maintaining diversity, says Vandana Shiva, scientist and activist against corporate agricultural practices, is our "cosmic obligation to keep a larger balance in place."[14]

We can also read the beetle's symbolic association with rebirth and regeneration as a sign to return to the soil what has been taken out in the process of growing food plants. All sustainable agricultural techniques return nutrients to the soil while conserving topsoil, water, and energy. Mixed planting (polycropping), crop rotation, trickle irrigation, and nitrogen-fixing crops are just a few techniques that reintroduce balance back into the fields.

Alternatives like permaculture and biointensive and biodynamic agriculture, which demonstrate the ability to produce equal or higher yields of more nutritious food, also build soil fertility while conserving natural resources. And there are sustainable agricultural technologies that could replace conventional methods right now with only a brief transitional drop in yields.[15]

Rewarding those who switch to sustainable methods would hasten the process of weaning American agriculture from its dependence on monocultures, petrochemical transfusions of fertilizers and pesticides, and megafarms. A special commission appointed in 1997 by the United States Secretary of Agriculture published a study that reiterates what we once knew: small farms are not less efficient than large farms—and small farms foster biological and social diversity.

As sustainable practices of growing food are reintroduced into the culture, it is important to stay alert to the assumption of insect as adversary. Some methods, although proposed as environmentally friendly, are still insect hostile, while others view the insects as messengers of the environment—an important distinction. True environmental friendliness, deeply rooted in a belief in the divinity of the natural world, will also include a respect—and even a fondness—for insects. The gardens of Scotland's Findhorn Community and Machaelle Small Wright's Perelandra in Virginia, for example, are uncommonly friendly to insects, and their founders have eliminated the influence of aggressiveness in the gardens with surprising results.

Findhorn Garden

Few gardeners are unfamiliar with Findhorn, a small gardening community in Scotland that gained world attention in the 1960s for its abnormally large fruits and vegetables. When pestered by reporters to reveal their "secret," the founders of Findhorn, Peter and Eileen Caddy and Dorothy McClean, attributed the success of the garden to a cooperative effort by people and plant "devas"—angelic presences or energies that oversee the growth of plants everywhere.

For the Findhorn community, gardening was a way to experience, work with, and learn about the essential oneness behind nature's diverse forms. Cultivating a state of attunement and love, the members of the community sensed, articulated, and met the needs of the garden.

In their book *The Findhorn Garden: Pioneering a New Vision of Man and Nature in Cooperation*, the community had this to say about insects:

> Insects...should be approached with an attitude of love to reach a solution that is for the highest good of all concerned. It is normal for plants to live with a balanced population of insects around them. We have observed that healthy plants are not harmed by the insects they attract, just as healthy bodies are not infected by the germs they come in contact with.[16]

Interestingly, in yet another instance of science catching up with intuitive wisdom, a recent study published in the journal *Science* supports this idea. In fact, plants that are "attacked" by caterpillar larvae have fewer problems with predators later and actually do better. Interaction with insects makes a plant strong and hardy.

At Findhorn, when a clear imbalance in the insect population did occur, it prompted members of Findhorn to examine their thoughts and behavior, a response similar to that of native people when bitten by an insect. Sometimes they discovered that the plants had not been receiving what they needed physically for their optimal development. Other times, the imbalance occurred when certain offensive methods of gardening were practiced—methods contrary to the spirit of cooperation that was necessary for balance. And still other times the problem originated from the general use of insecticides in the country, a use that often poisoned insect-eating birds in the area or in some other way ruptured the web of interrelationship and balance between species.

To discover why a particular insect was out of balance in the garden, Findhorn gardeners would usually contact the deva of the insect—what native societies called the overlighting spirit of the species, or group soul. Cofounder Dorothy McClean, who contacts the devic realm by feeling into the essence of a plant and then trying to harmonize herself with it, was the community's initial contact person for the devas. In one communication, she was told that everyone's feelings, thoughts, words, and actions around the garden had a great effect on its balance and the health of the individual plants and animals. And to correct a situation when insects appeared out of balance, the devas suggested that she visualize the plants as strong and healthy, an action, they said, that adds to the plant's life force and helps them withstand the attack. Removing the insects by hand or using an organic spray on the insects was also an acceptable course of action as long as the insects were warned beforehand and disposed of with an attitude of love and recognition.

Perelandra

For small communities and individuals with a penchant for gardening, working with the forces of nature has always been the primary method for growing food. In the wooded foothills of Virginia's Blue Ridge Mountains, Machaelle Small Wright has also followed a method of gardening based on collaborating with nature spirits.

Wright believes that a willingness to cooperate with these natural forces is the only way we will be able to reverse the damage done to the planet. It is a message that has been preached by many others from many different disciplines.

Wright has discovered firsthand that insects are messengers that quickly direct her attention to problem areas. When she sees a plant or a row of plants overwhelmed by insects, she opens to the appropriate deva and asks if the plant balance is off. Infinitely practical, the overlighting spirit of the plant often just communicates that the plant needs more mulch or water. Sometimes what is "off" and in need of correction is Wright's thoughts or intentions or those of the community connected to the garden. Sudden shifts in the emotional climate of nearby people affect the garden and its balance. Wright explains:

> When it comes to ungrounded, raw, emotional energy released by humans, nature functions in the role of absorber. Even though emotional energy is invisible, it is not less tangible in its effect on the world of form than insects, heavy rain, or drought."[17]

A case in point is the year Wright had a particularly gruelling schedule and was increasingly anxious about meeting it. The insects came one day and attacked the Brussels sprout plants to let her know that she was becoming an unbalanced and disruptive influence on her garden.

Giving Back. Wright also believes in giving back in gratitude to the Earth community and the soil what has been given in food plants. It's an ancient idea in accord with the beetle's message of regeneration, and it is also a central principle guiding sustainable societies.

One of many forms of regeneration that Wright practices is tithing ten percent of the garden back to nature—although she says nature has never fully taken that much. One time when cabbage worms infested her cabbage, broccoli, cauliflower, and Brussels sprouts, she gave one plant at the end of each of the four rows to the cabbage worms and requested that the worms remove themselves from all the other plants, except for the designated four. The next morning the plants in all four rows were free of cabbage worms except for the ones at the end of each row. And interestingly, even the end plants didn't get annihilated by the cabbage worms. They had only the number of worms they could comfortably support. The rest had simply disappeared or had been eaten by birds, wasps, and various other creatures who favor them as food.

Beetles and Roses. Wright didn't always feel friendly toward insects. Like most of us, she thought that given a chance insects would take as much as they could, which is a common projection. Maybe we only have to stop the war against them in our gardens to experience a different reality. After watching beetles take one or two flowers on a rose bush and leave the other ten for her, Wright gradually changed her attitude about insects. Today she doesn't focus on what they attack, weaken, damage, or destroy. She focuses instead on the gift each insect offers to the countless other members of the garden and encourages their health as she would anything else in the garden.

Wright's nonmanipulative approach to gardening also gives us an alternative to buying ladybugs at our local nurseries. One season the nature spirits that were stewarding her and her garden told her to plant costmary in the herb garden, although it was an herb she didn't use. Each spring, as she tended the herb, she saw that it became completely covered with aphids—thousands of them. About a week later, almost the same number of ladybugs appeared on the costmary. They ate all the aphids and then scattered throughout the garden. Wright decided that the costmary in the Perelandra garden was obviously meant to be a breeding ground for the ladybugs, an invitation, and a welcoming feast.

Japanese Beetles and Corn Crops. Another time, when Japanese beetles ravaged her corn crop, eating the pollen and demolishing the silk, Wright decided to contact the overlighting spirit of the beetle and request that the beetles leave most of the corn plants alone.

To her amazement, she touched into an energy that she could only describe as that of an abused and battered child. She says, "It was an energy of defeat, of being beaten into submission. Yet it still had mixed in with it anger and a manic desire to fight for its life."[18]

Sobered by the contact, when Wright again tried to contact the deva of the beetle, this time she made contact and was informed that her original connection had been made with the consciousness of an actual Japanese beetle. She was permitted to sense its consciousness so she would better appreciate what people's hatred had already done to it before she made any requests of its deva concerning her corn plants.

Wright was aware that the beetle, introduced accidently into the United States by an insect collector, had multiplied rapidly. It not only liked a wide variety of plants, but it also had no natural predators in the country. Consequently, Japanese beetles, like other exotic insects imported unintentionally, have been on the receiving end of a massive chemical attack that continues to this day.

Under the circumstances and still sensing its pain, Wright didn't feel it was appropriate to ask anything of it. She decided instead to ask the species to recognize Perelandra as a sanctuary, inviting it to join the rest of the garden community so that it could begin its own healing journey. To reinforce her message, she left an area of tall grass adjacent to the garden, a favorite haunt of the beetle, unmowed.

In the years that have followed since making the agreement with the Japanese beetle, Wright has noticed that the beetles have grown more calm. For a couple of years they only damaged the roses. Although tempted to knock them off the plants, she refrained. Eventually she adjusted her thinking so she could actually invite them to enjoy the roses. In time, they no longer clustered on the roses, although occasionally she sees one or two of them on them. Wright says, "Because of their shifts and changes, I've found that I have not had reasons to request anything special from them. Their presence here is in balance."[19]

The thought of the blind anger and incalculable pain of the Japanese beetle—my niece's Christmas beetle—has always touched me deeply. What would our imagined god of beetles think of our war against its chosen ones? And what aspect of our beetleness have we battered and abused in our inability to accord them messenger status and understand that we cannot declare war on any member of the Earth community as though we were separate and outside the circle of life?

When we can acknowledge our role as abuser and the damage our hatred has wrought on certain species, we have tackled part of the required shadow work that in turn will help us reconnect to the assaulted parts inside ourselves. And if we can forego the denial and justifications and undertake this work, we might be able to touch our greatly imperiled natural identity and coax our wounded "insect self" out of the dark corners of our inner zoo, reassuring it that its long exile is over. Reunited with our beetleness and infused with its vital energies, we might gain access to the power and will necessary to end the silence and subvert those who would hold us to our current self-destructive path.

The Chinese solar calendar designates a day in February as *Jing-Zhe*, or "Waking of the Insects,"[20] which tells the farmer that spring is coming. Maybe we can designate a day to celebrate the Waking of the Humans—the day we can see the beetle once again as Lord of the Sun, symbol of light, truth, and regeneration, and be reborn, emerging at last with the wisdom to create a sustainable life and the capacity to enjoy a perfect summer life.

9

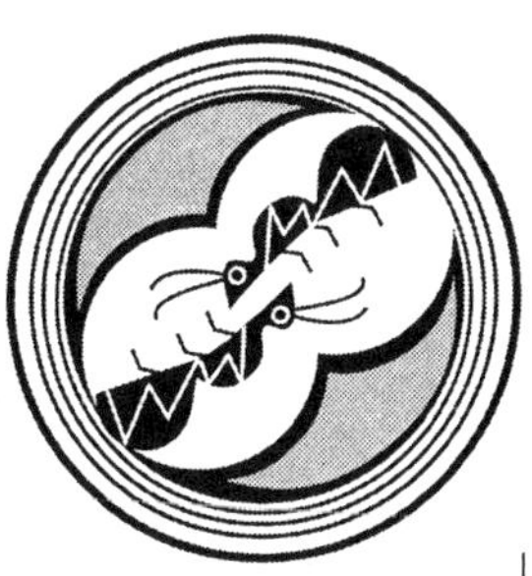

First Born

> Go to the ant, consider her ways, and be wise.
> —Proverbs 6:6

The discovery of a new ant species by a nonspecialist is a rare occurrence, but it happens. In the early nineties World Wildlife Fund (WWF) President Kathryn Fuller found ants in a palm plant in her Washington, D.C., office. She called on Edward Wilson, an ant specialist (or myrmecologist), to identify them, but he hadn't seen them before. It happened that they were a new species—later named in honor of Fuller. The news coverage gave WWF the opportunity to underscore its mission to conserve a biological diversity that has yet to be discovered.

Most of us (myself included) wouldn't recognize a new species of ant from a previously identified one—even in our own backyards. We know very little about ants, yet they cover our planet in unimaginable numbers—a conservative estimate from Wilson is a million billion. And while over 9,500 species of ants have been identified, it is thought that twice that number exist. Ants live everywhere except in the extreme arctic regions and taken collectively, as Wilson cheerfully reminds us, they may be the dominant life-form on the planet, rivaling our claim.

Not only do ants outnumber us, but their services are absolutely vital to the functioning of life on Earth. Since prehistoric days ants have served creation, circulating vast amounts of nutrients in the soil and pollinating plants by dispersing seeds. They also prey on other insects and spiders and dispose of ninety percent of all small creature corpses. If they were to disappear, species extinction would increase significantly over the current alarming rate. Land ecosystems would also diminish rapidly without their services. Not so if human beings were to disappear. Without people, the Earth would recover rapidly from its overburdened condition and flourish once again.

We are participants in an unequal partnership, but we don't have the knowledge or the good sense to appreciate the industry and organization of ants or our dependence on their services. In fact, we rarely notice them unless they enter our homes or show up at our picnics, and then we're just annoyed by their appearance.

Native people studied ants, reflecting on their ways. We let our specialists do that, publishing their observations in books and articles for other specialists. Maybe we believe that the separation between ants and humans is too great and that the manner in which they live their lives has no corresponding points to the way we live ours. But if we believe that, we've forgotten that every species has important lessons to share with our species and that size and appearance are not valid criteria for assessing importance. Native people knew that even an ant could impart a revelation and that within and about every aspect of nature were the universal patterns that drive all life.

White Ants. Termites are often linked in people's minds with ants and are sometimes called "white ants." In California, termites and ants are considered two of the state's top insect pests, sharing the least-wanted list with cockroaches, kitchen moths, and carpet beetles. To an entomologist the similarities between termites and ants are largely superficial. They classify them in different orders and consider termites social cockroaches. Since laypeople group them together, and because termites, like ants, live in sophisticated communities, we'll take a chance on annoying those who favor entomological classifications and include some termite behavior and lore in this chapter—while keeping our sight primarily on ants.

Termites are older than ants and other insects like wasps and bees that entomologists call eusocial,* meaning truly social. In fact, recently discovered fossilized termite nests in New Mexico are thought to be 155 million years old, whereas the oldest ants found in amber are about 92 million years old.

Like ants, termites live on every continent except Antarctica. Of the two thousand identified species of termites, only about forty live in the United States. Most live in the tropics, because they thrive on heat. And in rain forest areas, ants and termite colonies together compose nearly a third of the animal biomass.

Top Pests of California

Termites are ranked number one on California's list of pest species because they eat wood. Both wood-dwelling termites and subterranean termites cause millions of dollars of damage in wooden structures, and vast amounts of money are spent on pesticide services to control or eliminate them.

The differences between wood-dwelling and ground-dwelling termites determine where they live and what and how they eat, and to a lesser extent what method we use to kill them. Wood-dwelling termite colonies are relatively small—a few hundred insects—since colony size is limited by the size of the

---

* All eusocial insects have three things in common: adults who care for their young, two or more generations of adults living in the same colony, and members that are divided into those who reproduce and those who don't.

wood. These species exist entirely on wood, which they digest with the aid of special microscopic one-celled animals that live in their intestines. Ground-dwelling termite societies can hold millions of termites and build the famous structures most of us have seen in television shows about Africa and Australia where termite mounds are scattered across the landscape. These species are more advanced and less dependent on the activities of one-celled animals. They forage for food and eat a variety of things that don't need to be broken down.

The superior wood-destroying abilities of all termites, which have made them so disliked in our cities, are vital to our forests. Termites are directly responsible for keeping forests clear of fallen timber and tree stumps. They also enrich the soil by releasing nutrients from dead wood. Of the thousands of species that destroy wood, only eight groups do major damage in our cities.

To destroy them, we have traditionally fumigated soil and structures with the ozone-depleting, highly toxic chemical methyl bromide. Despite studies that have shown that methyl bromide drifts as much as 100 feet and poses a real health threat to people and animals in the area where it is used, in 1995 almost 600,000 pounds were used to fumigate about 10,000 California homes and countless businesses. This usage occurred even after the state enacted a ban on the chemical in 1984—a ban postponed three times after heavy lobbying by chemical and agricultural interests.

New methods to eliminate established termite colonies use elevated heat and liquid nitrogen, but prevention is the best course of action—although not a profitable one for the termite control industry. Recent studies of termites reveal that termites like air that contains about one percent carbon dioxide. When we insulate our homes with foam panels, which emit carbon dioxide, we may inadvertently be inviting termites to dine—and then feel like the victim of an invasion. Also, since termites thrive in humid conditions, many colonies are easily deterred at the onset by adding vents in the crawl spaces of buildings and fixing leaks when they first occur. Complicating the situation, however, are reports that some species have adapted to modern cities, having resisted our focused extermination efforts. These new termite species have expanded their food repertoire and are reputed to eat through reinforced concrete, copper, iron, and aluminum.

Ant Crimes. The crimes of ants seem mild compared with the dollar count against city termites. Ants are third on California's pest list. Small- and large-scale gardeners consider them pests because ants "guard" white flies, aphids, and other insects from predators, in exchange for a sweet substance these creatures produce called honeydew. In citrus groves, for example, the presence of Argentina ants allows scale insects (which suck plant juices) to flourish because the predators of scale will avoid any plant these ants occupy.

Since most of us are unaware of ant guards, our reasons for calling them pests revolve around their periodic "invasions" into our living spaces. When annual rains flood their nests, for example, many ants seek refuge in dry places like our kitchens. With the bountiful El Niño rains, ants are already being called the number one insect problem because they are seeking shelter in our buildings (interestingly, since termites can hold their breath for three days they can usually survive until the water has receded). Before El Niño, during California's drought, ants annoyed us as well when they entered our homes, probably searching for water. And sometimes in new subdivisions they'll enter a recently constructed home because it has been built on top of one of their ancient trails.

Whether it is raining or not, ants still enter and leave homes in established neighborhoods for reasons unknown—a scientific mystery currently being studied at Stanford University. Whatever the reason, however, we don't like trespassers, and insect control services for ants in home and garden are a billion-dollar industry.

The Wisdom and Ingenuity of Ants

It has only been in the last three hundred years, and primarily in places where industrialized society has a foothold, that ants and termites have not been accorded respect. In many parts of Africa and Australia, for instance, termites were consulted as oracles because their exceptional powers of communication (evidenced by their ability to work and live together in a harmonious and cohesive manner) were thought to include the gift of knowing at a distance. The Azande in West Africa, for one, considered the termite oracle a highly reliable source of information. Members of this tribe thought that termites listen selectively for questions put to them. And they consulted certain species of termites more than others because some species had the reputation of deliberately giving false information.

Ants have played an even greater role in myth and legend probably because they are more visible than termites. In ancient Thessaly, for example, people worshiped ants, and in the Harranian mysteries ants were grouped with dogs and ravens as the brothers and sisters of humanity. Hindu holy writings teach that ants are divine, the "First Born" of the world. Ants were also examples of the transitory nature of existence and practicing Hindus fed ants on certain occasions associated with the dead.

The tribes of Benin, West Africa, believed that ants served as messengers for the Serpent God. Among other African tribes anthills and termite mounds were related to cosmogonic ideas and sometimes linked to fertility, because if a woman sat on either, it was thought to make her fertile. In the creation myth of the Dogon of West Africa, for example, the god Amma makes Earth's feminine body from a lump of clay with an anthill for her sexual organ and a termite hill for her clitoris.

In the Middle East, the ancient Moslems believed that ants embodied wisdom and so placed them among the constellations as earthly teachers of Solomon.

Besides wisdom, ants were linked in the minds and imaginations of ancient people with ingenuity, the often underestimated power of the small. Stories from many traditions tell of the ant who outwits the larger, stronger adversary.

Traditional Japanese culture also held ants in high esteem for many reasons. In fact the Japanese word for ant is *ari*, and it is represented by the character for insect combined with a character that signifies unselfishness, moral virtue, justice, and courtesy.

Ants appear in many creation stories like the Navajo creation myth mentioned in Chapter 3 where Red Ant and Black Ant reside in the First World. In an Andamanese legend, the first man emerged from the buttressing root of a tree and cohabited directly with the inhabitants of an ants' nest. Another version tells of how he molded the first woman out of clay from an ants' nest.

The myths of the Pima Indians of South Arizona tell of an ant creator who divided the tribe into two primary moieties: Red Ants and White Ants, with Black Ants comprising a third tribe. Another Pima myth tells of an Earth Doctor who made the world from dust and created some black insects to make black gum on the creosote bush. Then he made termites to work on the small beginning until it was the size of our present Earth.

Hopi First Born. The Hopi tribe, like the followers of ancient Hinduism, believed that the first people were ants. These Ant People obeyed the laws of creation and gave a few pious human beings a place to live and food to eat when the world was destroyed to eliminate people who violated creation's laws.

The Hopi's belief in ants as "first born" has parallels to the legends of the Cere, who also believed that the first people were ants. A famous Greek legend recounts the origin of the Ant Men, originally "an Ant clan subject to the Goddess."[1] The Greeks called this race of Ant Men the Myrmidons and believed they were once ants transformed by Zeus into highly skilled warriors.

Becoming Antlike

Many cultures believed that under certain circumstances other species could transmit their qualities to humans. For example, the ancient Arabs, who saw wisdom and skill in ants, placed an ant in the hand of a newborn baby with a ritual prayer that the baby would be graced with antlike qualities.

The transmission of attributes from other species to humans was most often accomplished by eating the creature, being near it, or by being bitten or stung by it. In parts of Africa and India, for instance, childless women believed that eating termite queens would allow them to take on the insect queen's tremendous fertility. Since a queen in one African termite species can lay thirty thousand eggs a day and live for many years, it is easy to see how she could be viewed as the embodiment of fertility.

The Arawak Indians of the Pomeroon District of Guiana welcome the bite of the local black ant or *munirikuti*. An Arawak mother will even place this black ant on her newborn child, believing that the ant's bite will stimulate her offspring to walk quickly. So when a black ant bites a member of this tribe unexpectedly, the ant is not harmed, because they believe the bite is an omen that something good and satisfying is being bestowed on the one bitten.

The ant is so much an integral part of the Arawak culture that part of the preparation for a hunt involves enduring the bite of these ants. It is also part of the Arawak puberty initiation rites. Before young people of this tribe can marry, they have to undergo a trial by ants. Being bitten is a test of fortitude that bestows on the girl the strength and willingness to work and makes the boy skillful, clever, and industrious.

These kinds of practices are foreign to us, yet their existence across diverse cultures suggests they emerge from a universal pattern common to all people that links human beings with other species in deep and profound ways. In fact, in the domain of the shaman, all trials and tribulations presented by another species result in the transmission of certain qualities from that creature to the shaman. If we look at this phenomenon from a psychological perspective, it is likely that these practices assist and mark the death of a certain aspect of self and the awakening of another, more expansive identity. This new identity would reflect the favorable attributes associated with the creature while giving the individual new rights and responsibilities.

Nightmares About Ants

We have little encouragement from modern culture for coexisting with ants, much less seeking them out for their bite in order to become a beneficiary of their admirable traits. Jim Nollman, who wrote of his confrontation with yellow jackets, also writes candidly about his struggles with ants:

> One part of my mind fears that it is about to be taken over...What if a sudden stiff gust of wind should pick me up bodily and drop me onto the mound? What irony. Even as I dream a nightmare of ants as the incarnation of mindless terror, so that same nightmare prompts me to void my observed realization of essentially helpful ants. The queen's nonnegotiable demand reasserts itself [she demands that Nollman surrender to her his nightmares about ants and in return coexistence would soon follow].[2]

Although he realizes that he must relinquish his fear (as the queen ant demands), he had not as of that writing done so. Instead, following anger and fear's mandate after his daughter is bitten by an ant, Nollman pours gasoline down the anthill by his home and sets it on fire.

> I am not unlike a sleepwalker, bound to perpetrate a waking nightmare holocaust upon ants who, despite aggravating outward appearances, are trying to go about their business of living outside my dream. Unfortunately, they are never going to succeed as long as I live next door.[3]

Peaceful coexistence with other species is not always easy to accomplish when fears run wild and anger goes unchecked. If Nollman had been aware of the Arawak's beloved black ant, he might have comforted his daughter until the pain of the ant bite subsided and then celebrated the knowledge that the ant had favored his child, bestowing many fine qualities through its bite.

As mentioned in our discussion of the *sawgimay*, the discomfort of being bitten did not provoke outrage in aboriginal people. They did not insist that life be painfree. In fact, for the Arawak and other tribes, enduring pain was an integral part of becoming an adult. Pain matured an individual and prepared him or her for greater responsibility. Adults of the South American Orinoco tribe placed inch-long Bolla ants on the bodies of Orinoco boys on the threshold of adulthood. To prove they were ready to be adults, the adolescents had to withstand several stings, each one powerful enough to temporarily disable a person.

In the West we brand any ant that stings a pest, including the imported red fire ant from South America. Yet, the fire ants in their native habitat of Brazil are not considered pests but are looked on as helpful predators—the way we look on ladybugs. Their reputation as pests in the United States stems from alarmist news reporting and not because they pose any real life threat to crops, farmworkers, or livestock. In fact, red ants appear most reluctant to attack any large creature unless it ventures into their nest.

Our fear of being overrun and stung by large groups of ants has primed us to believe the worst about fire ants, especially when we hear reports from the United States Department of Agriculture's public information specialists. Capitalizing on our readiness to dislike a species that stings, a group of administrators in the 1950s (accused later of wanting to make a name for themselves) launched a massive eradication campaign against fire ants. Not only was the chemical war costly, it did great damage to other wildlife.

Ironically, our assaults and the fire ant's defenses only helped the species increase its range throughout the South. And since that time many entomologists have changed their view about fire ants. Studies show that whenever fire ants are abundant and chemicals are restricted, cotton boll weevil populations are checked (and in the absence of fire ants, in a balanced ecological system, native ants are also effective predators of this weevil).

Army Ants

It is not to say that taking special care isn't necessary around certain ant species. The army ants of South America—a favorite subject of journalists who typically embellish their reports on these creatures to sell more copy—are creatures that require some precaution on our part.

All 270 to 300 known species of army ants hunt, capture, and retrieve prey cooperatively, carrying out attacks in well-organized columns. They also emigrate periodically from one nesting site to another. A small colony may have ten thousand ants and a large one up to thirty million. Most species are blind and live underground. The few that live on the surface of the soil are the ones scientists have studied the most.

It's a myth that army ants move through the jungle like a tidal wave, consuming every living creature in their path—but it makes a good story. Army ants feed almost entirely on insects. Their most frequent prey is insect larvae, including maggots, grubs, and caterpillars, but they also dine on earthworms, spiders, scorpions, centipedes, millipedes, and slugs and have a profound, although unmeasured, effect on the dynamics of their tropical ecosystems.

When searching for food, army ants typically march with military precision in four parallel columns—six to eight ants across with guards on the sides. If an insect, small bird, or snake is in their path, they overwhelm it by their sheer numbers and then eat it. Consequently, most creatures get out of the way when army ants are marching, even the giant anteater of South America who can snare thousands of ants at a time with its long sticky tongue. Some creatures depend on army ants, including 250 species of birds (called antbirds) who live by eating the insects flushed out of the bush by the advancing ants.

If you coexist in an area with army ants, you need to be alert and cautious when they are on the move—much as you would if you lived in a city and had to maneuver around traffic. When these ants are headed toward a village, the local people pack up all their food and move out of their houses until they've passed. Most of these people agree that the benefits of army ants outweigh the inconvenience of an occasional evacuation, because when the ants are gone, there is not a snake, rat, or insect left in the house.

The driver ants of Africa are also army ants. Like their South American kin, driver ants hunt, capture, and retrieve prey cooperatively and periodically move en masse from one nesting site to another. Greatly respected and appreciated by African tribes, driver ants keep their general nesting and raiding area—usually the farmer's field—free of deadly snakes, including pythons. Since pythons are easy targets for driver ants after they have eaten and are temporarily immobilized, the smart ones won't eat if they detect the presence of these ants and usually leave the area.

Image-Making

Beyond taking precautions, we would benefit from studying ants. Within us is a natural curiosity about them, a part of our rejected wilderness self and an impulse that was once energized by the beliefs, practices, and rituals of all cultures prior to the industrialized age. By eliminating our biases and adding back information to our image-hungry minds—a primary objective of this book—we invite the natural self back into our consciousness, letting it guide our responses to other kingdoms—and in this case to eusocial insects.

Guardians of the Soil. Ants and termites and other creatures who live above and below the ground were once studied because they were thought to be bearers of special wisdom and in touch with the secrets of the soil. Ants in particular were regarded as guardians of what was in the ground. An Australian aboriginal myth tells of a nest of luminous green eggs guarded by the Ant People. Any disturbance to this nest of eggs was thought to anger the Ant People and cause great changes to the planet as a whole. Today we know these luminous eggs as uranium. And as the activities of the uranium mining companies exploit and threaten the life of the Australian Martujarra people and their land, prophecies foretell a dire outcome.

Ants were also believed to know about items buried in the ground by people, so they have a long association with buried treasure. In a Chinese story, a humble man's devotion is rewarded by the goddess he had worshiped faithfully for many years. She places a special ointment on his ears and tells him to listen to the ants. He does, and hears one ant complaining to another about the ground being cold, since an underground treasure prevents the sun from warming the spot. The ants move on to warmer ground, and the man runs for a shovel. By digging in the vicinity where he heard the ants talking, he soon finds many large jars full of gold coins and becomes a rich man.

Legends of Huge Ants. Legends from China, Persia, India, and Greece tell of huge ants that guard underground treasure. For instance, a Chinese source, Shan Hai Ching, which incorporates the shamanistic lore of the Yangtze Valley culture, refers to giant wasps and giant ants to the east.

The Greeks also mention huge ants in their descriptions of India. They called the giant ants "Arimasps" (also mentioned in Goethe's *Faust*), and said that they brought gold up out of the Earth. In fact, the Greeks believed that the Arimasps were the secret of India's great wealth.

European works of natural history and travel lore also featured giant ants until at least 1536, although some descriptions suggest that the creatures were red marmots, not ants. But even so, it does not explain the Chinese legends. The search for the giant ants continues and may underlie in some unmeasured way our modern fascination with giant ants in horror films.

Psychological Treasure. The psychological significance of these legends has also been lost today. From the perspective of depth psychologists, these tales refer to the universal process of psychological growth. Ants and other insects that make appearances in our psyches are aligned with beneficial forces and are willing to assist the human being in this process of retrieving the inner gold. Gold in this context is an alchemical symbol of the enduring essence of life within an individual, or the soul, or "higher self."

Scholars report that the symbolism of the ant at the time that these Indian legends were told was a complete parallel to the Egyptian scarab symbolism. An ant, instead of a beetle, resurrected the sun by pushing it over the horizon every morning.

Buried Wealth

Today only a small percentage of specialists are aware of the ants' historical connection to treasure or their psychic or symbolic connection to the treasures buried in our psyches. Yet these industrious species are still turning over inner and outer soil. Several years ago a news article even credited ants for the discovery of diamonds in Botswana. Geologists found diamonds in areas where ants, in their search for water, had brought to the surface particles of kimberlite, the soft volcanic rock in which diamonds are found.

In Rhodesia, geologists routinely analyze the mineral-rich soil of termite mounds to determine whether or not prospecting for gold and other valuable minerals like copper, nickel, and zinc is likely to be profitable. Like the desert ants of Botswana, Rhodesian termites bring up soil from deep beneath the surface as they tunnel in search of water. And because the soil from termite mounds has a heavy mineral composition, it has long been used in Africa as material for road paving, pottery, and bricks, and in South America as a dietary mineral supplement.

Turquoise Harvesters. Ants are also connected to turquoise and in New Mexico, certain ants harvest it. National Park Service archaeologist Thomas Winde has studied the role of harvester ants as turquoise collectors at Chaco Canyon. By mapping ruin sites and ant nests in one large site, he found that harvester ants have a definite affinity for blue and green stones—the color of turquoise—perhaps because of the stone's thermal properties. The fragments found at the ant nests cover the outside of the structure's dome and may help regulate nest temperature.

For the Lakota tribe, the stones ants collect are sacred and are used in traditional ceremonies. Healers like medicine man John (Fire) Lame Deer seek out these anthills for the tiny rocks and place 405 of these sacred stones in their gourds and rattles for ceremonial use. The stones represent the 405 tree species native to their ancestral land.

Messages from Ants

In keeping with their ancient ties to the ground, ants are also credited with forecasting earthquakes. In his recent book on American Indian symbols, Bobby Lake-Thom tells the story of when he and his wife were in Northern California to do a presentation, and the sponsor took them and a few other speakers to lunch at an expensive restaurant in San Francisco. While engaged in a lively discussion about rituals and symbolism, a large black ant slowly walked across the white tablecloth. All conversation at the table stopped, and all eyes went to the ant as though directed by an unseen power. The ant walked over to where Lake-Thom was sitting and stopped in front of him. Then it danced in a circle four times, and jumped two times as Lake-Thom talked to it in his native language. After he thanked the ant, the insect traveled south over the edge of the table and out of sight.

They all knew something unusual had happened. Insects were not usually seen in plush restaurants. Another visitor at the table voiced what they were all thinking. The ant had appeared out of nowhere and had communicated something deliberately to Lake-Thom.

Lake-Thom explained that "it was a sign…and messenger of Nature."[4] The ant told him that a big earthquake was coming to South San Francisco in four days. It would be preceded by two small warning quakes. The people at the table laughed. Lake-Thom and his wife canceled their plans to do some sightseeing over the next few days and returned to Washington. Four days later, a large earthquake hit Hayward, a city southeast of San Francisco, knocking down bridges and buildings and even killing and injuring people.

Ants and the Harvest Mother

Since ants were frequently observed carrying grains and seeds, ancient people also linked them with the Harvest Mother, one of the many personifications of the goddess. This connection to grains and seeds is frequently seen in symbolic stories where ants appear, like flies and other insects, as helpers. In one of the best known myths, the Greek myth of Psyche and Eros, ants play the role of helper to Psyche (who symbolizes the feminine soul) during an archetypal rite of initiation.

In the myth, Aphrodite tests Psyche by giving her three impossible tasks. For one task Psyche must separate, before morning, a great quantity of wheat, barley, millet, poppyseed, peas, lentils, and beans all mixed together in a heap—a task chosen so Psyche will fail. The ants take pity on Psyche and divide the grain for her while she sleeps.

In a Chinese folk tale, a young man is also faced with a series of impossible tasks by a witch trying to kill him. When he is told he must collect all the linseed he had sowed in a field, thousands of ants perform the deed for him. In another version, a swarm of ants is being carried by flood waters past a boat with the young man and his mother in it. The compassionate mother uses a sieve and manages to get the ants safely into the boat. Later, when the young man is faced with an impossible task by the witch, the ants come to collect seeds for him in payment for their lives.

In both tales, which are master teachings of the dynamics of the psyche, ants represent energies aligned with the innate push toward wholeness. They come during these periodic rites of passage to assist those who are ready to advance to the next level of consciousness, helping them outwit those who guard the gates to these zones.

Ants and Plants. The link between ants and the Harvest Mother is also played out on the physical level where relationships between ants and plants takes many forms. In fact, harvesting by ants dramatically changes the number and local distribution of flowering plants and reduces both the vegetative mass of plants and their ability to reproduce.

Many ants have mutually beneficial, symbiotic relationships with plants that secrete a solution rich in the sugars and amino acids that ants love. Ants guard these plants from herbivorous insects in return for a ready supply of this sweet substance. The Rocky Mountain aspen sunflower is one such plant that depends on ant guards to protect it from three species of flies who try to lay their eggs in its flower heads. If the flies are successful, their larvae will eat all the seeds. Ants collecting nectar on the sunflower's buds act as a deterrent by charging and trying to catch female flies as they lay their eggs.

Ants are also in symbiotic relationships with animals as noted earlier when discussing how the Argentina ant guards scale and other insects from their predators. In fact, according to a recent tally, ants have this type of "farming" relationship with over 580 different kinds of creatures. The best known example is that between ants and aphids, a relationship that invokes many a gardener's wrath.

Another mutually beneficial relationship, that usually goes unnoticed by all but entomologists, is between ants and the caterpillars of certain butterflies. The metalmark caterpillar's guardian, for example, is a species of wood ant that protects the caterpillar by forcing it down into an underground chamber during the day. The ants simply pester it until it moves in the desired direction. At night, using the same method, they coax it back up to feed in safety. In exchange the caterpillar secretes honeydew, which the ants eat. The relationship continues through the pupa stage until the butterfly emerges.

Plant-Insect Relationships in Native Myth

Native people knew of these plant-insect relationships, because they lived in intimate relationship with the natural world and observed them closely. They also learned about these relationships from their shamans, whose visions frequently revealed intricate connections between certain plants and certain insects. Among the *vegetalistas* or "plant-inspired shamans from Peru,"[5] for instance, certain plants brought insect spirit helpers. One such vegetalista under the influence of two hallucinogenic plants—*ayahuasca* and *chakruna*—not only communicated with a large ant, an "ant of knowledge,"[6] but was invited by the ant to ride on its back to the shaman's home (being helped by an animal or riding on its back was a way to assume that animal's qualities). When this shaman reached his home, a tiny chakruna plant emerged from the abdomen of the ant, and the dust and pollen that clung to a sticky substance secreted by the ant's body turned into the *ayahuasca* vine.

The Red Ant and the Manioc Plant. Indigenous myths often reveal plant and insect relationships like the myth included in *Wisdom of the Elders*, by David

Suzuki and Peter Knudtson, which describes the relationship between the women of the Amazon Kayapo tribe and a tropical red ant they consider guardian of their fields, and hence a friend and relative.

In Kayapo myth, both ant and people share a reverence for and a responsibility to the manioc plant, a sought-after source of nectar for the ants and a precious source of food for the Kayapo. The ant and manioc plant are in a symbiotic relationship or, as Kayapo myth describes it, they are bound by deep bonds of friendship and kinship. Drawn by the promise of the manioc's nectar, these ants frequent vegetable gardens where Kayapo women cultivate manioc and other domestic food plants. To reach the young manioc plants the insects cut trails through the tangle of bean vines that might otherwise choke the plants. This activity doesn't kill the bean vines, but encourages them to redirect themselves to nearby corn plants. So ant, manioc, and Kayapo women exist in a mutually satisfying relationship. Their shared reverence and responsibility for the manioc plant unite the women and the ants, and to honor the connection, they care for one another.

What is striking about this relationship is the feelings the Kayapo women have toward these ants. They cherish them and seek to be more antlike. Each symbiotic relationship in their lives is evidence of all relatedness in a sacred world in which they know themselves to be an integral part.

Our specialists don't lack for information about symbiotic relationships, but neither expert nor layperson holds dear what is known. The information we might read about this phenomenon exists outside of our lives and speaks of a world we no longer know from direct experience. Consequently, it has lost its power to inform and enrich us. Complicating the situation, we're discouraged from translating the wealth of available facts into anything real and immediate for fear of anthropomorphizing. Risking it now, consider that anthropomorphizing allows us to relate to other species. We can understand symbiotic relationships because we have many of our own, and healthy relationships cross species lines. They're the ones that are mutually beneficial to both parties—you help me and I'll help you.

The miniature world in our backyards would come alive if we paid attention to it and learned who's who with a genuinely friendly and curious attitude. And if we knew what relationships exist between other species, we could honor them. For example, we might not pluck the caterpillar off its plant and away from the insects pestering it. We might instead call it "tough love" and encourage the caterpillar to submit to the ant's insistence, telling it that it must go underground for a few hours for its own good.

Ants and Termites as Food and Medicine

Not only have we placed ourselves outside the web of relationships upon which our life depends, but we are also divorced from our sources of food and medicine, unable or unwilling to acknowledge and give thanks for the daily sacrifice of all manner of plants, insects and other animals. The idea of eating ants and termites fills most of us with disgust, but until recent times, they, like other insects, were prized and praised for their nutritional and medicinal value.

Tribal cultures depended on local species for food and medicine. They understood that a creature's body was animated by the overlighting spirit of its species and that the spirit itself determined whether or not the body was available to the hunter or gatherer. The current insistence by animal advocates on vegetarianism would mystify them, for in tribal cultures plants were every bit as aware as animals. The inhumane and disrespectful treatment of food animals and food plants, however, would sicken them. Appropriate rituals to maintain contact with the spirit world and an attitude of respect and gratitude were the primary conditions for benefiting from the physical bodies of plants or animals.

Eating insects (called entomophagy) is a current fad in many high-school biology classes and college entomology classes in the United States, but it is not a viable option for most people. Although gaining in popularity through the efforts of the entomology department at the University of Wisconsin at Madison, which publishes *The Food Insects Newsletter*, the manner in which insects are occasionally eaten is still stained with our approach to the eating of any other species—that is, we demonstrate a total lack of regard for the creature. Often, for instance, an insect life is taken on a dare or is part of losing a bet. A high-school principle in a Midwestern state, for example, made a wager with the children of his school—if they would read a certain number of books over summer vacation, he would eat a mealworm. The children did their part and the principle ate the insect, taking its life publicly, surrounded by the press who considered the event hilarious. And in other instances when the insect consumed is considered a pest, the act of eating it is often viewed and declared an act of vengeance.

Ants as Food. Almost all Amazon Indians as well as the tribes of the Mosquito Coast of Honduras used leafcutter ants as food, collecting them at the start of the rainy season as the winged ants emerge from the nest.

The Bushmen of South Africa relish ant eggs and all over Africa, ant larvae are called "Bushman rice."[7] Central American and Australian tribes prize the honey from honey-ants, a species that stores honey in their abdomens. To honor the honey-ant as food and ensure the continuation of the species, the Northern Aranda of Australia perform a honey-ant ritual that involves an elaborate ground painting. The painting represents the feminine and fertile Earth as the source of all honey-ant life and as a sacred gateway to the Other World. And at the conclusion of the ceremony the participants lie down on the ground and rub their bodies in the painting. It is an act of respectful identification with the ants and is believed to ensure the ants' well-being.

Termites are another popular food. In Africa some tribes pour water over termite nests or hit them rhythmically with sticks in imitation of rain in an attempt to trick the winged ones into swarming out. In Uganda termite mounds are individually owned. Anyone caught collecting termites from a mound that he or she doesn't own is treated as a thief. In South America some tribes break open the termite mounds and wave large leaves in the opening, hoping to entice soldier termites to attack the leaves. When they do, they pull the leaves out and gather the soldier ants, who remain fastened to the leaves in their dedicated defense of the colony.

Recent studies have substantiated the native wisdom of eating insects. Most tend to be rich in B vitamins, iron, and zinc, as well as the amino acid lysine. Dried, they have a concentrated protein content that is higher than equivalent portions of beef or chicken. Ants and termites in particular contain many nutrients needed by our human bodies and are reputed to bolster the human immune system. And ants have a high concentration of zinc, which children need to grow and develop.

Ant and Termite Medicine. Ants also have medicinal value—and not just for our species. Bears, for instance, are reputed to eat ants to cure themselves when they are sick. Some birds use ants to help them get rid of parasites in a process called "anting," although other experts say the purpose is to soothe and stimulate the body and produce a feeling of well-being. Anting occurs when a bird picks up an ant and pushes the annoyed insect into its feathers. The ant sprays formic acid, which kills the bird's parasites at that spot. Sometimes the bird moves the ant around to different locations on its body to repeat the process. And birds are not the only ones to appreciate the power of the formic acid[8] produced by ants. It was the model for the first human anesthetic—chloroform.

Ants were also used by indigenous people to close and "stitch" wounds because of their powerful mandibles or jaws. And ant bodies were mixed with alcohol and the mix used to treat rheumatics. Zaire healers used another red ant mixture for respiration problems and mixed termite nests with the red bark of a local tree to stop internal hemorrhaging. In other native cultures, crushed honey-ants were applied as a poultice to wounds, and the Lakota tribes use red ants in certain medicines to accelerate wound healing.

In China the weaver or black mountain ant is used both as food and medicine. Ant medicine in this culture is particularly important in reducing inflammations and pain, slowing the aging process, controlling convulsions, and treating tumors, asthma, and insomnia. Medicine made from ants also plays a role in protecting the liver and is used by practitioners of Chinese medicine to treat patients suffering from chronic hepatitis.

The Chinese also used termites for food and medicine. In fact they prescribed marinated termite queens for everything from colds to cancer. Members of some tropical tribes also eat termite queens as a tonic. Eating termite queens is reputed to energize elderly people, and scientists, who suspect that these queens may provide vitamins lacking in ordinary diets, have also discovered that some queens contain an antibiotic substance.

Looking Out for the Community

Since most of us don't eat ants or termites, our contact with them is infrequent—unless they enter or, in the case of termites, eat our homes. In the popular culture, termites, when mentioned at all outside the services used to exterminate them, are victims of our one-dimensional imaginings. In 1993, for example, reports of a rabid termite whose bite makes people devour wood was the lead story in a popular tabloid.

Ants don't fare much better than termites—although recently in the 1998 bestselling novel *Empire of the Ants*, author Bernard Werber takes us into an ant colony and lets us experience the lives and struggles of a fellow species. Traditionally, however, while we may extol their industry, as illustrated in the popular fable of the grasshopper and the ant, we rarely consider ant or termite society as a model for community organization and efficiency. Nor do we admire these creatures for the cooperation, flexibility, and resourcefulness that gives them a competitive advantage in their world.

Learning that they communicate with each other hasn't altered our attitudes either. It was known as early as the 1900s, through a discovery by myrmecologist William Morton Wheeler, that ants communicate with each other. This evidence of intelligence, although fascinating to specialists, did not get translated into the mainstream culture in any positive way. Popular stories and films continued to capitalize on our fears by portraying ants as a robotic mass of bodies ruled by a ravenous hunger. The 1954 classic film *THEM*, for example, features ants exposed to radiation from an atomic-bomb site. The radiation transforms them into giants with humanoid eyes, a bellowing war cry, and an insatiable appetite for humans.

With the later discovery that ants communicate primarily through chemicals or pheromones emitted by the colony's queen, filmmakers used the knowledge in another classic film called *Empire of the Ants* (no connection to the book). They endow the same mutated ants from *THEM* with an intelligence superior to humans. The colossal insects capture people and expose them to the queen's pheromones. Under the influence of this chemical, the terrified prisoners become servants, without personal will or the energy to resist.

Without reflection, we are destined to remain uncomfortable with the strange appearance of ants, their vast numbers, and their devotion to the colony. When on occasion we notice ant life and sense the colony's unified purpose, for example, we are at best confused by it. Its solidarity exists in sharp contrast to the "looking out for number one" mentality that permeates our society. In fact *ANTZ*, a 1998 animated ant movie, is sure to be a hit because it features an ant that wants to be an individual and leads a revolution against her colony.

A persistent bid for individual success and recognition in Western culture leaves us, as its citizens, without a context for understanding the evolution of a species whose service to the colony and self-interest are one and the same. Unaware of our own chemical signals, we are quick to judge these insects as mechanical creatures absolutely ruled by the chemicals of the colony's queen. We typically assume an oppressive, even despotic, arrangement is involved that prevents all individuality. It makes our discomfort more manageable and lets us pity them, while fearing their numbers and appetites. But contrary to popular opinion and horror-movie theatrics, research has shown that chemical communication is not an exact mechanical science as much as it is an expressive art, and there is surprising room for individual interpretation and response.

The Individuality of Ants

Communication by odors or pheromones is not the only means of communicating that goes on in an ant colony. Bert Holldobler and Edward Wilson have demonstrated that ants use many modes of communication including "tappings, stridulations, strokings, graspings, nudgings, antennations, tasting, and puffings and streakings of chemicals."[9]

Termites, like ants, also communicate in a variety of ways—through sound, touch, sharing food, and specific pheromones. And both ants and termites interpret the meaning of a pheromone differently. Its quantity and whether it is alone or part of other odors influence its meaning.

That ants are individuals that cooperate with other members of the colony and behave sensibly and individually in the context of their lives is old news in research circles, but few laypersons realize just how individual ants can be. Edward Wilson says that in at least some ant species "personality differences are strongly marked even with single castes."[10] Different ants, for example, will perform the same task at different speeds with some worker ants consistently fast and others consistently slow in their responses. Worker ants of the same age and species can also respond differently to the same situation. For example, when a group of ants was presented with a caterpillar known to be a source of food for this species, individual worker ants behaved differently. Only a fourth of them attacked the caterpillar and fought it until it was dead, while another fourth did not attack it at all. A third portion fought the caterpillar but not to its death, and the remaining ants retreated from it as if afraid. Apparently the evolutionary process involves mechanisms for producing and maintaining individual variations within each species—even within eusocial insects.

Different colonies of the same ant species can also behave very differently. In one species, for example, one colony was found to live on dead insects, another on food stolen from ants of another species, a third on sprouted corn kernels, and a fourth on leftover food in a human kitchen.

Learning in Ants. Besides individuality, ants have shown considerable learning abilities and routinely alter their environment artificially to reach their goals. They also demonstrate great innovation and flexibility when put in a new or different situation and can learn difficult mazes, some having up to ten blind alleys, and remember the route when tested four days later. And like many types of birds, most species of ants learn to use the sun as a compass by making connections between its position in the sky, geographic directions, and the passage of time. In short, all the ways in which an ant meets and overcomes new difficulties and adapts itself to conditions entirely foreign to its experience show a considerable amount of independent thinking.

Focusing on Ants and Termites

The ways of ants and termites invite our inspection and reflection. We would be surprised to know that we have an innate, and surprising, emotional affiliation for these creatures. This propensity to focus on them is part of a larger tendency to focus on nature and its processes—Edward Wilson's "biophilia." Not only do we depend on the natural world and its inhabitants for food and shelter, says Wilson, but we also depend on them for aesthetic, intellectual, emotional, and even spiritual meaning.

Part of this innate attraction to the natural world is connected to the timeless and universal patterns encoded within it and within each species, patterns that illuminate our human nature and which are being revealed almost daily by discoveries in such diverse fields as physics, mathematics, biology, psychology, and economics. The macrocosm is in the microcosm. This verity is also the central message of indigenous wisdom: human nature is a reflection of the nature of the universe, and we can learn about ourselves by paying attention to the natural world.

Living on the Edge of Chaos. To entertain the idea of ant life as a reflection of an aspect of the macrocosm, recall how household cockroaches, who instinctively seek out the cracks and crevices in our homes, signify the spirit of boundaries, those potent transition places at the edge of chaos where change is possible. In these boundary places, between stability and chaos, enough coherency and stability is lost to permit change, transformation, and reorganization—without falling into complete chaos. What does all this have to do with ants? New evidence reveals that ants and other eusocial insects live on the edge of chaos and actually seek out this transition place.

The finding comes out of new research on ant colonies where the behavior of individual ants was revealed as chaotic (as opposed to rhythmic or random) until enough ants were interacting or communicating with each other to shift them all into a rhythmic, orderly state of being. This shift to the "superorganism" state, a unified state long sensed by ant aficionados, depends on density, that is, on the size of the territory the colony inhabits in relation to the number of ants in the colony. Ants purposively regulate the density in order to live near this transition point—between order and stability on the one side and chaos on the other.

Biologist Brian Goodwin believes that if we knew why ants always try to live near the edge of chaos, we might have a theorem about life—about everything that is complex and nonlinear, which scientists tell us is nearly everything. He suggests that complex systems that behave in ways that cannot be explained as the simple sum of its parts, and that can evolve—like brains, economics, high temperature superconductors, beehives, and ant and termite colonies—are "expressions of this emergent order and agents of higher levels of emergence."[11] In other words, within these systems is the next evolutionary phase waiting to emerge—given enough complexity or numbers and level of interaction. Ant colonies, by adjusting their density, somehow know intuitively to seek out this edge of chaos, where optimal life for the colony exists.

It appears that optimal life is at the edge of chaos in all open complex systems, and that an ability to shift, to advance creatively, is an essential requirement for the emergence of new levels of evolutions. It means we could follow the ants and also live on this edge, in readiness for that next creative step into a new order. This model, anchored as it is in eusocial biology, points to new possibilities in the functioning of human open systems.

Author Margaret Wheatley, president of a research foundation working on the design of new organizations, has applied the idea of optimal life at the edge of chaos to organizations. She believes that generating enough information and interaction between people (and weathering the state of confusion that it brings initially) is the state to be in if the members of an organization want to be open to new thoughts. "You can't get there without going through this period of letting go and confusion."[12] In other words, confusion or chaos appears to be an essential part of a deep process of organization that, given time and enough interaction, can arise from a stew of ideas and people.

The Global Brain. What the ants are demonstrating is also evident in the vision of physicist-futurist Peter Russell. Russell has taken chaos theory, quantum physics (which says that at the quantum level all is relationship), and the science of living systems and proposed that humanity as a complex living system is moving toward an evolutionary shift into the next level of complexity and order. Russell calls that next level the "global brain" and says each of us is a "part of a rapidly integrating global network, the nerve cells of an awakened global brain."[13] In this model, described in detail in his book *The Global Brain Awakens*, as our population increases to stabilize at around 10 billion in the year 2020, and as communication technology continues to link us and permit us to interact more frequently with each other, conditions are right for a new order to emerge and shift us into the next evolutionary phase.

Although some might call this vision of how the information revolution is shifting consciousness and creating planetary citizenship an overly optimistic perspective, the fact that it is deeply established in the physical world of eusocial insects lends it added strength. The pattern already exists. And moving toward a more rhythmic and more integrated society would ease our sense of isolation and alienation. Individual activity, held within such a stable matrix, would increase, and new patterns of order would be available to the individual to try out, with abundant creative energy to go with them.

And there is still more to be gleaned by looking at ants. Ant species like the previously mentioned Argentina ant (a species which is fast spreading across the globe) also indicate the possibility of a "super, super organism."[14] Consider that although most ants will fight any strange ant, even one of the same species that lives in a nest a few hundred yards away, the Argentina ant recognizes as family *all* others of its species, even across international boundaries. And this ant will move from nest to nest joining forces without conflict. As humans continue to identify themselves as planetary citizens, regardless of race, nationality, or religion, perhaps they will follow the biological model provided by the Argentina ants and cease fighting with each other.

The Pleasure of Unselfish Action

Although the fact that Argentina ants extend their circle of community beyond their specific colony is a new discovery, the fact that ants display many so-called human emotions and engage in altruism has long been known. William Morton Wheeler believed ants resembled people to a degree not often recognized by the average person. He reports that, like our species, ants display a range of emotions including pain, anger, fear, depression, elation, and affection. They also demonstrate empathy when they help crippled and distressed sister ants, another behavior of human beings. A remarkable piece of film produced in 1973 by a Russian entomologist and mentioned in Stephanie Laland's *Peaceful Kingdom: Random Acts of Kindness for Animals* showed an Amazonian ant extracting a splinter from the side of another ant. Other ants in the community formed a circle around the "patient" and "doctor" ant and preserved a clear space around them until the procedure was completed.

What is not, however, a predominant human behavior is the suppression of the sex drive in the majority of ants and their consistent altruistic behavior. In certain advanced forms of ant life, sex totally disappears in the majority of individuals—a fact sure to evoke ambivalence in us. More startling than that biological fact is that this practical suppression or regulation of the sex drive appears to be voluntary. Ants have learned how to develop or arrest the development of sex in their young by nutrition.

Rigidly restraining all sex to within the limits necessary to ensure survival is one of many vital practices of ant species. The female worker does not associate with males. Her femininity is expressed in every way but sexually, for she has all the tenderness, patience, and foresight that we consider maternal.

The Altruistic Society. Those who study ants have noted that ants don't seem to have any individuality that is purely selfish. Some scientists go so far as to acknowledge that in regard to social evolution, ants have advanced beyond us. Scientist David Sharp notes that ants in many regards have perfected, better than the human species, the art of living together in societies and have acquired some of the industries and arts that greatly facilitate social life.

Scientist Herbert Spencer points out that the life of an ant is entirely devoted to altruistic ends and he believes that ants are ethically, as well as economically, in advance of human beings.

In ant society, the will of the individual and the well-being of the community are a seamless whole, so that the only possible pleasure is the pleasure of unselfish action—a stance similar to some native people like the altruistic society of the Hodi of south-central Venezuela.

Ants attend to their individual life only so far as it is necessary for the life of their society. In other words, the individual takes only the food and rest that are needed to maintain its vigor. No ant takes more, and no ant sleeps longer than necessary to keep its nervous system in good order. Each ant works without stopping, and workers keep themselves and their inner living areas neat. By modifying their physiology, they have apparently repressed every capacity for individual pleasure except to the degree to which that pleasure directly or indirectly helps the community.

It is not that ants (or bees or termites) have a sense of duty or constantly sacrifice themselves for the colony. In fact, the concept of duty is meaningless. Instinctive morality replaces any need for an ethical code—it is simply the nature of eusocial insects. They have a biological disposition to pursue altruistic ends instead of egoistic ends, and as their relations we have a biological basis for community and service as well.

Herbert Spencer believes that someday humanity will reach a state of civilization ethically comparable with that of the ant. If we turn our imaginations in the right direction, we might entertain the idea of a global society in which a regard for others produces so much pleasure that it overrides the pleasure derived from direct gratification of personal desires. Spiritual traditions teach the rewards of service whereby taking care of yourself means taking care of others. Eventually then, there may come a time in humanity's evolution when egoism and altruism are so in agreement that they become one and the same.

Relating to Ants

As we've seen, ants have been models of industry and resourcefulness in cultures all over the world. The fourteenth-century ruler Tamerlane is said to have been inspired by an ant trying to carrying something heavy up a wall. The ant failed sixty-nine times before succeeding on its seventieth attempt. Afterward, Tamerlane resolved to go out and conquer Asia and acquired an empire that stretched from Russia to Arabia and from Turkey to India.

Those who study ants find fresh insights into human society and a vision beyond the subject of their observations. Auguste Forel published a monograph on ants in Switzerland and then went on to become a psychiatrist and social reformer. A Jesuit scholar from the Netherlands made many studies of ants and their parasites and was able to see in them manifestations of divine power. The best contemporary example of an individual who, focusing on ants, found all the mysteries of life and stimuli for many insights is Edward Wilson, whose contributions appear throughout this book and beyond these topics.

Compassion for ants is found in the stories of religions that placed great emphasis on cultivating harmlessness and right relations with other species. A Sufi legend tells of a man who traveled several hundred miles every month to purchase supplies. After one trip, when he returned home, he discovered a colony of ants in the cardamon seeds he had purchased. He carefully packed the seeds up again and walked back across the desert to the merchant from whom he had bought them. His intent was not to exchange the seeds, but to return the ants to their home.

Talking to Ants. A relationship with ants often springs from their appearance in our homes. When ants invaded his kitchen, J. Allen Boone felt angry and wanted to eliminate them. Before he could do so, however, he felt guilty and decided to try and contact them. He couldn't find their leader, so he tried to broadcast a message to all of them. When he began his communication, he reprimanded them for spoiling his dinner and questioned their right to be in his house. They ignored him completely. Then he remembered that in his other attempts at interspecies communication, he had discovered that all creatures like to be appreciated.

He sent the ants his admiration for their keen intelligence, their energy, their focused attention on their tasks, their harmonious relationships with each other. He asked for their understanding and cooperation. Then he left the room, feeling like he hadn't made contact. Later, he went out for the evening, returning just before midnight. When he entered the kitchen, there wasn't an ant in sight. The food that had attracted them earlier was still there, but not one ant was visible. He was never bothered again in his home, or when he traveled, although ants lived outside his door in great numbers and regularly invaded his neighbors' homes, much to their annoyance.

A Swiss businesswoman Kathia Haug also had success in communicating with the ants visiting her home each summer. Although in previous summers she had tried to solve her "ant problem" by killing them, she had never been completely successful. One year, she embraced a new spiritual discipline that brought a change of heart and prompted her to try talking to the ants mentally.

She told them that she wished them a long and happy life, but that she wanted them to stay out of her house. In one day, the ants were gone from the house, although they still passed across the doorstep from her kitchen into the garden.

In Rebecca Hall's exploration of the psychic connection between humans and animals, she says that to successfully communicate with ants (or any other species) requires a combination of compassion, positive expectation, and the ability to hold a mental picture. When it doesn't work, usually the person attempting to communicate is in some kind of mental turmoil, or is feeling ill-at-ease, or impatient. She tells of one woman who wanted to get ants out of her kitchen and lectured them crossly saying, "Get out, you load of little communists."[15] The ants ignored her until she realized she had to be more respectful.

Household Spirits

In *The Spell of the Sensuous*, David Abram tells of the time he was in Bali and a guest in a magic practitioner or balian's household. Each morning the balian's wife brought Abram a bowl of fruit. She also carried a tray containing many two- or three-inch boat-shaped platters woven from a section of palm frond and filled with white rice. After handing him his fruit, the woman disappeared from his view behind the other buildings. When she returned for his empty bowl, the other tray was always empty. He asked her what the rice platters were for, and she explained that they were offerings for the household spirits, that is, gifts for the spirits of the family compound.

Curious, one morning Abram followed her and saw her set one platter at each corner of the other buildings. Later Abram walked back behind the building where she had set the platters down. The rice was gone. The next morning after the woman picked up his empty bowl, he went back behind the buildings again. As he looked at the platters of rice that she had set out, he saw one of the rice kernels moving. He knelt down to look more closely and saw a line of tiny black ants winding through the dirt to the offering. At the second offering he saw another line of ants carrying away the white kernels of rice.

He returned to his building amused. Here his host and hostess had taken great pains to appease the household spirits with gifts only to have them stolen by ants. Then a strange thought slipped past his Western orientation and entered his mind. What if the ants *were* the household spirits? What if the offerings were made with them in mind? The family compound was constructed in the vicinity of several ant colonies and was, therefore, vulnerable to infestations by the sizable ant populations, especially where there was food. Maybe the daily gifts of rice prevented such an attack by keeping the ant colonies well fed. Perhaps the offerings also established a certain boundary between the ants and the humans since the platters were placed in regular locations at the corners of the compound's buildings. By marking and honoring this boundary with gifts, the balian's wife may have intended to enlist the cooperation of the insects to respect the boundary and not enter the building.

His encounter with ants was the first of many experiences suggesting that the "spirits" of an indigenous culture are those modes of intelligence or awareness that are in a nonhuman form. Instead of waging war on them, these people acted in ways that we would do well to emulate.

Interestingly, more esoteric sources reaffirm the wisdom of both Boone's approach and that of the balian's wife. In *Awakening to the Animal Kingdom*, trance channel Robert Shapiro and assistant Julie Rapkin offer messages received from the oversoul or guiding spirit of various animals. When they contacted the overlighting spirit of the insect kingdom, they received the following message about making contact and requests.

> Respect us if you would, for we are in fact an idea of God/Goddess/All That Is.... Be willing to come outside...and speak to us where we live. If we are inside your dwellings, speak to us there, out loud or in your thoughts. Speak to us with honor and respect and ask us if we would be willing to live outside of your dwelling space. Be willing to share your foods with us, for at times we do require it. Sometimes man will pave over the surface of the ground which denies us access to that surface. It is at times like this that we will emerge in your structures to remind you that we are creations that need to exist in love and harmony with physical sustenance for food.[16]

The insect spirit from our less-than-orthodox source also asks that for three days people bring insects gifts of food to a safe outdoor location to ensure their prosperity and continuance. And it directs us to be respectful and thank the insects for their willingness to cooperate with our lessons. The suggestions are reminiscent of how native people brought gifts to other species to show respect and enlist their cooperation. Respect is necessary for harmonious coexistence. Without respect, the communication channel is closed and we stand outside our own community, looking in and unable to unlock the door and enter.

Returning to the ground of science and ant enthusiast Edward Wilson, we find that Wilson also encourages us to respect the ants. The most frequent question he is asked is what to do about ants in the kitchen. He always answers: "Watch where you step. Be careful of little lives. Feed them crumbs of coffeecake. They also like bits of tuna and whipped cream."[17]

Kindness is always appropriate and the springboard from which we can best observe and learn from ants. Seeing in the ant colony a level of synergy and cooperation that we might want for our own species and a unifying spirit that makes each member a part of a cohesive whole—greater than the sum of its individuals—we might return to our human community inspired. We have gone to the ant, considered her ways, and are the wiser for doing so.

10

Birds of the Muses

> Last night, as I was sleeping, I dreamt—marvelous error! that
> I had a beehive here inside my heart. And the golden bees were
> making white combs and sweet honey from my old failures.
> —Antonio Machado

More has been written about bees and the life of the hive than about any other insect. Since prehistoric times bees and honey have assumed a sacred role in the mythology of cultures worldwide. Cave paintings from fifty thousand years ago depict strange beelike creatures, epiphanies of the Goddess, and Chinese legends speak of a giant race of bees who live in the K'unolun Mountains. The Pankararé tribe of Brazil considered bees enchanted creatures protected from human exploitation by the guardian spirits of plants and animals.

Ancient cultures venerated bees, believing them to be endowed with divine gifts and mysterious powers. Some scholars say that wherever people worshiped the Earth Mother or Great-Mother—the goddess of fertility, wildlife, and agriculture—bees also had a sacred status.

Western culture views bees primarily as resources, commodities of commerce. After all, the honey that bees produce each year is worth millions of dollars. Of greater value still is their pollination services. In the United States bees pollinate more than one hundred agricultural crops worth about ten billion dollars. Add in the milk and meat from farm animals who feed on bee-pollinated crops and the figures more than double. Some vegetables are also grown from bee-pollinated seed, and homegrown vegetables pollinated by honeybees are worth about 1.4 billion dollars a year.

The Value of Bee Products. Bee venom, bee pollen, royal jelly, propoplis (bee glue), and beeswax are also sought-after commodities. Bee venom, for instance, is valuable because of its healing properties, although the beneficial effects of venom that beekeepers and others claim have still not been investigated thoroughly. Stories abound, though, of bee venom's effectiveness in treating arthritis[1] and other ailments.

Bee venom is also thought by some to help multiple sclerosis, a disease that degenerates the central nervous system. A neurologist who heads the Multiple Sclerosis (MS) Center at Georgetown University Hospital says a certain protein in the venom, apimin, might be helping MS patients by improving the conductivity of nerve sheaths.

The venom from the sting of the honeybee is standard in homeopathic first-aid kits for healing swollen and inflamed tissues from wounds or burns. Internally it is an excellent anti-allergy remedy and a natural antihistamine—without any side effects. And by rapidly reducing water retention in the tissues, it also helps our kidneys excrete excess fluids.

A chemical analysis of honeybee venom reveals that it contains a substance that acts favorably to minimize rheumatic pain. Scientists say that therapeutic effects of bee venom come primarily from its being a natural specific stimulant, that is, it mobilizes the defense forces of the organism, provoking a general nonspecific reaction. Although many people believe they are allergic to bee venom, some studies show that reactions to these stings are often due to a protein hypersensitivity and not to the bee venom itself. Beekeepers who suffer regular stings have better than average immunization and a resistance to certain sicknesses.

Collecting venom for any kind of "sting therapy" is a process where our view of bees as commodities only, without rights or privileges, is clearly evident. It is most often done with a charged grid that rests in or near the hive. A thin synthetic material stretches under this frame, and when bees land on it, they get a mild shock that makes them sting through the material. The venom collects on the underside of the material. It takes ten thousand bees to produce one gram of pure venom. Happily, bees' stingers rarely get caught in the material, and so ninety-nine percent of the insects survive the process.

In the Company of Bees

When we only consider a species as a resource—whether for honey, pollination, venom, or other products—it hinders our ability to care deeply for them, regardless of their importance. Dollars don't add up to meaning. If we aren't beekeepers, we're left outside the circle of bee and human, a circle that has fed the imaginations of people since the recorded history of humanity.

In *The Queen Must Die*, contemporary beekeeper William Longgood's knowledge of bees comes from his feelings and intuition and what a long intimacy with these insects has taught him. He tells of bees grieving over the loss of a queen, making war cries, or humming with contentment. He describes them as angry, fierce, calm, playful, and aggressive and distinguishes their happy sounds from their distressed ones. Not good science, perhaps, but good sustenance anyway for our minds and hearts. We can relate to what he says. We know about grief and distress calls. We hum too when we are content and emit all manner of sounds to express how we feel as we go about our day.

If we kept company with bees, we would know that there are correspondences between our species and a community of bees. Those who do spend time with bees report that they are calming to be around and invoke peacefulness. The bees' contentment is apparently contagious. Perhaps it stirs some neglected part of ourselves, our wilderness self who honors the bees' selfless devotion in making honey—a substance that all mythologies, save ours, have considered the essence of wisdom.

The strong sense of order and purpose that dominates a community of bees and their hum of confident industry, so in accord with Nature, has long inspired beekeepers. In his classic book on bees, *The Golden Throng*, Edwin Way Teale calls the beehive "one of the great living philosophies of the world."[2] This lover of bees believes the solidarity of the swarm reassures beekeepers that life has a grand purpose and helps these men and women find courage and faith in eternal ways.

Before the industrialized age, before we silenced our imaginations to prevent ourselves from falling into superstition, people all over the world linked bees to peace, harmony, propriety, renewal, fertility, industry, and eloquence.

Bees' historical association with peace, harmony, and propriety, for instance, is so strong that people believed that in times of war bees would sicken and die, and that a hive would not do well if it were stolen. It was also believed that bees would react to the immorality of their beekeeper with a stinging fury, and this notion of honeybees as guardians of morals is still common in France.

When stinging bees disrupted a female circumcision ceremony (a practice known as genital mutilation and defended as an essential ritual to ensure chastity) in Freetown, Africa, in January 1998, those who know about the bee's historic concern with propriety might have taken their stinging actions as a commentary on the brutal act—nature's way of speaking out against it. Viewing the bee's actions this way is an imaginative truth far more interesting than the tired old assumption of bees stinging from mindless aggression.

Living outside the circle of bee specialists and bee symbologists, most of us know too little about bees to draw any scientific or imaginative conclusions about their actions (except as we said by pronouncing them aggressive if they sting us). We don't reflect on our correspondences or allow ourselves to be inspired by bees either. The division between us and bees has existed for too long for us to bridge unless some incident with bees forces us to reconsider them.

Many who have read the rich lore of bees find that even when popular beliefs were based on wrong information about the functioning of the hive and other notions that scientists have long since disproved, knowing how the bee was thought of in ancient cultures is rewarding. The mythology of ancient cultures seeds our imaginations and rejuvenates the image-making creative pattern within us that is a kind of "hieroglyphic of the spirit."[3] In time, we might even find ourselves reenchanted with the world, more able to understand our own mythologies, and allow new discoveries to lead us further into life's mysteries.

The Sacred Bee

To tell the history of people's involvement with bees would require its own volume, so we will limit ourselves to a few prominent ideas. For one, bees have long been linked to death and the human soul. The Aztecs, for example, believed that the human soul becomes an insect, and referred frequently to a Bee God which leads experts to think that they may have believed in a bee-soul.

For the ancient Egyptians the bee also symbolized the human soul and was associated with the solar cult of Ra. Myths say that when Ra wept, his tears were honeybees—a symbolic link between compassion and its ability to be transformed into something beneficial and greater than itself.

The bee has appeared in sacred books of many traditions from the Rig-Veda, written in Sanskrit between 2000 and 3000 B.C., to the Koran and the Book of Mormon. And Mohammed taught that the bee is the only creature ever spoken to directly by God.

Among the Hindus in India, Vishnu, one of the great nectar-born triad—Brahma, Vishnu, and Siva—was represented as a blue bee on a lotus blossom. Krishna and Indra were also frequently portrayed as a bee on a lotus flower, and Krishna wears a blue bee on his forehead. The Indian god of love, Kama, carried a bow whose string was formed by a chain of bees, and the Indian god Prana, who personified the universal life force, is sometimes portrayed encircled by bees.

The Greek deity Zeus was called the Bee Man because as an infant he was hidden in a cave and guarded by bees who nourished him with honey. To drown out his cries, the Curetes, warrior-priests of the Great Mother, were said to beat their spears and shields together, and in one version of the myth, the bees came to the cave because they were attracted to the noise of the clashing weapons.

This myth may underlie the still-prevalent belief in rural areas that "tanging" or banging on old pots and kettles will attract a swarm of bees to an empty hive. The practice persists even though we now know bees cannot hear sounds in the way we do. They can only sense sounds as vibrations traveling along a surface with which they are in contact.

Bees were also linked to fertility and its seasonal cycles. Wild hives tucked away in the hollows of ancient trees or in the crevices in rock faces linked the bee to the secretiveness of generation in the womb and the secrets of life before incarnation. The fact that bees disappeared in the winter and reappeared in the spring tied them specifically to the Earth Mother's annual renewal of fertility.

Christian bee symbolism followed that of the Earth-Mother religions. Bees symbolized resurrection and immortality. Their disappearance in the darkness of the hive during the three months of winter was like the three days that Christ's body was hidden in the tomb. And the insect's reappearance in the spring symbolized Christ's resurrection.

Christian saints associated with bees have strong parallels with the ancient gods and goddess. Many experts believe they were assigned to older legends to legitimize the stories and cancel out the Earth-Mother images. St. Sossima, for example, the patron saint of beekeeping in the Ukraine, who is said to have brought bees from Egypt in a hollow reed, resembles Zosim, a pre-Christian Russian bee god. And a Brittany legend that corresponds to the myth of the Egyptian sun-god Ra tells how bees were created from the tears Christ shed on the cross.

Birds of the Muses. In the Far East bees were called the "Birds of the Muses." The idea that bees have the power to impart eloquence to a child of their choosing has a long history in many parts of the world. People believed that bees had swarmed on the mouths of Plato, Sophocles, and Xenophon when they were babies, giving them their eloquent style of teaching. And when the poet Pindar was an infant, bees were said to have fed him their finest nectar.

Among the Hebrews, the bee was related to the idea of language, and its name, *dbure*, and the Hebraic root *dbr* means word or speech. In Christianity, St. John Chrysostom was called "Golden-Mouthed." It was believed that the sweetness of his preaching came from the swarm of bees hovering around his mouth when he was born. The same story was told about the famous preachers St. Bernard of Clairvaux and St. Ambrose. In fact the legend of the child Plato and the bees was probably transposed to St. Ambrose.

There are relatively few American Indian stories about honeybees, because these insects are not indigenous to the Americas but were brought over by the colonists as an agricultural aid. Stingless bees did, however, inhabit tropical and subtropical areas of the Americas. The Mayans domesticated the species found in their area and made an intoxicating drink from the wild honey.

The Symbolism of Honey. Honey has as long and as complex a history as its creator. Rock paintings dating from the earliest era of human life show humans stealing honey from wild bees. The Incas of Peru offered honey as a sacrifice to the sun. Babylonians and Sumerians erected their temples on ground consecrated with honey and offered honey to their gods at religious ceremonies.

In ancient cultures eating honey was a sacrament with as great a meaning as the sacrament in the Catholic tradition in which bread and wine are believed to be transubstantiated into the living flesh and blood of Jesus Christ. Honey was believed to change into a mystical substance after being consumed. This substance of spirit, then, had a beneficial power on the person eating it. Symbolically, honey was both wisdom and the sweetness which comes out of the mouths of the wise and the strong.

Interestingly, when Sharon Callahan, a practicing interspecies communicator in northern California, was asked by a beekeeper friend to contact the overlighting spirit of the bees, she discovered (without knowing anything about honey's ancient role) that as the bees tell it honey still should be regarded as a Sacrament, a holy communication between the one partaking of it and God. She learned from the bees that honey is transformed inside the body of those who eat it. Once digested, its properties act upon the DNA structure, and honey created from pollen gathered in different regions carries different energetic patterns or information—so we would do well to eat honey created from our own region.

Alchemic Symbolism of Bees and Honey. Bees and honey also figure prominently in the alchemic symbolism of certain spiritual traditions. Bulgarian spiritual teacher Omraam Mikhaël Aïvanhov says, "the only thing that the great Initiates, those true alchemists, teach is how to become a bee...."[4] The highly advanced disciple or Initiate, for instance, who takes the pure and divine elements from the people and prepares a food in his or her heart for the angels is a bee.

In the secret order of the Freemasons, the bee's ability to collect pollen from a flower was also a metaphor for humanity's ability to extract wisdom from daily life. In these teachings the hive was a reminder that diligence and labor for a common good brings happiness and prosperity.

The Spirit of the Hive

Like ant and termite colonies, a hive of bees has long been thought of as a single living organism, inspiring many analogies. The Desnan, a Columbian tribe, drew a sophisticated and detailed comparison between the meticulously organized beehive (or termite nest) and the human brain—a comparison that those inspired by the possibility of a global brain will appreciate. The bishop of Milan, St. Ambrose, patron saint of beekeepers, compared the beehive to the Church and the bee to a diligent Christian. A contemporary observer likens the hive with its core of liquid honey to the Earth with its molten core. Others have compared the hive to the human body with the queen serving as the brain and the individual bees performing like cells.

What was until early in this century the unifying "soul" or "spirit" of the hive (and ant and termite colony) has been hypothesized by some biologists to be a self-organizing and self-evolving field intimately linked to the queen. This ordering field is believed to contain blueprints or patterns that organize and coordinate the individual insects much as a magnetic field organizes iron filings.

As we saw in our brief discussion of chaos theory and the new findings about ant colonies living on the edge of chaos, a growing number of biologists believe the coordinated activity of bees is an emergent property of the hive itself and not an outside field. In this view the beehive, like the ant colony, follows the same principles as other complex nonlinear systems—moving from chaos to rhythmic order as the level of interaction between a specific number of individuals increases.

Theory aside, from the time of the ancient Egyptians, who were the first to start keeping bees in a methodical way, beekeepers have found the community's unity and division of labor extraordinary and not a little mysterious. Some of these mysteries are tied to their language, others perhaps to the effect of this hypothesized unifying field or emergent order. Yet, even with recent discoveries, few who keep company with bees are not moved to admiration by the bees' organizational genius and ability to work together to accomplish their goal.

The Popular Culture's View of Bees. Outside the field of entomology and the big business of agriculture, the popular culture's jaded view of bees focuses on their ability to sting us. The fact that bees can sting plugs them into an unimaginative track in mainstream society where other creatures with the potential to harm us also reside. In the past few years we have been put on the alert about Africanized bees, popularly called "killer bees." The media's portrayal of killer bees more than satisfies our unconscious need to find and fight insects as enemies. Sensational news reporting about them has left us panicky about all bees, and Hollywood has lost little time in adding to our fears by depicting killer bees as creatures that maliciously seek out victims to kill. So great has been the uproar about killer bees that much of this chapter will be devoted to investigating the charges against them and finding a way to coexist. First, however, some background on bees is required to debunk the popular notion that bees are little more than robots wired to respond in fixed, unyielding ways.

The Dance Language of Bees

When we reduced bees to resources, we assumed they were mechanically programmed to find and pollinate vegetation and produce honey. But science has verified what ancient people knew intuitively: honeybees, like other social insects, respond flexibly to their environment, communicate with each other, and have a great deal of intelligent awareness. Consider that honeybee scouts don't proceed robotically to report their new food discoveries. They report them only when the colony needs additional food sources, when the new source's distance from the hive is not too great, and when its quality and quantity is worth mentioning.

Karl von Frisch, an Austrian zoologist and Nobel laureate who studied bees for fifty years, discovered that bees communicate by dancing. The "waggle" dances and "round" dances of bees are part of a complex social communication system that goes on continuously in a community of bees. The dances occur when something that has been in short supply has been discovered—like water, waxy building materials, or food.

A bee will perform a dance only where it has an audience of other bees to watch (perhaps bees that dance without an audience are like people talking to themselves—slightly suspect). Sometimes, when the dance is performed in the darkness of a hive, the dancer makes faint sounds. The audience also makes sounds, a kind of begging signal to the dancer to stop and regurgitate food samples.

A bee performs a round dance when she has discovered food relatively close to the hive. Waggle dances are performed to communicate information about things located at a greater distance from the hive. Both convey the direction and distance to the food source using a kind of geometrical symbolism. The bees also communicate the quality of what has been discovered—the more intensely the bee dances, for instance, the better the quality.

Although many researchers have tried to disprove that an insect is capable of symbolic communication—perhaps wanting it to be a unique human capability—James Gould, a professor at Princeton University, has proved that bees really do use the information coded in the dance. Gould says that the dance language of bees is definitely symbolic communication, that is, it is language. And it accomplishes what human language does—it describes something removed in time and space.

A lot of communication among bees goes on when a part of the bee colony with a queen breaks off from the main group and flies off looking for another home—a phenomenon called swarming. Studies show that honeybees "plan" to swarm, for swarming involves a lot of preparation including gorging on honey so the departing bees and their queen will have enough energy to make it to a new location.

A swarm will often cluster on the branch of a tree a few yards away from their old home. From there, scouts go out to look for a suitable location for a new hive and "report back" to the other bees. Waggle dances performed in front of a swarm lead to a group decision about where the bees should move to establish a new colony. After many hours of exchanges of information, when, finally, all of the scouts indicate by their dance that they favor the same hive site, the swarm flies off to their new home. This consensus depends wholly on the "speaking" and "listening" interactions between bees.

After seventy-plus years of studying symbolic communication among honeybees, researchers have discovered that the dance language has different dialects in different parts of the world. Other findings reveal that bees other than honeybees also communicate symbolically. Several stingless species, for instance, use signals comparable to the Morse code to convey distances.

Bees and Quarks. The most far-reaching research, and research that promises to join mathematics and biology, has been conducted by a mathematician at the University of Rochester, Barbara Shipman. She has described all the different forms of the honeybee dance using a single coherent mathematical or geometric structure (flag manifold). And interestingly, this structure is also the one that is used in the geometry of quarks, those tiny building blocks of protons and neutrons. From this and technical evidence too complex to present for our purposes, Shipman speculates that the bees are sensitive to or interacting with quantum fields of quarks.

Researchers have already established that bees are sensitive to the planet's magnetic field, but they have always attributed it to the presence of a mineral in the bee's abdomen. Shipman's research indicates that the bees perceive these fields through some kind of quantum mechanical interaction between the quantum fields and the atoms in the membranes of certain cells. Shipman says simply, "The mathematics implies that bees are doing something with quarks."[5] If Shipman is correct and bees can "touch" the quantum world of quarks (without "breaking" it as we do when we try to detect a quark), scientists say it would revolutionize biology, and physicists would have to reinterpret quantum mechanics as well.

The Intelligent Awareness of Bees

The more James Gould works with bees, the more he is convinced that they have a great deal of intelligent awareness, including a navigational intelligence that surpasses ours. They learn early to read compass directions from the sun by observing and remembering the sun's apparent motions across the sky and relating these motions to the passage of time and to geographical distance.

Their awareness of what is happening during an experiment is also obvious. In a simple test, for instance, when Gould and his colleagues moved food systematically at regularly increasing distances from the hive, some of the bees began to anticipate their movements. These clever bees flew past the training station and waited at the next food location for the researchers to bring food.

Although Gould does not believe that bees are self-aware (and says he would have to end his research on them if he knew they were aware because some die in his studies) perhaps he suspects it, for he goes to great lengths to see that his bees survive as often as possible.

Gould tags his bees with numbers for easy identification, and in his long intimate association with them, has discovered that, like people, they have both distinctive characteristics that make them individuals and predictable traits that are common to them all. He is always impressed by the chaos in the hive, because the behavior of bees is not reliable statistically except when taken as a group—a fact that suggests that bees, like ants, live at the edge of chaos where life and possibility abound.

Each bee, in fact, is so individual in the way it approaches its duties, Gould and his colleagues can recognize them by their differences. One is brief and goal-oriented at the food dish while another appears to feed at leisure, even taking time to fly around the dish looking for drips.

> Gould could often predict with his eyes closed which insect was coming in; this was the one who always circled twice, or the one with the very throaty buzz. New recruits were even easier to identify—they flew in to the food source low to the ground, hovering and hesitant, with a slightly different buzzing tone.[6]

Since bees only live six weeks, Gould has even seen them age. Suddenly they are not quite as good at foraging. During the last few days of their life, they are also apt to be blown off the feeder by the wind, and Gould would know that his "old friend" was not long for this world.

Telling the Bees

The intelligent awareness of bees was recognized long before it was demonstrated in our experiments. This observed awareness is probably one of the reasons that "telling the bees" is still a time-honored custom in rural areas.

Throughout Europe it was generally believed that bees wouldn't thrive if important news was withheld from them—especially the news of when their keeper died. The custom has deep roots in the bee's association with the human soul and their habit of frequenting caves. Caves were thought of as semi-sacred entrances to the underworld and bees were seen as the souls of the dead either returning to Earth or en route to the next world. In fact, the myths of ancient Germanic tribes say that bees came from an underground paradise where they dwelt with the Fates.

So when bees were told of important events like births, deaths, and marriages, it was a way of conveying news to souls no longer in a human body. Telling the bees also arose from the typically close relationship between the bees and the beekeeper and the fear that if the bees weren't told, they would take offense and leave.

Telling the bees that their keeper had died was done in different ways. Sometimes relatives of the deceased would place black crepe or a small black piece of wood on the hive for the period of mourning. In other places they turned the hives as the coffin left the house.

Sam Rodgers's Bees. Today, telling the bees is still done and has given rise to at least one recorded mysterious event. It occurred in Shropshire, England, in the mid 1960s. In the small county of Myddle lived a beekeeper named Sam Rodgers, who was a postman, cobbler, and handyman. Although he liked his work and had many friends, his great love was for his bees. Each day he would go to the hives and care for them. He would always talk tenderly to them as if they could understand his every word and every affectionate gesture. Stories about Rodgers say that he would sit in a chair in front of his hive and ring a small bell. The bees would swarm out and cluster all over him "in the greatest delight, loving him as much as if he were a queen bee."[7] Then one day the benevolent beekeeper, who was eighty-three years old, did not come to the hives. He had died.

His family, aware of the old customs of the area, knew that someone must tell the bees of his death or risk the chance that the bees would leave the hives and never return. Two of Sam's children walked down the path to the hives and solemnly told the bees of Sam's passing.

Sam was buried a few days later in the local cemetery behind the town's church. The Sunday after the funeral, parishioners called the parson outside to witness what was happening. The Reverend John Ayling reported later that he saw "bees coming from the direction of Mr. Sam Rodgers's hives which were a mile away."[8] They formed a long line and headed straight for Sam's grave. Some witnesses said the line of bees reminded them of a funeral procession. Once the bees had circled Sam's gravestone, they flew directly back to their hives. Friends and members of Sam Rodgers's family were mystified by the strange behavior of his bees; so were qualified bee experts in the area. No one could explain logically what they had witnessed, for it seemed as if the bees had gone to the grave site to pay their last respects and bid Sam good-bye.

Building Bridges

The folklore of the bee and its rich symbolism through the history of humanity has little relevance for us today; it is no more than an interesting curiosity—if that. Ask anyone what the bee means to them and they are likely to answer honey production or stinging pain. Never has a culture had more facts about the natural world than we have today, but instead of adding to our lives, our knowledge has stripped our imagination of images that once brought inspiration and meaning.

In our fear of succumbing to superstition and fancy, we have elevated the myth of objectivity to the point of severing (through denial) our relationship with the natural world. There are those who would advocate throwing the methodology out and replacing it with anthropomorphizing, if only to honor the obvious commonalities between people and other species. As laypeople, we could start by assuming that we are connected to bees and have many behaviors in common. Once comfortable with that idea, we could use our skills of observation to highlight our similarities and differences.

As it is, contemporary beekeepers, if they feel "known" by their bees, are likely to keep it a secret or dismiss it as fancy. A New York beekeeper and philosopher wrote in *Bee Culture* that people always ask him if his bees recognize him. Although he admits that it would seem so since he can move about his bee yard at certain times of the year without a veil, and if a stranger suddenly appears in the yard, he or she is likely to be stung, he answers, "No," that his bees have no idea who is he. It only seems as though they do, he decides, because he has learned how to move about a bee yard without rousing their defenses.

Although this beekeeper would receive an approving nod from traditional scientists for avoiding the trap of anthropomorphizing, bee specialist James Gould (and others) would argue with his conclusion. Gould discovered that just as the experimenters could tell the bees apart, the bees soon learned to tell the humans apart. He thinks the bees do it by smell. If one of his observers, for example, who had been watching the bees feed at a feeding station, walked off into the bushes to relieve himself, his bees, noticing his absence, would usually come to him in the bushes. And foraging bees that Gould had been working with would also come looking for Gould first thing in the morning. If a stranger or other researcher came out onto the field instead of Gould, the bees would fly to him or her, hover for a moment (as though taking a sniff), and then leave.

The desire to be known, recognized, even loved by another species is great within us. It emerges out of the needs and hunger of the instinctive, natural self that we have abandoned. Its cry to be known, as it once was known by the sentient world, is felt today as a cry of loneliness and alienation that no amount of technological toys and scientific reasonableness can eliminate. Our insistence on a scientific methodology that separates the observer from the observed and our devotion to numbers and abstraction has cost us our feeling link to the world. It has also distanced us from the transcendent creative pattern that operates behind all form and that could rejuvenate us and return us to a loving interdependence with other species.

As we saw in our look at flies, cockroaches, and ants, our facts are often contaminated by unexamined biases and assumptions. Common evaluation of other species involves only two questions: Do they serve us and can they hurt us? With bees, the answer to both is, "Yes," throwing us into a state of ambivalence. Bees so obviously serve us and all creation, yet we don't know how to be in relationship to those who can also hurt us.

Killer Bees

The thought of killer bees frightens us. Although the sting of a killer bee is no more toxic than the sting of the normal honeybee, these bees are reputed to be easily agitated, and their "angry nature" makes multiple stings more likely. We are not accustomed to being careful around other creatures. We are used to seeing them at a safe distance—on television or in zoos. Killer bees are autonomous beings, who live out their lives with little regard for our million-dollar honey industry and our rent-a-hive crop pollination enterprises.

Killer bees are a relatively new strain of bees created by crossing two subspecies: the European honeybee and an African honeybee. The result of the crossing has been a more aggressive bee that is harder to manage and raises the possibility of a rapid genetic takeover of our domesticated honeybee—a phenomenon that has already occurred in South America.

To understand the issues involved and separate media invention from fact, we need to know a few more details about bees in general and the European honeybee and African bee in particular.

Of some twenty to thirty thousand bee species, about six are honeybees. These six probably originated in southern Asia over millions of years and are considered more advanced than bees that live alone. The honeybee species commonly found today in Europe, Africa, the Middle East, and the Americas is *Apis mellifera*, which means honey carrier. There are several subspecies of this honeybee ranging from the gentle and easy-to-manage Italian bee (*Apis mellifera ligustica*) to the larger, more aggressive German bee or "dark" bee (*Apis mellifera mellifera*). The bee that started the whole killer bee scare is *Apis mellifera scutellata* from eastern and southern Africa. Since all these honeybees are of the same species, bees from one subspecies can mate with bees from another subspecies and create even more variation within the honeybee tribe.

Bees are shaped by their environment. The temperament and habits of European bees, for instance, evolved out of interactions with their environment. Year after year, European bees were faced with changes of seasons that included a long winter. To survive they had to store enough honey to last through the cold months and develop ways to keep warm. During the warm months, abundant foraging opportunities, plentiful water, and few predators let them maximize their honey storage to make it through the winter. And to keep warm they learned to conserve heat by forming a tight ball around their queen.

African bees had different challenges. They had a warm climate year round with frequent droughts and unpredictable weather patterns. They also had to contend with a large number and variety of predators who would attack and destroy any colony that was not fiercely defended. During times of drought, when vegetation diminished or when their nest was threatened, African bees developed a survival tactic called absconding, that is, leaving their old nest and finding a new one.

Knowing even this little bit about African bees, it is easier to see how the current situation evolved. To survive, they have learned to respond to any perceived threat to their colony with quick and efficient aggressive tactics, and, to the dismay of those who would "manage" them, they are quick to leave the area when threatened by lack of food or predators. Their aggressive tactics, overstated and exaggerated by sensational news reports, emerge from their will to survive.

The African Bee Comes to Brazil

The killer bee scare originated in 1956 when the African bee (*Apis mellifera scutellata*) was brought to Brazil by Warwick E. Kerr, a Brazilian university professor. Kerr was acting under the authority of the Brazilian Ministry of Agriculture who wanted to strengthen beekeeping in the warmer parts of Brazil. Since the African species was already used successfully for commercial honey production in South Africa, Kerr believed it would prove to be of immense value to Brazilian beekeepers.

The trouble began in 1964 when military forces took over the government and Kerr, an outspoken human-rights advocate, criticized the military government. He was jailed twice. To discredit Kerr as a scientist, military leaders are suspected to have played upon people's general fear of stinging insects. News leaked out about a few African queen bees escaping from the school's hives and taking over colonies already established in Brazil. From that point on, when anyone was stung, whether by wasps or bees, government officials blamed the incident on Kerr's killer bees.

Cornell University bee expert Roger Morse is one who thinks that the Brazilian military government deliberately manufactured much of the hysteria about Africanized bees to undermine Kerr's human rights activism. So does beekeeper and author Sue Hubbell, from whose illuminating account of these bees in *Broadsides From the Other Orders: A Book of Bugs* this chapter borrows heavily. She says the term "killer bees" was probably a government-instigated media invention picked up and spread in news reports about people being stung to death—by any kind of stinging insect. The original name for these bees was "assassin bees" because on rare occasions they would raid the hives of other bees and kill the queen. The term "killer bees," however, went deeply into our imaginations. Our fear of stinging insects in general had already prepared a place for it.

Controlling Killer Bees

In 1986, to stop Africanized bees from moving north, the United States Department of Agriculture and scientists in Southern Mexico joined forces in a $1.5 million joint program to build a string of traplines in a region called the "Bee Regulated Zone" (BRZ). When the first natural swarm of killer bees crossed the Rio Grande into Texas in October 1990, they were prepared. United States agency officials killed all the bees within hours (so who are the angry and easily agitated ones?). As predicted by beekeepers, however, the BRZ did not hold. The bees showed up in Arizona in 1993. Two fatalities from killer bees were subsequently reported in Texas, although one thousand deaths have been attributed to killer bees throughout South America. Both Texans were elderly men. The first was stung at least forty times when he tried to eradicate a colony with a flaming torch and had an allergic reaction to the stings and died. The second man, who was ninety-six years old, was sitting on his porch when a nearby gardener disturbed a colony. The bees mistook him for the threat and stung him repeatedly. In 1995 a pair of tree trimmers were stung in the small California river town of Blythe. One man received twenty-five stings and the other fifteen—unpleasant but not life-threatening by any stretch of the imagination. Healthy adults can sustain as many as fifteen hundred stings and survive, and they can outrun bees (which fly at between 10 and 15 mph) and find safety in a car or other enclosed space.

About forty people die each year in the United States from the stings of venomous insects. Ordinary European honeybees (probably the German bee, which dominates our hives in North America) are responsible for about half the deaths—usually from an allergic reaction. Since killer bees are hard to identify, any bee that stings a lot, including the German bee, is assumed to be an Africanized bee.

The temperament of bees, like people, is not a fixed trait. It varies. Most beekeepers agree that climate and altitude influence the temperament of bees. Hubbell points out that bees in low-altitude areas, which are noted to be very ill-tempered, lose much of their testiness when they are at an altitude over 6,500 feet. Apparently the cooler, drier climate at that altitude has a soothing effect on them. And California beekeeper Andy Nachbaur has found that when he moves honeybees from Northern California to the desert regions in the southwest, they become "violent," something he also attributes to environment and weather.

Bravo Bees

The caution that we need to exercise around Africanized bees is matched by the caution we need to bring to what we read and hear about them or any species the public has decided is an unqualified threat to human life. Most of the stories about killer bees are products of the active imaginations of journalists unfamiliar with bees. The old adage "when a pickpocket meets a saint, all he sees are pockets" is true for journalists and bees. Journalists are looking for action, or better yet, threatening behavior. Hubbell reports that a friend who keeps two thousand Africanized hives in Costa Rica allowed an American television producer who wanted to film the bees for a story to accompany him on his bee rounds. When the bees showed no evidence of being ferocious, the producer offered to pay him to stir the bees up, explaining it would make a better story. Hubbell's friend refused.

One step we could take toward promoting coexistence is getting rid of the term "killer bees." Sue Hubbell calls them "bravo bees," or "fierce bees," following the example of some South American beekeepers who have learned to work with the bees' natural inclinations. We'll adopt the former term throughout the rest of this section as we sift through the information on these bees to assess the bravo bee's real threat and distinguish it from media hype.

Opinions differ, but Marla Spivak, who has studied bravo bees throughout Latin America and is coeditor of the book *The African Honey Bee*, thinks the arrival of the bravo bees may eventually be the best thing that has ever happened to beekeeping in the Americas.

Bret Adee, another beekeeper friend of Hubbell and co-owner with his father of the largest commercial honey operation in North America, thinks the bravo bee's aggressiveness has been overstated by scientists who compared bravo bees with Italian bees or other bees more gentle than the dark, or German, bees that commercial beekeepers in North America are already used to handling. After talking to a number of Central American beekeepers who work with bravo bees, Adee discovered that they preferred these Africanized bees because they are more productive and disease-resistant than other honeybees they have worked with.

Consider that before Kerr brought these bees over from Africa to breed with their domesticated honeybees, Brazil ranked forty-seventh in world honey production. Now, after the bravo bee takeover and a generation of learning new beekeeping methods, Brazil ranks second in Latin American honey production and seventh in the world.

Living with Bravo Bees. To assume the role of beekeeper to bravo bees requires caution. Bravo bees will defend an area up to two or three city blocks in size, so beekeepers must move their beehives away from houses, barns, and farm animals who might inadvertently bump their hive and be attacked for the action. Beekeepers must also contend with the tendency of bravo bees to abscond when the blossoms in one area cease flowering or when the hive is threatened. Minimizing threats and feeding them sugar syrup when flowers disappear has already persuaded many hives to stay put.

As laypeople, if we are still balking at the idea of living with bravo bees, we need to understand that honeybee researchers say that we can't destroy bravo bees without destroying all bees, so we have no choice but to learn to live with them. The key to sorting out irrational fears from reasonable concerns is knowledge. If we are afraid of bees or any stinging insects, it behooves us to learn something about them. For example, the best defense is not to swat at the bees—which makes them more likely to attack—but to run for shelter. Bees don't look for trouble, nor do they attack without provocation. And any bee, even a bravo bee, will not sting while happily collecting pollen or nectar unless she is stepped on or otherwise threatened. She can only sting once and dies afterward, so she will save that action for the defense of her hive, its honey, and her sister bees.

Knowledge about swarming bees also prevents undue panic. A swarm may look dangerous to us, but they actually may be more docile than normal because they don't have a home to defend and are busy "discussing" new possible nesting sites. What's more, they are usually so full of honey that they can't bend their abdominal area to insert their stinger even if they wanted to.

The Likes and Dislikes of Bees. A fact commonly known for ages, and verified today by scientific studies, is that honeybees are sometimes aroused by human breath. In fact, experts advise that we not breathe directly on nearby bees.

Bees are also sensitive to bright colors. The man who discovered the dance language of bees also discovered early in his career that honeybees have excellent color vision. In fact, the sensitivity of bee color vision precisely matches the hues in flowers, and flowers have adapted their coloration to the bee visual spectrum as though wanting to be noticed and found attractive. So when we wear clothes that simulate the bright hues of flowers, we must expect bees to notice us and find us attractive.

If you will recall, James Gould noticed that bees seem to distinguish people by their smell. This sensitivity extends to a variety of smells. People who wear colognes and perfumes outdoors, for instance, will find themselves under close scrutiny by bees. Be aware too that bees don't like the body odors that come from uncleanliness. A person who has not bathed for some time might find himself or herself the focus of an attack in a bee yard, while other people are left alone.

Like their relatives the wasps, honeybees like sweets, especially liquid sweets in the form of open bottles of soft drinks. They gather around eating areas at open-air events like fairs and carnivals and crawl around on the straws and bottle caps, so it's always advisable to examine the inside of your soda can or bottle before taking a swig.

Replacing myths with facts will also help gardeners coexist safely with bees, for all bees are beneficial to gardens. What is not a good idea, however, is providing bees with appealing nesting sites in our gardens like empty boxes, cans, buckets, upturned flowerpots, tires, open pipes, or overgrown shrubs and trees. Experts also recommend that we seal any opening larger than one-eighth inch, such as pipe entrances on walls or where stucco meets wood or brick, and repair or replace damaged vent screens on foundations and eaves.

As we become more comfortable with the idea of coexisting with bees—even bravo bees—and as beekeepers adapt their management techniques to the ways of these bees, I suspect that we are going to have an opportunity to learn that there are always unexpected benefits in situations where our customary manipulations have led us to believe that we are in control of the forces of nature. Consider that the same life force that allowed African bees to survive in an environment with many dangers appears to make bravo bees more productive and resistant to disease. Their hardiness may prove to be the single most important reason to welcome them into the United States, for our wild and domestic honeybee populations are facing extinction.

Our Declining Honeybee Populations

In the 1980s and 1990s two varieties of mites have been wiping out honeybee populations in Europe and every state in the U.S. except Hawaii. A 1997 report says that an alarming ninety-five percent of the United State's wild honeybees have died during the past two or three years. Before the mite epidemic there were 5.2 million bee colonies in the United States—half of which were wild. Now there are only about a quarter of a million wild colonies. The mites are also taking a toll on our domesticated honeybee. Tennessee's domesticated bee population, for example, dropped from 36,000 in 1986 to 4,000 in 1995.

With the declining populations, rental costs for agricultural pollination have climbed more than 50 percent in the past few years, and other pollinators are being studied. An unexpected benefit is that protection has increased for wild pollinators. Farmers are being encouraged to keep the edges of their fields wild to provide nesting areas for wild pollinators and to refrain from applying pesticides. A downside, however, is that many are writing off honeybees rather than investigating more deeply the reasons why our honeybees are so vulnerable to the mites.

If we consider the possibility that the mites are *not* the real problem but are messengers, what might their overwhelming presence be telling us about honeybees or the larger environmental picture? A growing number of beekeepers believe that honeybees have been so weakened by years of manipulation that they cannot ward off the parasites that a healthy bee population could.

Gunther Hauk, beekeeper and director of The Pfeiffer Center, a biodynamic gardening and environmental program in New York, has worked with bees for over twenty years. He and others like him believe that the mites and other diseases that have affected honeybees since the 1960s are all symptoms of a weakened immune system in the bees and an overall degenerated state of health—a result of our own limited insights into nature and an attempt to improve on one of nature's miracles. According to Hauk, one of the most serious factors in our manipulative management of bee colonies is the way we breed queens. He explains:

> We have taken an emergency situation in a colony where, upon the loss of a queen, the workers raise a new queen out of the egg or larva of a worker bee, and...raised it to a standard method of propagation. That's like driving around on a thin spare tire instead of a full tire.[9]

To understand why this practice has been so detrimental to bees requires an understanding of the impact of form on the development of embryonic life—for in nature queens are raised naturally in round cells and workers and drones are raised in hexagonal cells. To the public and community of professional beekeepers unfamiliar with the influence of form upon life (what is called "sacred geometry" in esoteric traditions), it doesn't make sense. So our "perfected" technique of artificially raising queens with artificial insemination has not been recognized as one of the reasons for the weakened health and vitality of our bees.

Hauk points out that when Rudolph Steiner warned people in 1923 that the honeybee might not survive the end of the century, he was referring to the way we raise bees and how our best methods contradict the requirements of nature to create and sustain a healthy hive. As the plight of the bee grows more serious each season, more beekeepers are willing to explore what those requirements are—something Hauk teaches in his "Bees in Crisis" workshops—and we can only hope that it won't be too late to save the honeybee.

A Hidden Killer. The saturation of our environment with pesticides has also contributed to the ill health and in many circumstances the death of both wild and domesticated bees. We already know that pesticides like Sevin, which was used to kill the gypsy moth after DDT was banned, is very toxic to bees, destroying both wild and domesticated colonies. Encapsulated pesticides like Penncap also pose a real danger to bees since they are carried back to the hives by the bees and lodged with their stores of honey. Death comes later in the winter when the bees use the stored food.

A Colorado county bee inspector reported in 1997 that he thinks encapsulated pesticides are the real culprit in decimating the honeybee populations in his area—a replication of losses in 1970 that were traced back to Penncap. As evidence he offers the fact that despite rigorous mite control procedures (including pesticide strips in the hive), honeybees are still dying and honey production has dropped dramatically.

Transgenic Crops. Another problem facing honeybees is of unknown magnitude, and one that using bravo bees and returning to natural methods of beekeeping may not be able to amend. Bees are picking up mutant pollen from "transgenic" crops—the crops genetically engineered with foreign genes. Early studies have shown that rape-seed oil plants, for example, engineered to produce the insecticide Bt, not only killed the caterpillars and the beetles it was engineered to kill, it killed bees as well. Another study indicated that more than 30 percent of bees who visited a new Bt cotton hybrid died, although no further assessment was done on whether their death was actually a result of the altered cotton. And honey produced by bees who take pollen from plants with this insecticide gene may also be poisonous, or cause severe allergic reactions in humans.[10]

If we consider all the factors contributing to the plight of our honeybees, we might want to not just put up with bravo bees, but start welcoming them into North America. They are hardier and more efficient than our European bees, and their queens lay more eggs and at a faster rate than European queens. The eggs of bravo bees also develop into adult bees faster than European eggs, which means bravo bees can work earlier in the season and dominate the food sources. They even forage late into the night when the moon is bright.

To soften the takeover, Texas researchers are breeding bravo bees with the domesticated European bees, which appears to reduce the bravo bees' aggressiveness. And in managed hives, experts are replacing the bravo queen bees with European queens to keep the Africanized queen bees from propagating. If those practices continue, however, we may, out of our fear (and a misplaced faith in our own cleverness), compromise the hardiness of bravo bees by our manipulations and sabotage their ability to resist mites and thrive in our pesticide-polluted environment.

Addressing the Real Threat. All in all the so-called threat of bravo bees has not materialized. Although bravo bees are in North America, they tend to confine themselves to a year-round existence in southern Texas. The grim possibility that honeybees will be extinct by the turn of the century, however, is the more pressing threat and if that happens, many believe that the consequences would be dire and far-reaching, for agriculture as well as for nature's intricate interrelationships.

In Sharon Callahan's 1998 communication with the overlighting spirit of the honeybee, (in which she was told about the potency of honey), she also learned that bees play a critical part in the re-patterning of the Earth and planetary grids as well as struggling to hold in place what they call the "new pulse" of energy from the Creator—struggling because of their weakened constitution. Within this energy is all the information needed for all species and the Earth to move in harmony into the new millennium. "Natural selection, comb shape, the speed of vibration of the bees themselves are all vital to the bees' ability to keep up with instructions from the Creator [on how]...to build the vibrational matrix necessary for...[all to move] into the next phase of becoming or spiritual evolution."[11]

Perhaps it is these energetic instructions that the honeybees are receiving from their mysterious interaction with quarks in quantum fields. And perhaps in this area science is meeting the insights provided by "spiritual sight." Both are needed if we are to help restore the bee to health, enlist its continued cooperation in pollinating our food plants, and help all bees—if only by not interfering—hold the new energetic pattern emerging in our midst.

Communicating with Bees

For the nonbeekeeper, contact with bees may be limited to watching them visit our flower beds and gardens. But for the adventuresome, and for those times when there are no other options, communicating with bees is also a possibility. A classic story of communicating with these insects is found in Doug Boyd's account of intertribal medicine man Rolling Thunder, who teaches that all fear comes from misunderstanding. When a friend of his was gathering herbs, she reached for some horehound, but drew back suddenly when she saw that bees were swarming all over the plants. She stood up abruptly, pale with fright. Rolling Thunder told her that she wasn't really afraid of any living thing, she only believed she was. He reminded her of the loving experiences she had had as a child with other species and urged her to talk to the bees. He advised her to tell the bees that she wouldn't hurt them and to ask them to share the plants with her.

> I did as he said, and...the bees actually understood me, and they moved! I just can't describe how I felt. All the bees on the plant I was looking at moved. They all moved together to the back of the plant....[12]

Then Rolling Thunder came up to her and told her that the response from the bees was a gift from the Great Spirit. When we can set aside our fears and self-consciousness, we open ancient pathways that have always linked us to other species.

Chiquinho of the Bees. The link between people and other species, even potentially dangerous ones, is sometimes revealed in special human beings, people who have a "gift" so unusual and outside the norm that their behavior is rarely thought of as a latent human potential. One such special person is a young man named Francisco Vicente Duarte, called "Chiquinho of the Bees,"[13] who was born in a small Brazilian city of farm-worker parents and grew up in the company of nine siblings. Since the age of three, Chiquinho has demonstrated a rare ability to understand other species. Poisonous snakes don't bite him, dangerous spiders become his friends and confidants, and bees (no doubt bravo bees) and wasps land amiably on his face.

Chiquinho's affinity with other species and his preference for dangerous invertebrates have given him quite a following in the city where he lives. Born with an affliction that makes him look more like a ten-year-old than the thirty-plus-year-old man that he is, Chiquinho is also believed to be slightly retarded. Perhaps these disabilities have provided the permission he needed to explore his other gifts, instead of spending all of his time laboring in the fields. Today he sometimes helps his family by performing in the city park. He also helps by responding to calls from people having trouble with bees or wasps in their homes. He goes to the home and calls the insects. They land on his body and stay with him as he leaves the house, heading for his home and an improvised bee yard.

Parapsychologist Alvaro Fernandes, who investigated Chiquinho as a boy of twelve, says, "It may seem hard to swallow but Chiquinho can really talk to animals and they really obey his requests."[14] Although skeptical at first, American parapsychologist Gary D. Richman accompanied Chiquinho on his excursions into nearby farms to catch animals and was amazed at what he observed. No creature, fish, spider, or snake appeared to be scared when the boy approached them. They remained quiet, as though waiting for him to catch them. He could reach into a nest of snakes, for example, and calmly bring up a cascavel or a jararaca or any other poisonous snake coiled around his hand as if it were hypnotized.

In a 1990 article published in a Brazilian magazine, Chiquinho says of his rapport with other species, "The animals have always understood me and they even obey my commands." He doesn't spend any time analyzing his affinity to animals but simply accepts it. "I talk to the animals and they talk to me. I can understand everything they say. My talent is a gift from God."[15]

A friend of mine, Silvia Jorge, had an opportunity to speak with Chiquinho at his home and to film him with his five poisonous snakes and later with a nest of wasps and a wild beehive. Jorge said Chiquinho's mother told her that when he was about three years old, she found her son in their backyard intertwined with a large poisonous snake. She feared greatly for his life before realizing suddenly that the boy was playing with the snake. She says that is when the gift emerged.

Since that time his mother admits that she has had to protect her son from researchers who wanted to take him away and study his abilities. One group even offered to pay her for the right to take Chiquinho to Europe to study him and dissect his brain when he died. She refused.

The day Jorge visited Chiquinho in the summer of 1997, she found him personable and eager to demonstrate his ability. She watched as he talked softly and patiently to the poisonous snakes that live with him, who would lick his face in unmistakable affection. Then she snapped pictures as Chiquinho lifted his hand to a large wasp nest hanging from a nearby building and called the wasps to his hand. Finally the neighborhood boys located a wild beehive and

Chiquinho and Jorge went to have a look. Demonstrating his affinity with bees, he talked to them in a soft, affectionate tone as he put his hand into the hive. During his demonstration, a few bees flew over to Jorge, landing on her as she was videotaping. She called out to Chiquinho and told him that she was very nervous. He called to the bees telling them not to sting Jorge. He told them that she only wanted to take their picture because they were "very great beings" and to fly back to him. The bees left Jorge and went to Chiquinho.

After witnessing Chiquinho, it was apparent to Jorge that his communication with these species is a natural, even casual, behavior on his part. Fernandes didn't view Chiquinho's abilities as paranormal either, because Chiquinho's accomplishments are not achieved through any great mental concentration power. Rather it seems that what occurs between Chiquinho and the animals happens at an instinctive level, a body-to-body exchange of energy. Barbara Brennan, former NASA scientist and pioneer in the exploration of the human energy field, says that the communication of feeling between people and other species is an energy field interaction and that our intentionality influences our human energy field. So Chiquinho could be broadcasting to other species—via his energy field—his deep accord and intention not to harm long before he interacts with them physically.

A Brazilian professor, Dino Vissoto, believes that Chiquinho doesn't pose a threat to other species because he isn't afraid. When people are afraid, he explains, they produce and give off some kind of secretion. Rolling Thunder talks about this scent of fear in terms of vibrations (perhaps Brennan's energy field interaction) and teaches that we can learn to control our vibrations and so affect the way we interact with other species.

J. Allen Boone believes that when we are taught to despise a creature, our feelings are a kind of mental poison that is somehow communicated to the creature, which then reacts against it. Chiquinho loves and trusts other creatures—including sometimes dangerous invertebrates like bees. That in itself sets him apart from most other people raised in industrialized society. Whatever the mechanism, these feelings appear to be communicated to the creatures as he approaches them, for they display great trust in him.

Special people like Chiquinho hint at the great latent potential within all of us to communicate with other species. Interspecies communicators Sharon Callahan and Penelope Smith maintain that these abilities are our heritage. We have only to pay attention and cultivate them to enter the community of other species where we belong. In *Life Song*, Dr. Bill Schul says communication with other creatures may not only be important, but it may also be critical to our survival. "Interspecies communication becomes a matter of the parts recognizing their kinship with the other parts and the whole…all life is a creation of and with God, the All Conscious."[16]

11

All Our Relations

> Perhaps everything terrible is in its deepest
> being something helpless that wants help from us.
> —Rainer Maria Rilke

Bacteria are news. Pick up any newspaper and chances are it will have an article or two about a bacterium implicated in an infectious disease or a new "magic bullet" antibiotic that promises to kill a drug-resistant strain. Another popular topic is antibacterial soaps and disinfectants and products with an antibacterial coating. Market analysts have noted our growing fear of germs and our obsession with germ-killing products. Advertising is already playing on our fears as manufacturers scramble to meet the increasing demand.

We can now buy sports bras, pillows, kitchen towels, baby blankets, toys, pens, carpeting, socks, and even toothpaste that all claim to wipe out bacteria on our bodies or in our homes. About thirty percent of the $2.1 billion hand- and bath-soap market is already comprised of antibacterial products, and 150 antibacterial products were introduced in just the first nine months of 1997.

Car companies are following suit. Honda Motor Company is offering its Japanese market a germ-killing steering wheel in its Life minicar. So is Nissan Motor Company, along with antibacterial-coated gearshift knobs. They call it part of the move to build "healthier" cars.

The chemical that goes into many antibacterial formulas is a pesticide called triclosan which prevents the growth of bacteria. When we use products with triclosan, our upper levels of skin absorb a small amount of the chemical and it stays there even after we rinse with water. The agent damages the cell walls of the remaining bacteria, slowing their growth and ability to multiply. The usual concerns about its effectiveness and safety over the long run have been downplayed.

Yet many microbiologists who promote the use of antibacterial products are not unbiased advocates, especially those whose research is funded by the companies that manufacture these products. The notion of the disinterested, unattached academic researcher who produces pure data is a romantic one. The

Krimsky study, published in the journal *Science and Engineering Ethics* in 1996, indicated that about one-third of the papers that appeared in fourteen prominent scientific and medical journals had "at least one author with some financial interest in the research published."[1]

Helping Bacteria to Mutate

As a preventative measure, using antibacterial products seems like a small but effective action we can take against a world overrun by mutant and potentially deadly bacteria. But like our war against insects whom we have branded pests, it's an ill-conceived strategy born of fear and a lack of knowledge about bacteria.

Specialists on bacteria and infectious diseases are warning us that by washing our hands with antibacterial soaps and using antibacterial disinfectant cleaners in our homes, we are probably helping bacteria mutate into other more potent strains. Think about it—they have to. We are killing them indiscriminately, most often in their benign forms, and they are fighting for their lives. We need to realize that bacteria are noted for their ability to rapidly adapt to changing environmental conditions—it's one of their strengths. Changing our home environments from friendly or tolerable to deadly is sure to trigger their survival mechanisms. In fact, in hospitals it already has. Drug-resistant strains of bacteria abound in these institutions and are known for causing major infections that are difficult to heal.

Using antibacterial housecleaning products is particularly dangerous when family members are hospitalized and then discharged early to complete their recovery at home. They return home to an environment of bacteria that have survived the antibacterial products and that can probably resist conventional drugs. Microbiologists concerned with this trend say that if we would relax and use basic soap to clean our bathrooms, floors, and bodies, we could save antibacterial cleaners for those times when we or other family members are particularly vulnerable to infections.

Understanding Bacteria. A little bit of knowledge about bacteria could go a long way toward easing our fears. It might even catapult us into a totally different relationship to these life-forms. Since we have already discussed bacteria in several other chapters, we'll build on that information and highlight the ways in which bacteria are helping us. Our emphasis, however, in light of the germ-fueled terror that permeates our culture, will be on the war against disease-producing bacteria and those strains now resistant to antibiotics.

To begin with, we will define a few terms. Bacteria are microbes, and the term microbe is actually an all-inclusive category of microscopic organisms that refers to any of thousands of species. We call microbes "germs," "pathogens," "bugs," "bacteria," or "protozoans," depending on the context. Bacteria are the most numerous microbes. Germs are the least numerous. Only one microbe out of a thousand is a germ.

Our relationship to bacteria is a remarkable one. As mentioned in the chapter on flies, there is a growing body of evidence that supports the idea that all life is the offspring of bacteria, recombinations of the metabolic processes of these life-forms that appeared some two thousand million years ago. Every plant and animal on Earth seems to have come from and be sustained by bacteria.[2]

Microbes were the ones that invented all of life's essential chemical systems, including putting oxygen into the atmosphere. And at some point, through symbiosis (a major source of evolutionary change on Earth), they combined with other microorganisms, forming permanent alliances, and created us. The products of some of the first mergers between ancient bacteria invented human digestion, movement, and our tactile and visual systems. They also created speciation, cannibalism, and genes organized on chromosomes and eventually even our kind of cell-fusing sexuality.

Mitochondria, descendants of these first mergers, live inside our cells, reproducing at different times and in different ways from the rest of our body's cells. As hosts, we provide them with food and shelter. In turn, they give us waste disposal and oxygen-derived energy. In fact, ten percent of our own dry weight consists of bacteria, many of which, like those bacteria that live in our intestines and produce vitamin B-12, we can't live without.

Since microbes can constantly and rapidly adapt to environmental conditions, they support the entire biosphere—primarily through networking. As a part of that intricate network, we are dependent on bacteria. The implications of having bacteria as ancestors and intimates are far-reaching. Although a radical notion for some, our link to bacteria may be the humbling we need in order to give up our inflated ideas about our position in the circle of life. Maybe we could learn to take pride in the fact that we may be the offspring of a supercreative life-form operating on this planet, for as we will see in the next section, our tiny ancestors and companions are the embodiment of creativity and vitality.

Strange and Wonderful Appetites

In a teaspoon of soil as many as ten million bacteria and other small organisms live. As nature's most efficient recyclers, these underground creatures routinely break down organic matter and release plant nutrients. To help us clean up the environment, bacteria are indispensable allies because they have an astonishing range of appetites, including a taste for all kinds of waste products.

Microbiologists are excited about their appetite for contaminants. Once a particular strain of bacteria has been identified as liking a particular by-product, these specialists encourage their growth in a contaminated area or introduce new bacteria with similar tastes. It is still being argued whether it is more effective to apply microbes to a spill or stimulate those already in the soil or water. Many believe that introduced bacteria can't compete with hardy indigenous soil bacteria, and only time and experimentation will settle the issue.

Some bacteria even break down chemicals from which they appear to derive no nourishment or other clear benefit. Perhaps they do it for the sheer love of the process or for the challenge. There are even bacteria that love lethal substances like arsenic. One that actually thrives on arsenic has been found in a watershed north of Boston where toxic waste had been dumped for years. This "new" bacterium dubbed MIT-13 is anaerobic (lives without oxygen) and depends on arsenic, not oxygen, to release energy from food.

Alongside these active types of bacteria are others in the environment that have previously adapted to environmental change and then go into a state of suspended animation. Stresses such as climactic changes or the appearance of chemical pollutants activate these quiescent microbes. Once aroused, they impart a memory to the ecosystem of other changes that have been successfully adapted to and they help guide the total system in coping with the new stresses.

Bioremediation. The accomplishments of bacteria that feed on organic and chemical waste may seem commonplace after realizing their importance in creating and sustaining life on Earth, but it is still a cause for celebration. Once treated with skepticism, the abilities of bacteria are now well documented. In environmental circles one of the most frequently recounted stories tells of how bacteria cleaned up the lethal cyanide from a gold processing plant that had poisoned a South Dakota creek.

In a Maryland sewage-treatment facility, bacteria also take center stage. The wastes, both human and industrial, feed four million pounds of microorganisms, and the result enters the Chesapeake Bay. Microbes also make possible flushless toilets for vacation homes and campsites and turn smelly sewage into odorless compost.

New technologies have sprung up, capitalizing on the abilities of microbes. Bioremediation is one. At its heart are bacteria that break down toxic compounds like pesticides and leave water, carbon dioxide, and other harmless products in their wake. It's a technology used by over fifty cleanup companies in the United States whereby microbes work on everything from gasoline-soaked soil at gas stations to the sites with the worst carcinogens.

Industrial Waste Clean-Up. Sites whose soil and water are contaminated with grease and paint have posed some of the most difficult challenges. A microbiologist sorted through twenty thousand mutants of a single strain of bacteria and isolated the most effective form known to degrade trichloroethylene (TCE), a contaminant found in solvents for removing grease and paint. Today this bacterium is used with another strain of bacteria and a fungus at many paint-stripping sites to remove the paint from metal. Once it is removed, a third kind of bacteria eats the paint.

Some substances at contaminated sites actually glue the soil together, preventing nutrients, pollutants, and bacteria from interacting. Fortunately, a microbiologist has identified a bacterium that can make these soils more accessible. Another microbiologist, digging through some abandoned wood-preservative sites in Florida, discovered other bacteria that degrade carcinogenic PAH, a contaminant found at coal gas sites that gums up the soil along with creosote and coal tar. A bacterium that feeds on quinoline, a poisonous byproduct of oil shale and coal processing, has also been found.

Extremophiles. In the spotlight as well are extremophiles—microbes that thrive under extreme conditions and that promise new strategies for cleaning up pollutants and creating new medicines and products. Many communities of extremophiles live beneath the surface of the Earth. A search for these subsurface microbes was fueled in 1987 by the fear that microbes could disrupt the integrity of sealed chambers containing radioactive waste. The results of this quest revealed that subsurface bacteria are ubiquitous. They live in most sedimentary rocks, nourished by a supply of organic compounds that were originally produced by plants and later buried and consolidated into solid rock. Microbes also thrive in igneous rock, which makes up most of the continental crust of the planet, and may capture energy from inorganic chemical reactions. Other extremophiles include microbial communities that populate the Antarctic sea ice, microbes that like highly acidic conditions, and ones that favor intensely saline environments like the natural salt lakes.

In 1978 a scientist found that extremophiles thriving in Yellowstone National Park's hot springs contain enzymes that enable replication of the bacteria's hereditary machinery at high temperatures. The discovery was a critical piece of information for molecular biologists searching for a way to synthesize DNA fragments, and today these enzymes are the key to the polymerase chain reaction which replicates DNA fragments.

With its thousands of steam vents, geysers, mud volcanoes, and hot springs, Yellowstone contains eighty percent of the Earth's terrestrial geysers and more than half of its thermal features. Scientists now believe that the microscopic biodiversity in Yellowstone may rival or even surpass that of rain forests. While the citizens of the United States "own" the Yellowstone microbe, what is making current news is the debate about who owns the results of genetically manipulated microbes found in Yellowstone and other places. Patents given on these genetically engineered (GE) microbes have angered many who feel you can't own a life-form—plant or animal—or a life process.[3]

Patenting Life

The controversy over who is going to profit from owning these life-forms heated up in 1997 when Yellowstone officials signed a controversial and revolutionary "bioprospecting" deal. The deal transferred proprietary rights of a microbe found in Yellowstone to a California company in return for fees and royalties from any commercial sales. Since the market value of the resulting products is estimated at $15 billion annually and the royalty returned to taxpayers from the deal may be as small as half a percent, the deal obviously favors the company and sets a dangerous precedent which threatens the resources in all our parks.

Some people, like Indian physicist and social activist Vandana Shiva, consider this kind of deal piracy, claiming that all such contracts only serve the interests of the company holding the patent and are not in the best interests of the public. The patent system has given private rights to what many think should remain a biological commons. Shiva, who is currently organizing farmers in India to resist efforts of companies to take, alter, and attempt to patent native seeds, identifies the issue as one of human rights. She and a growing number of others are demanding freedom from patents on life, whether it be on seeds or life-forms and life processes. And many agree with her. In early 1998 a lawsuit was filed against the National Park Service, accusing the organization of illegally selling off federal resources in secret contracts with biotech researchers.

The problems surrounding GE plants and animals are in many ways a natural extension of the problems inherent in our entire ownership system. Chief Seattle is said to have asked, "How can you buy or sell the sky, the warmth of the land?...If we do not own the freshness of the air and the sparkle of the water, how can you buy them?"[4] Like others since, he was questioning our basic assumptions about life. If the natural world and its processes and inhabitants can be reduced to their DNA, then maybe we can own them if we sequence and then change their DNA. But if, as a new biological science of qualities tells us, organisms have an emergent order greater than the sum of their parts, and if they have their own kind of consciousness, trying to own them will mean little more than buying the right to manipulate and move them around and charge others for their natural abilities. And without critical knowledge of the role and importance of different bacteria in the larger health and evolution of the planet, these manipulations at best will be ineffective and at worst may result in a crisis of unparalleled magnitude.

The Patent Controversy. Patents come out of a long-standing political pattern in which commercial interests dominate other interests. In the case of GM bacteria, plants, and animals, those commercial interests take precedence over human concerns.[5] Those who argue for the patenting of altered bacteria maintain that patents will motivate researchers to find new treatments for genetically based diseases. They also point to the state of the environment and say patents will provide fair reward and incentives for companies to clean up polluted sites. A case in point is a Texas firm that has built a worldwide business applying its oil-eating microbes to petroleum-polluted sites. This company claims that a proprietary catalyst speeds reproduction of microbes and helps sustain the bacteria as well. And biologists at Clark Atlanta University have patented a system of "microbial mats" made from bacteria and grass clippings.

The mats float on polluted water and remove metals, oils, acids, pesticides, TNT, uranium, and rocket fuel.

Perhaps it is not the technology that is the problem but the way it is thrust on us and the motivation behind all the manipulations. If the same factors that drive genetic manipulations of crop plants are operating in the field of medicine and ecology (and there is little reason to think they are not), there is cause for grave concern. One solution might be to impose limits on the amount of money that companies can make. Appropriate limits would cull out those driven by greed alone and leave the field to those seeking honest solutions to some of the challenges that face us. Slowing down the development process and requiring long-term testing would also weed out many entrepreneurs looking to make a quick fortune and would prevent short-term solutions from impacting us or the environment.

Without regulations that harness economic objectives to human concerns and values, we stand little chance of shaping the way these new technologies will be used. Already a hybrid bacteria that produces ethanol fuel[6] from crop wastes and trees is in production even though preliminary tests indicate there are serious problems with it. When one of these altered soil bacteria was added to soil in a test plot of wheat, all the plants in one soil type died from alcohol poisoning. What's more, the same altered bacteria reduced the soil's normal microbial bacteria by about fifty percent, thereby impeding the natural composting cycle. Companies claim they can fix natural cycles by other means (a familiar agricultural stance), like implanting fungi with a gene that helps it to resist the effects of the other hybrid. But piling one unpredictable fix on another only compounds the problem.

It would serve us, as the public affected by these developments, to get very clear about the potential dangers of genetic engineering and stop assuming that scientists, for all their technical brilliance, have developed enough of a sense of ethics to set the course. Slowing down the rush to implement new developments, often a result of one company trying to beat another to the market, and insisting on adequate testing by independent researchers are two measures that might help protect our health and the health of the environment.

The Power of Bacteria

Bacteria wield a great deal of power. It is probably what attracts certain people to biotechnology—that and the thrill of discovery. The bacterial power that has launched a full-scale search for microbes is the same power that we fear will be used against us in biological warfare.

Despite an international agreement signed at the Biological Weapons Convention in the early 1970s, some experts estimate that between ten and twelve countries are currently pursuing or already have an extensive biowarfare capability, including Iraq, China, Libya, North Korea, and the United States, which has recently allocated millions of dollars to some of the top researchers in the country as part of a scientific "counteroffensive" against biowarfare.

Although it is fairly certain that biological weapons have been used, there is little documentation. Japan may have used the plague and other bacteria against China in the 1930s and 1940s. A recent article on biological terrorism says that Iraq once put deadly anthrax spores (*Bacillus anthracis*) into bombs and Scud missile warheads, but fortunately didn't use the weapons. Closer to home, a former top aide to deported cult leader Bhagwan Shree Rajneesh was convicted of unleashing salmonella bacteria in an Oregon community, which resulted in hundreds of people getting sick. An Ohio white supremacist also caused an uproar when he mail-ordered three vials of freeze-dried bubonic plague microbes from a Maryland laboratory.

Enzyme Magic. A different and more desirable kind of power is wielded by bacterial enzymes. In fact, one of the ways microbes work their magic is by secreting enzymes. We all produce these proteins that serve as catalysts for the chemical processes in our bodies. Enzymes produced by fungi and bacteria have given rise to the ancient art of fermentation, and the Japanese have long used these microorganisms to ferment sake and beer, miso, a bean dish called natto, and soy sauce.

An enzyme that penetrates the dirt-holding niches in cotton fabrics is the key to the effectiveness of the first bacteria-based laundry product, Attack. Developed by Japanese scientists, the product uses enzymes from a bacterium found in a rice field. This bacterium can survive alkalinity lethal to most other microbes. Also being developed for commercial use by two biotechnology companies from India and Australia is a bleach-boosting enzyme that reduces bleach chemicals in wood pulp while increasing paper brightness.

Other Microbe Contributions

Microbes have also played an important role in mining since Roman times, leaching copper out of rock. Gold-mining bacteria, however, are a relatively new development. When high-grade ores began to dwindle in the early 1980s, miners had to mine lower-grade mineral deposits that contain lower proportions of gold. Besides being more difficult to extract, lower grades of ore had to be roasted at high pressure to burn off sulfides, a process that has contributed greatly to the acid rain problem.

Today, using a bacterial process called biooxidation, gold miners in Brazil, Australia, and South Africa treat the raw ore with microbes before final processing with cyanide. The bacteria feed on sulfides and other chemicals while releasing an acid that leaches out metal from the ore, so the roasting stage is eliminated. And with help from bacteria, miners can now extract from seventy to ninety-five percent more gold.

Biodegradable plastic is another gift from microbes. So is cement. A new process was developed whereby bacteria fill fissures in cement from the inside out (unlike conventional sealants), meshing completely with the existing material. This bacterial builder may be used to seal up cracks and fissures in concrete buildings and other structures, and its creators have suggested that Mt. Rushmore would greatly benefit from a bacterial facial.

To complete this partial list of the results of people working with microbes, an electrochemist has created a living battery in a little plastic chamber of microbes. He thinks the energy released by these bacteria during digestion could potentially power everything from watches to vehicles. And it might even bring inexpensive electricity to places too remote or too poor to be served by conventional electrical generating plants.

The talents of bacteria are so varied there appears to be no end to the ways they can work and the ways creative minds can enlist their help. How that help is enlisted, however, is critical. On every front we are being asked to decide what values will determine political and economic policy as the technological prowess of the human race continues to supersede its overall emotional and spiritual maturity. The current situation also requires that we take steps to protect all of the planet's microscopic life-forms, many of which may disappear from the Earth before being uncovered or are being monopolized by companies whose priorities are not aligned to anything greater than their profit margins.

Germs

The adaptability of bacteria that has made them the workhorses of the biotechnology industry is the same ability that has allowed a small percentage to mutate when faced with inhospitable or hostile environments. As we noted in our discussion of malaria, by the late 1800s the discoveries that linked illnesses to microbes were being made almost monthly. Since that time the list has continued to grow, and microbes have become the new enemy, a formidable foe over which science and medicine must persevere and triumph.

The war we are waging against bacteria has strong parallels to the war we waged against the mosquito. Just like our initial successes and the invention of DDT led us to think we could eradicate the mosquito, the advent of antibiotics led us to believe that eliminating infectious diseases was simply a matter of time. Everyone now agrees our optimism was not well founded.

As laypeople who have invested a great deal of power in the scientific and medical professions, expecting them to protect us, we had no idea that bacteria, after being assaulted at every turn by antibiotics, would find the means to resist them. But they have, stirring a fear in our society that threatens to erupt into widescale panic.

The War Against Germs. Although not insects or arachnids, microbes invoke similar reactions in people. News articles about them have an uncanny resemblance to news articles about insects. The use of military language, for instance, is common to both. We think that we are being realistic about germs, but we are actually becoming paranoid. Psychologist Sam Keen says that the way to detect hidden paranoia is to pay attention to words and metaphors.

The vocabulary of paranoia is organized around the words war, battle strategy, tactics, struggle, contest, competition, winning, enemies, opponents, they, defenses, security, maneuver, objective, power, command, control, willpower, and assault.[7]

Now consider a sample of titles about bacteria taken from recent magazine articles and news reports: "Germ Warfare," "Killer Viruses: Concern Proves to be Contagious," "News from the Front Line of the Battle Against Bacteria," "It's Us vs. Them...and They Have a Head Start," and "New Weapons in Development." Paranoia revealed.

Predictably, books and movies are adding to our fears. *The Andromeda Strain*, a 1969 film, was one of the first. Based on a novel by Michael Crichton, *Andromeda Strain* features extraterrestrial bacteria that devastate a city and threaten the rest of the world. More current films and books present outbreaks of certain viruses in sensational and gruesome detail. *Outbreak*, for one, was a recent movie thriller about a fatal virus that spread by coughing. Despite its unlikely plot, the movie drew record-breaking crowds. Its popularity, like the draw of all horror stories, is connected to our projecting an evil or demonic nature onto germs. Connie Zweig, coeditor of a timely book on confronting the shadow, says, "Through a vicarious enactment of the shadow side, our evil impulses can be stimulated and perhaps relieved in the safety of the book or theater."[8] Small wonder that these movies are box-office hits.

The best-seller *The Hot Zone* by Robert Preston recounts an actual 1989 flare-up of Ebola sickness among a group of macaques in a quarantine facility in Reston, Virginia, but the grisly descriptions he employs arouse and entertain more than inform. Another book, *The Coming Plague* by Pulitzer Prize winner Laurie Garrett, chronicles history's worst epidemics. Although the book provides a wealth of information, underlying Garrett's account and the views of the hundreds of specialists she quotes is the belief that we are at war and that if only we had paid more attention to microbes and had not taken our attention off "our predators," we would not be in the danger we are today with the rise of infectious diseases.

On every front it looks like we are at war, fueled by our fear of the great potential of bacteria and viruses to harm us. But do we have our sight on the right enemy or just the most convenient one? How long will the focus on microbes as our enemies and our rhetoric on battling evil prevent us from looking at our cultural shadow and the complexities of our relationship to microbes? And how long will the search to find a magic-bullet drug prevent us from focusing on the determining factors and issues underlying the health of all the inhabitants of the Earth?

An Evolutionary View of Infectious Diseases

Outbreaks of deadly exotic types of bacteria have always taken a toll on human life. In the West, as global travel has brought every village into our living rooms, we can only expect that bacteria capable of causing illness and even death live among us already. Yet experts tell us that, particularly in North America, the odds are against our contracting a deadly illness from an exotic bacteria. One reason is that severe strains like the one that erupted in Zaire simply cannot survive very long, because they kill the very hosts they need in order to survive, and their transmission requires that a new host be exposed to infected blood products or be in very close contact with a sick individual. When transmission between hosts is difficult, deadly viruses must evolve into milder forms, or they die out entirely.

Another reason that it is unlikely for an outbreak from an exotic microbe to occur in North America is the fact that the conditions from which they emerge—for instance, jungles—do not exist in these parts. Scientists say that the encroachment of new populations on once remote countryside and wilderness is what releases a virus like Ebola. New bacterial and viral diseases occur after an upset in the natural balance of existing ecosystems, and an upset, more often than not, that stems from our activities.

Blissfully ignorant or with a wanton disregard for the existing balance, we continue to open new habitats abruptly, barging into ecosystems and disrupting the relationships between organisms that have evolved over time. When an ecosystem changes rapidly, mild viruses that may have cycled through animals and isolated human populations for years erupt into new severe strains, such as the new fly-carried disease that occurred in Brazil after the Amazon highway was built.

Once a few people are infected, infectious agents spread via our global transportation system. Overpopulated slums filled with poor, malnourished, and susceptible people also permit severe strains to spread easily to new hosts through air, water, and food, as well as through shared hypodermic needles and sexual intercourse. Preston maintains that "we're looking at a biological Internet; every household in America is wired up with every household in Southeast Asia, Africa and South America. It's a disease superhighway."[9]

A more realistic threat, experts say, are the less brutal, slower types of bacteria that are mutating to antibiotic-resistant forms. To counteract them means taking a long overdue stand against the way our food is produced. Feeding antibiotics to poultry and cattle and spraying crops with antibiotics, for instance, speeds up the mutation process in bacteria. Constantly bombarded, only those bacteria that can resist these assaults survive, and they are the ones that multiply and move into new hosts or transfer their resistance to benign bacteria.

Bacteria and Medicine

We are starting to shift from thinking that antibiotics are the cure-all for every symptom to an understanding that we can no longer depend on these drugs for our health. It is the overuse and incorrect use of these drugs, like the overuse of antibacterial products, that is giving rise to new strains of bacteria, and thus the shift in our thinking can't happen quickly enough. Using our buying power to purchase antibiotic-free meat and produce is one way to catch the industry's attention. If we did that *en masse*, the industry would race to meet the new demand.

Meanwhile, we must deal with the fact that many infections are now antibiotic resistant. More than a quarter of all new gonorrhea cases, for instance, are now resistant to penicillin. Many lung-related infections are also resistant to penicillin and other commonly used antibiotics that once were almost infallible in killing these bacteria. And according to a 1997 report, an antibiotic-resistant, deadly strain of tuberculosis has erupted in at least four continents.

Resistant Staph Infections. Staph (*Staphylococcus aureus*), a leading cause of hospital infections, has also developed a strain resistant to several antibiotics. Although staph bacteria usually live peacefully within the human body, they can become dangerous when present in an open wound or sore, and staph is known for causing hospital infections.

From the 1940s to the late 1970s, we exhausted penicillin- and methicillin-based antibiotics as different strains developed resistance to them. Vancomycin was the "silver bullet" for staph, the last drug that could kill all of its strains, and so it was used sparingly for thirty years. But in May 1997, a staph infection resisted vancomycin for the first time in a Japanese infant who had undergone heart surgery. Three months later, the vancomycin-resistant staph germ showed up in the United States in a Michigan man.

Enterococci, another gastrointestinal bacterium once considered benign, has also mutated into a new form. The new form, VRE, does not respond to antibiotics and has spread worldwide. In healthy people, VRE lives silently in the intestinal tract causing no illness, but in emergency rooms, intensive care units, and nurseries, it can be deadly. The more pressing concern is its proximity in the gastrointestinal tract to staph and the possibility that it will transfer its resistance gene to its staph bacteria roommate.

Bacteria Must Mutate or Die. When a bacterium mutates to resist succumbing to an antibiotic, it is not doing so because it is evil or malevolent—it is only trying to survive. After all, its life is being threatened. Recent research points to the fact that bacteria can alter their mutation rate when faced with any kind of stress that interferes with their ability to grow. Consider that in hospitable laboratories, where *E. coli* bacteria are not threatened, but are fed on a regular basis, only one in a thousand mutate. But when the same bacteria are trying to adapt and survive in a stressful or hostile setting, the number that mutate becomes more than one in a hundred—and in some strains five or six in every hundred. Any bacteria threatened by an antibiotic (or an antibacterial product) can quickly make the mutation they need to survive. The presence of the antibiotic triggers the bacteria to start recombining DNA.

Many lines of research are working to prevent the return and increase of infectious diseases and treat them effectively when they do occur. Research has commenced on bacteriophages,[10] or "bacteria eaters," for example, that prey on other microbes and can kill bacterial cells by disrupting their cell walls, picking up where studies left off in the 1940s when antibiotics arrived on the scene.

Unlocking the secrets of viruses[11] (a microbe much smaller than a bacterium) is another area of research with both promises and problems. Scientists hope that virus research[12] including a new DNA vaccine[13] will lead to new approaches in the treatment of certain afflictions, while critics fear our ability to manipulate viruses will backfire and cause serious health problems.

Another new direction focuses on using an enzyme of a bacterium to block the genes that provide resistance to specific antibiotics. And still another development promises new antibiotics and perhaps even anticancer agents—and all from a soil-dwelling microscopic worm.[14]

Holistic Medicine

All of these lines of research and countless others may be essential but insufficient solutions to the dilemma of bacterial infections and infectious disease. For example, the combination of deadly parasite plus human being does not necessarily equal disease. Studies on the human immune system and the role of susceptibility and the impact of an environment filled with toxic by-products all point to other factors in the equation.

Many health officials have acknowledged the folly of past approaches. Marc Micozzi, head of the National Museum of Health and Medicine in Washington, D.C., is one who believes that trying to find "silver bullets" to shoot down these pathogens is the wrong approach, given the fact that we are surrounded by bacteria and viruses. He says:

> I think the problem is we've looked at it as a war. It's not winning a battle, but achieving a balance. The question is not to eliminate all the microbes, but to understand we're able to heal ourselves.[15]

Another problem impeding progress is the narrow conceptual base of Western allopathic medicine. Its practitioners always search for the source of illness in a physical cause independent of the patient. Likewise, their treatments prescribe only an external agent. Yet, most physicians agree that there is a fundamental connection between emotional dysfunction and physical illness—it is the connection underlying what we call susceptibility. More factors are involved in disease than just the physical ones, and most of us know that. The rise of alternative holistic medicine is a response to our dissatisfaction with the reductionist approach of conventional allopathic medicine.

How we define a root cause of an illness defines the remedy. If scientists identify the cause as simply an isolated physical disease agent, like a bacterium or a gene, this eliminates the need to look at anything else. It's simple and it's comfortable. We are simply hapless victims—but are we? Sometimes being a victim is easier than looking deeply into our situation. Perhaps we fear the fundamental changes in our lives that might be required to heal ourselves.

A Soul's Approach. Marc Ian Barasch, author of *The Healing Path: A Soul Approach to Illness*, points out that if we insist on keeping our diseases outside the dysfunctional patterns of our lives and our economically based society, even "successful" treatments that result in remissions can do little more than support a personal or societal structure that needs major renovation.

To understand what makes us susceptible to infections and illness and what needs to be done not only to survive, but also to become well and thrive, requires that we look more deeply into how we live, love, work, and feel. Neurologist Dr. Candance Pert, who has studied neuropeptides—the chemicals triggered by our emotions—says the distinction we commonly make between mind and body is misleading and false. Our emotions, through these neuropeptides, reside in our cells and tissues. As Pert concluded on Bill Moyer's "Healing and the Mind" program, "the mind (and emotion) is in every cell of your body."[16]

In line with Pert's conclusion are the findings of psychoneuroimmunology (PNI), a relatively new field that studies the links between the immune system and the central nervous system. Chemical messengers operating in both brain and immune system are also the most dense in the neural areas that regulate emotion.

The emotion that science has clearly connected to the onset of sickness and the course of recovery is anxiety—the distress evoked by the pressures of life. When anxiety is irrational or triggered by a false sense of danger or when anxiety is our chronic response to situations over which we have no control, research demonstrates repeatedly that our immune system is affected and medical problems are triggered and exacerbated. In men chronic hostility and a tendency to react with anger to a wide variety of situations increases their risk of heart disease. And for women it seems to be anxiety coupled with fear.

It is not that emotions cause disease, but that they invite it and are one of many interacting factors. Barasch believes that taking the widest possible "soul approach" to our illness lets us start to eliminate the sources of pathology from our lives and our society rather than merely suppressing our illness's symptoms in our bodies. So symptoms tell us where to start looking. Psychologist James Hillman suggests that the area of affliction is the part of our body-mind where we feel the most vulnerable. That site then is a doorway, and inside the door is a potential for greater life. And in this approach to illness, symptoms are, paradoxically, both an attack on the body and the body's expression of what it needs.

Eastern and Western Medicine. Western medicine, which insists on linear causality and the validity of taking a reductionist approach to living organisms, directs Western physicians to search for a single disease agent. Once isolated, they then attempt to control or destroy it. Those schooled in Eastern medicine examine the individual as a functional and structural unity. Parts exist in relationship to each other and together express the nature of the person. This is a perspective in accord with our understanding of quantum fields and very different from the reductionistic model in which the parts of the body are assumed to be fashioned independently and then assembled like a machine.

In Eastern medicine, since body parts and symptoms arise as a result of interactions within the developing organism and within the matrix of interrelationships in the environment, practitioners search the individual's body, mind, and emotions, looking for where the pattern of harmony has been upset. This approach has gained recent support from biologists who have studied models of complex systems. Brian Goodman, an advocate of a new "science of qualities,"[17] says, "The emergent qualities that are expressed in biological form are directly linked to the nature of organisms as integrated wholes."[18]

On the other side of the relationship, but also within the boundaries of this broader field of interaction, is the disease agent, a living organism that seeks its own ideal circumstances for living. The virus or disease agent is not an impersonal object—it too will seek its own optimal conditions or pattern of balance. Many studies, for instance, demonstrate that a virus behaves differently in a person with a strong immune system than in a person who is stressed, poorly nourished, or fatigued.

Imbalances in the Culture. To add further complexity, a growing number of physicians and psychologists believe that our search for health should be widened to acknowledge the pattern of disharmony in our culture. Again the principles of wholeness and coherence in a generative field delineated for the individual apply to the society.

Psychologist Robert Sardello says our bodies reflect the Earth's body. Disharmony in one will involve the participation of the other, a view endorsed by deep ecologists and ecopsychologists. It means that when we harm the Earth, our own bodies must reflect in some manner the effects of that harm. It also means that our illnesses are not always a sign of personal dysfunction, but can be a sign of a larger dysfunction. In these cases—and it is important to explore illness from every perspective—taking a soul's approach may involve accepting that the illness may not be "personal" but may instead reflect imbalances in the culture or environment. The rising number of cancer patients along with the increase in toxic chemicals provides just one example—although there may also be personal elements in each cancer patient that influence the course of treatment.

A soul's perspective of these kinds of afflictions might even see environmentally determined illness as having been unconsciously taken on, although not consciously, to either demonstrate forcibly, and tragically, a cause of imbalance in the society (consider that Rachel Carson died of breast cancer), or perhaps to hone the personality and give it its life task. Ecologist Sandra Steingraber, author of *Living Downstream*, who was diagnosed with bladder cancer in her twenties, was sensitized by her illness to the many ways in which our environment is being contaminated and was subsequently motivated to do something about it.

All Is One

Theories aside, few debate the fact that our health and the health of the environment are deteriorating rapidly. A mad race is underway to develop new technological fixes—new drugs for better health, as well as new products for our ease and comfort, and new offenses and defenses against biological warfare. When a leading virologist was asked about emerging infectious diseases, he replied, "We're underestimating the enemy and the microbes are winning the race,"[19] voicing the popular view of microbes versus people. It is a simplistic military approach to disease that reduces the treatment of disease to a crusade against evil. Armed with such an outlook, the doctor sees him or herself as the heroic defender of what is good and right and divine, the last stronghold in the mortal struggle with inimical disease agents.

As we look at the history of our battle against bacteria, our manipulations, our limited successes, our defeats, and our rising fears, we find ourselves on a treadmill. We isolate a chemical or gene to resist a deadly strain and, in time, it mutates out of reach. We find another, and we're safe for a while, then it too mutates. The repeated cycles of success and defeat bear an uncomfortable resemblance to the lost war against the mosquito, which was based on viewing life as a battlefield and believing that we can and must eradicate anything that causes people to die. But we die anyway.

Psychologist Sam Keen says that a consensual paranoia in modern culture, which seems in this decade to have erupted into an all-out war on bacteria, is rooted in "the perennial struggle within us between basic trust and mistrust of life."[20] Either we trust the world or we don't; either we think it friendly, or we think it hostile or indifferent. Paranoia comes out of a perception of the world and others as hostile. If that belief is lurking behind the actions of those who are exploring bacteria, their solutions will reflect its influence and preclude other solutions.

As individuals, if we trust life and trust that there is a reason for being here at this critical stage in the evolution of humanity, perhaps we can move past paranoia and find the tools that will help us explore the current paradigmatic conflicts and align ourselves with an authentic view that supports real growth in ourselves and the culture.

Imbalances in Energy. One ancient view emerging in contemporary language diagnoses disease as an imbalance of energy and works to balance the individual's energy before it manifests itself as disease in the body. Medical intuitive Carolyn Myss is one of its leading proponents. She introduces people to a new way of understanding their human physiology, insisting, as ancient cultures have long known, that the human spirit is not only real, but also that it is the underlying force of life from which all else flows. She sees our current condition, the advance of physical diseases, and the poisoning and pollution of our environment, as an epidemic of the fragmentation of the human spiritual condition.

Myss and countless other health practitioners have learned that illness often follows certain patterns of stress or trauma emerging organically out of our lives. What resources we have available to cope with life are intimately linked with our quality of health.

Concerning susceptibility to harmful strains of bacteria, Myss believes our value system is a critical key to our being able to resist them. When we feel like victims of our culture or our families, she explains, we open ourselves up to group negative experiences—like viral and bacterial epidemics.

The predominant value system in the world at present is based on economics, so it should not be surprising then that this system has reduced millions of people and other species to a position of little or no value, since they are lacking financially, politically, or socially. A feeling of worthlessness is then translated into a feeling of victimization and can manifest as a susceptibility to particular diseases.

Here is where Myss's model of energy anatomy gets interesting in light of our fears about germs and the current rise of infectious diseases. The center of energy or power connected to the health of our immune system is the same center that determines positive and negative group experiences, and epidemics are a negative group experience. If our fears and attitudes are similar to those held in the culture at large, it seems as though we become energetically susceptible to an epidemic. In other words, all epidemics reflect our personal values, the current social issues of the culture, and the health of the society's "immune system."[21] And when an entire group becomes infected with fear, that energy extends to its children. Since children absorb their family's energy, they become as susceptible to these diseases as their parents.

Thus the perception that illnesses such as bacterial and viral infections occur as a result of being invaded by bacteria and viruses is only part of the story. We actually invite illness when emotional, psychological, or spiritual stresses overwhelm us and thereby weaken our bodies (in a manner reminiscent of how sickly plants invite insects to prey on them). Our personal energy level, and the collective energy of our society, family, or group, is directly related to all that we experience physically. So given our unbalanced mode of operation in the modern world, it is not surprising that challenges to our biological immunity have followed accordingly. And Myss reminds her audiences repeatedly that the sacred truth of this first energetic center, or tribal center, is that "All is One."[22]

The positive side to that maxim is that whatever we do as individuals affects the rest of the group, be it family, society, or the Earth community at large. When we sort through our inherited beliefs and discard the ones that violate our interconnectedness with others—like the world is a battlefield and we must be vigilant against insects and bacteria—we help our biological system and influence all life positively. We also withdraw our support from negative group beliefs and remove ourselves from their consequences. And because every belief, true or not, directs a measure of our energy into an act of creation, our best defense against bacterial epidemics and other diseases is simply to hold only those beliefs that support the kinds of experiences in which we wish to participate.

Relating to Bacteria

Finding the appropriate stance in relationship to bacteria, our tiny ancestors and associates, will help us disengage from the culture's fear and panic about bacteria's power to render useless our antibiotic weapons. On a practical note, we might serve ourselves and life by retaining a genuinely appreciative attitude to the invisible life-forms that surround us, encouraging them to evolve into benign forms. Those involved in research with bacteria are also urged to approach their subjects with feeling. A feeling for the organism, as McClintock so aptly put it, will allow these specialists to penetrate the secrets of this world so intimately interwoven with our own.

In *Kinship With All Life*, J. Allen Boone introduces us to a chemist named J. William Jean, whose beliefs about bacteria provide us with a model that we can use as a standard to measure and adjust our own beliefs. Jean has a reputation for solving "unsolvable" problems for companies, including the petroleum, rubber, and airplane industries. Boone visited him one day hoping to learn his secret. At that time Jean was working with microorganisms, and Boone soon discovered that Jean's fundamental belief that all things were "God's purpose in action"[23] was the primary reason his research was so successful. Jean believed in a universe that was well planned and managed—and innately good. Bacteria and other microorganisms were included in his universe and considered to be cooperating "fellow beings." Just as Jean lived and worked as an intelligent expression of this unifying life force, so too did the microbes function as intelligent expressions. Intuitive bridges built on his intent and belief directed their interactions.

It was easy for Jean to maintain an attitude of friendliness, respect, admiration, encouragement, and limitless expectancy toward bacteria. His feeling for them gave him an ability to understand and cooperate with them that, in turn, appeared to evoke a spontaneous reaction in them. Boone says about Jean:

> Here was a very wise man who had learned that the most effective way to achieve right relations with any living thing is to look for the best in it and

> then help that best in the fullest expression. As a result, millions of unicellular organisms, invisible to the naked eye, were turning to him with an enthusiasm that was as startling in its effectiveness as it was in its implications.[24]

The result of his approach has given Jean an understanding of microbes that few scientists attain. Boone believes this man knows their "attitudes toward life, their methods of doing things, their likes and dislikes, and even their ambitions. He knows what they need for their peace of mind and their fullest expression."[25] Taking great pains to make their living conditions ideal, Jean's consideration and integrity have resulted in all sorts of new and useful products that benefit people.

Jean would have found good company in Roman Vishniac, a microbiologist, microphotographer, and physician. In Scott McVay's essay on affinity or what he calls our "Siamese connexion" with other species, he says that Vishniac revered the microscopic planktonic organisms he gently ladled from a nearby pond. Watching their behavior under a microscope, Vishniac considered these one-celled animals "as friends and neighbors, deserving of civilized consideration and occupying a status equal to his own in the natural order of things."[26] Although in most laboratories these creatures usually die in a day or two, in Vishniac's care they thrived. After he was done observing and photographing them, he returned them to the exact spot in the pond from which he took them.

In previous chapters we have looked at people who have found in their interactions with insects an individuality in the creatures, and a consciousness that, however impenetrable, responds and thrives under our care and concern. Now we are witness to this same phenomenon in people like Jean and Vishniac. The latter was heard to say:

> Some people think microscopic animals are all pretty much alike—but oh, no! They have individualities that make them different from one another, just like human beings.[27]

These men's regard and patient attunement to microscopic organisms gives us a model of right relationship based on trust in a world infused with creativity. It also gives us a model for good science whereby our understanding of other organisms requires intimacy and a blurring of the boundaries that separate us. When we do this, our vision is enhanced, and we can rejoice in the fact that our intimate associates embody the creative forces of life and respond, as all life does, to our appreciation and respect. All is One.

PART III

The Eight-Legged People

- Spinners of Fate
- Insect Initiators

12

Spinners of Fate

Man did not weave the web of life, he is merely a strand in it. Whatever he does to the web, he does to himself.
—Chief Seattle

In 1988 British news was filled with reports of an invasion of foreign spiders imported in a shipment of grapes. Shoppers panicked, and to appease them health officials ordered huge quantities of grapes destroyed.

A few years earlier Japan went to war against the redback spider, a native of Australia. When a dragnet near Osaka captured over a thousand redbacks, health officials went searching for others—shining lights down into wells, turning over manhole covers, peering into cracks in gravestones, and perusing schoolyards and parks. The panic reached near hysteria when officials found another hundred on Japan's east coast. Emergency shipments of antitoxin were airlifted in from Australia, and urgent updates on the "infestation" were broadcast each night on the news.

Although no one had been bitten, the search widened to Tokyo, and the Japanese people rallied behind the detection and eradication efforts. The health official leading the offensive justified the war effort by explaining that the two-inch spider was a serious danger because its bite was life-threatening.

The people of Australia, meanwhile, were amused at Japan's panic. Australians consider the shy redback spiders as more nuisance than threat. Redbacks have a reputation for being timid and easily frightened. If disturbed, they often roll up in a ball and play dead. Every backyard has a few, and it's impossible to live in Australia very long without seeing these spiders.

For the record, the redback's bite does kill a few people around the world each year, but no one in Australia has died since an antitoxin was developed in the fifties. A bite makes most people sick for a few days, and then they recover.

The Huntsman Spider. A fear of spiders that is out of proportion to the danger of the situation is called arachnophobia. In areas with large spiders, like Australia, it usually has a particular spider as a focus—like the huntsman spider, a hand-sized, black, hairy spider. A few years ago a panicky Australian teenager tried to kill a huntsman spider by setting a can of insecticide spray on fire and hurling it at the creature. He burned down his family's home. No one knows if the spider escaped.

Each year the huntsman spider is also implicated in many disastrous encounters involving people and cars. When people find themselves enclosed in the car with this large ambling creature, they panic and swerve all over the road. Some even drive off the road, hit telephone poles, overturn their cars, and jump out of them while the car is still moving. Although a few people have been bitten in these circumstances, they didn't suffer from the bite. Their injuries were a result of their panic-driven responses to this spider.

The huntsman spider has huge fangs, and, like many primitive spiders, rears up on its hind legs when threatened, but it offers no real threat to people. Its venom is weak and its bite feels like a pinprick. Known as the giant crab spider because it can run sideways quickly, the huntsman is a predator that occupies trees or the cracks of rocks waiting for an unsuspecting insect to go by. Small places offer it protection and are one reason this spider ends up hiding in the air vents or sun visors of cars.

The Fear of Spiders. In British surveys in 1950 and 1988, spiders were the second most unpopular animal—below the snake in the first survey and the rat in the second. Interestingly, the dislike for and apprehension about spiders is not always a result of the fear of being bitten. Other research, also conducted in Britain, indicates that people primarily fear large spiders, black spiders, and long-legged spiders. The rapid movement of spiders was also frequently cited as a cause of aversion—but surprisingly, not the spider's bite. Perhaps in Brazil and Australia where being bitten by certain spiders can have real consequences, the responses would have been different. Yet the huntsman spider is large and black with relatively long legs, three reasons for any arachnophobe to respond hysterically.

According to biologist Ronald Rood, some fear is helpful, a safety device warning us not to go beyond our understanding. In contrast, an inordinate amount of fear that creates panic seldom serves us and, as we've seen in the incidents with the huntsman spider, may even place us in danger.

The distortion of healthy fear comes from misconceptions about the creature and the fear of a painful experience. When the focus of irrational fear is on spiders, scorpions, or any other venomous creature, it usually contains the assumption that creatures capable of inflicting pain or death spend their lives looking for humans to attack. One of the many distortions in the movie *Arachnophobia* was the portrayal of large, hairy spiders seeking humans to kill. The movie was a highly successful money-maker because it fed this widespread fear.

The belief that poisonous and venomous creatures want to harm us endures despite scientific evidence that the assortment of "weapons" these creatures possess is merely an important aspect of their survival strategy. Poisonous insects like blister beetles and certain caterpillars store toxins in their body tissues to dissuade predators from eating them. Venomous creatures like spiders, scorpions, bees, and wasps have evolved with venom apparatus to capture food and for defense.

Spider venom. Spiders, who have been around for about 300 million years, are built differently from insects. They have eight legs, not six, and their bodies are divided into two sections, not three. Entomologists put spiders in the class Arachnida along with mites, ticks, and scorpions, and only about 34,000 of an estimated 120,000 species have been described. Fewer than that have been studied. All spiders are venomous in the sense that all but one species possess a pair of poison glands. Since spiders use their jaws to employ their venom, they bite, jabbing their fangs into their prey while squeezing venom out from these glands (unlike scorpions, bees, and wasps which have stingers in their tails). Chemically, spider venom is a mixture of many different toxins and digestive enzymes. The Bushmen of South Africa took advantage of its deadly power and tipped their arrowheads with spider venom, as did some indigenous tribes of North and South America.

Researchers today are investigating venom as a medicine. Necrotic venom, which is found in spiders like the brown recluse and which causes tissue decay, might be helpful in dispersing blood clots that cause heart attacks, according to one study. Another recent study of the venom of the African Cameroon red tarantula revealed a compound (SNX-482) that blocks specific channels in nerve cells that are linked to brain function.

Spider-venom-derived medicines in homeopathy affect the nervous system, heart, and brain, with each species having its own particular accent. In fact, a spider is credited with helping Constantine Hering, who devised the Laws of Cures and continued the development of homeopathy in America, discover this method of treating disease. In 1898, Hering brought back a Cuban tarantula from his travels, and although it died en route, he produced a medicine from its dead body that became a great remedy for decayed tissue, badly infected boils, and puncture wounds.

Tarantulas. Only twenty to thirty spiders are potentially dangerous to us—although about five hundred can inflict a significant bite. Most people, especially after seeing movies like *Arachnophobia*, assume that the large and hairy tarantulas are one of those five hundred because they look dangerous. But even the South American varieties who stalk lizards and small birds and are ten inches across are not seriously venomous. And like the Australian huntsman, their bite is reputed to feel like a pinprick.

The true tarantula is from Italy. This spider was believed to have bitten the citizens of Taranto in southern Italy and caused its victims to dance vigorously for relief and to flush out the venom. Since victims danced in a kind of hysterical trance, the malady was called tarantism. When the phenomenon was investigated, scientists determined that the spider involved was not a tarantula at all but a Mediterranean black widow species.

The tarantulas in the western United States, the females of which can live twenty-five years and the males ten years, are actually large wolf spiders, and the bite of even the large ones isn't any more painful than a bee sting. More uncomfortable, but still nothing to worry about, is the bite from a dwarf spider. As adults, these tiny creatures are smaller than the head of a pin with a bite that feels like a wasp sting.

This is not to say that some spiders can't harm us, and under certain circumstances, even kill us. My purpose is not to attempt to substitute naiveté for life-serving caution or fear, but merely to dispel the irrational fear that binds us to a narrow track of behavior, filled with anxiety and trepidation. We need to know that most spiders will not bite even when they are threatened. And any fear-based misconceptions about being bitten we might harbor usually mask or ignore the fact that virtually all injuries and deaths attributed to venomous animals are directly related to human interference, albeit generally unintentional.

In North America there are three thousand species of spiders, but only two, the black widow and the brown recluse, are poisonous to people. And knowledge about them should greatly diminish our fear. Then we will know how to identify them and when to take judicious action.

Black Widows, Brown Recluses, and Redbacks

The black widow spider is one species that requires some caution on our part, although this spider rarely lives up to its reputation. If present in our area (and it has a huge geographical range), the first step is learning how to recognize it. Female black widows, the dangerous ones, are shiny black creatures with an hourglass pattern of red or yellow on their underside. Should we encounter one, it will ease our mind to know it is painfully shy. Although its venom is more deadly, measure for measure, than the venom of the rattlesnake, the chances of receiving even a drop are slim. One reason is that the body of a female black widow spider is only about the size of a green pea. And second, the black widow spends her life in hiding, preferring lumber piles and the nooks and crannies of old building foundations—although be aware, she also likes outhouses and clothes and shoes left outdoors.

If you disturb a black widow, her first reaction will be to curl into a little ball in the middle of her web. If threatened further, she might even leave her web and run away. Thus statistics on black widow spider bites are usually inflated. One study, conducted before antivenom was developed, revealed that only about fifty cases of human deaths from this spider could actually be verified amidst hundreds of claims. With antivenom, deaths are virtually unknown, and only one percent of untreated bites prove to be fatal.

Like other spiders, the black widow spider has a superior sense of touch and receives messages through the vibrations of her web—as when food is available. Rushing toward an insect that lands on her web, she throws silk at it with her hind legs.

Sometimes the vibrations are caused by a cautious male trying to determine if she is hungry and, therefore, dangerous. The smaller-sized male is banded and streaked with yellow, orange, or red. Males who are not eaten during or after mating are usually the ones that have a strategy for determining how hungry a potential mate is.

Male spiders of many species must deal with dangerous females. In one species, for example, the male chooses to mate just after the female has shed her skin, while she is still soft and helpless. In another species males tie their mate down with silk while she is in a receptive mood—it buys them time to get away after mating. The black widow male merely tweaks the web of a female. If she doesn't charge out, he takes it as an indication that she isn't starved, and the chances of his mating successfully with her and leaving after the act are good. Sometimes it works, other times it doesn't.

Reports on the black widow focus on the deadliness of her venom and her reputation for eating her mate. I suspect our fears and ambiguous feelings about sexuality contribute to our fascination and horror with the latter behavior. In Junichirö Tanizaki's famous short story "The Tattoo," written in 1910, the black widow is a symbol of cruelty and sexual aberration. The heroine of this story possesses a sensuous beauty and, as the reader discovers, a latent sadism. The story begins with a renowned tattoo artist whose artistic talent is tied to a sadistic desire to see his clients in pain while tattooing them. A shy young girl destined to be a geisha comes to him for a tattoo. He shows her a picture of a woman looking at corpses of men with an expression of pride and satisfaction on her face and tells her, "This painting symbolizes your future,...the men fallen on the ground are those who will lose their lives because of you."[1] Then he tattoos a large spider on her back, which in a mysterious act of satanic-like transfiguration brings out her spider-like sadistic leanings. When the tattoo is finished, she becomes cool and calculating, supposedly like the black widow spider, and tells the artist that he will be her first victim.

A short story called "The Spider" written by Hanns Heinz Ewers in 1921 also plays on the male fear of the erotic danger of women and features a seductive black-haired woman who spins each day in front of her window, luring the male occupants of the room opposite hers to their death. More recently, a made-for-television movie called *The Black Widow Murders* aired. It was based on the story of a real woman who drew men near her by using her power of seduction and then killed them. And a 1997 news article about a scam to rob elderly men of their money and then kill them is called the "black widow case."

Men versus Women. These tales reflect the popular theory that there is an inherent conflict between males and females, even in the world of creeping creatures, and in species where the female is bigger than the male and eats him during or after copulation, the hapless males are simply overpowered by the beautiful and insidious females. It's not a new idea. The Aztecs, who lived in an aggressive, male-dominated culture, also viewed spiders as malevolent. They interpreted the female's habit of eating the male during copulation as hostility in general and hostility toward men in particular. In their myths spiders represented the souls of warrior women (the Amazonian Fate-spinners) from the pre-Aztec matriarchy. At the end of the world, they thought these spider-women would descend from heaven on their silken threads and devour mortal men.

It is outside the scope of this book to speculate on the nature of this apparent conflict between men and women or its origin in patriarchal cultures in which men typically exert power over women. That the spider is intricately connected to the feminine is true, primarily because many spiders weave webs. Weaving is a symbol of creation—the creation of the world, of human beings, and of relationships. And as a symbol of relatedness, psychologists associate weaving with the essence of the feminine (not female) nature. Symbolically we could view the cannibalism of male spiders by the female not as hostility, but as an act that returns the male to the ground of creation—the source of being and life itself.

On the physical side, remembering that nature's creatures are not possessed of the motivations that drive and/or confuse us will keep us from judging them in a context that doesn't apply. In the insect world, cannibalism is commonplace, a survival strategy tied not to cruelty, but to energy resources and, as we will see, propagation strategies.

The Redback Male. The behavior of the male redback spider, the species that caused such a panic in Japan, gives us another take on cannibalism. Although the panic in Japan was caused by the fear of being bitten, the redback female's cannibalistic habits did not win her any points. During copulation, she eats the male redback about sixty-five percent of the time. What's particularly interesting is that spider researchers tell us that the male seemed "peculiarly, almost exuberantly, reconciled to his fate."[2] In fact, he actually flips into his mate's jaws while copulating.

If the Aztecs had investigated the redback spider, they might have had to change their myths to account for the male's eagerness. Modern researchers trying to explain the phenomenon are looking at what advantages there may be in being eaten. One idea is that being eaten allows the male to copulate longer (more than twice the time as males who don't get eaten) and thereby fertilize more eggs. It's an important objective for the male if he wants to pass his genetic material along, especially since the female redback sometimes takes two mates, one after the other. Supporting this theory is the fact that females who eat their first mate are about seventeen times less likely to accept a second mate than females who don't.

Whatever the motivation, judging them by human standards is inappropriate, and projecting malevolence onto the female spider or foolhardiness onto the male doesn't serve the human-spider relationship. What does serve it is letting certain mysteries work in the nonhostile, non-fearful imagination and displacing our anxiety about spiders with information that gives us strategies for coexisting with them.

The Brown Recluse. Another spider in the same family as the black widow, and one that also requires caution on our part, is the brown recluse spider. Sometimes called the fiddleback or the violin spider, the brown recluse is yellowish brown with a dark brown pattern resembling a tiny violin on its head. Recently the brown recluse, once found only in the southwest, has spread across North America in suitcases and trunks of clothes. In cold areas, it has had to make its home inside our buildings to survive. Although the brown recluse has only killed six people in a hundred years, it gained notoriety in the 1950s because of a number of bites that resulted in severe tissue damage.

Photographs of unhealed wounds always accompany a talk on the brown recluse. And while it is true that its bite may cause a wound that is stubborn to heal, many people have been bitten with no harm at all—a fact worth noting. With modern communication, we also hear about brown recluse bites more frequently, although we are still left to our own devices to manage our anxiety about them and avoid encountering them. If we remember that this spider is also doing its best to avoid us, we can trust that a good measure of attention when we are cleaning dark, hard-to-reach places will be sufficient to keep most of us out of harm's way.

Arachnophobia. Although learning which spiders require caution will serve the average person, those with arachnophobia consider all spiders, even the common harvestman spider (daddy longlegs), as threats. For these people avoidance is a key defense. But when the inevitable encounter with a spider occurs and triggers an immobilizing fear, friends and family must come to the rescue and deal with the spider by killing or removing it. Treatment for arachnophobia, like other phobias, involves aversion or desensitization therapy. The phobic patient is exposed gradually to words and spider-related objects and images while under a state of deep relaxation and while being given reassurance and encouragement.

In traditional Freudian psychology, phobias like arachnophobia were thought to be related to underlying unresolved issues with the individual's mother or father. Freud proposed, for example, that the fear of being bitten by a spider was really a fear of punishment by one's father. Other psychoanalysts have viewed the spider in dreams as a malevolent symbol representing the devouring or castrating mother, an interpretation probably based on our anxiety about the female's cannibalistic habits. In depth psychology, the spider is a symbol of the Self for women, although Jungians also see a spider in terms of the negative Self in dreams in which the spider makes an appearance as a symbol of one's dread of the unconscious forces of integration. Today, although many of these theories have been discarded, psychologists still talk about underlying causes of the fear of spiders like a fear of the unknown.

Fear and Fascination of Spiders

In Paul Hillyard's *The Book of the Spider: From Arachnophobia to the Love of Spiders*, he points out that people either inherit a tendency to fear spiders within a family or, as in most cases, are conditioned in childhood to fear them. Patience Muffet, the real girl behind Mother Goose's "Little Miss Muffet," apparently suffered from a lifelong fear of spiders because of negative experiences with spiders as a child. Those experiences originated with her father, who loved and studied spiders. In his enthusiasm he insisted his daughter swallow live spiders as a cure for any number of minor ailments.

A recent survey in the United States showed that biology classes have frequently awakened animal phobias. Unfortunately, once awakened, the fears of the student are usually laughed at or ignored unless they seriously limit the individual. Sometimes those same classes can be the cauldron where the intense fear of an arachnophobe is transformed into an equally intense fascination. Consider George Uetz, who was terrified of spiders from the age of five when he put his hand in a bush and a large spider ran up his arm. In college, Uetz enrolled in a biology course, unaware that its focus was spiders. Rather than bolting, he stayed. Learning about spiders turned his fear into fascination, and that fascination led him to become the president of the American Arachnological Society and a leading spider researcher.

Perhaps it was fate that led George Uetz to spiders—first through fear and then through fascination. The spider's connection with fate is an ancient one. In the oldest myths, the spider is associated with the triune Great Goddess, as spinner, measurer, and cutter of the threads of life. In Hindu mythology, the spider represented Maya, virgin aspect of the Triple Goddess, who was portrayed as the Spinner sitting at the hub of the Wheel of Fate. As the weaver of the web of illusion—maya—she also created life from her own substance.

Spider Woman and Grandmother Spider, both weavers of the fate of humans and animals, plants, and rocks, are portrayed with a wise and knowing nature and figure prominently in many indigenous mythologies. Like the Three Fates in the East—Clotho, Lachesis, and Atropos—these female spider deities control by their weaving who lives and who dies—an art that we have already noted as being linked to acts of creation, and one we will discuss further in the section below on spider webs.

The belief that spiders have an intimate knowledge of future events (because they create them) is widespread. The men of the Mambila tribe, who live on the Cameroon-Nigerian border, try to capitalize on that attribute and use the large hairy burrowing spiders as a way of divining the future. They receive answers to important questions through a process called *ngam* that involves letting the spider move around leaf cards and then reading the position of the cards.

The Spider Trickster

The belief that fate could enter and influence a person's life was generally accepted in ancient cultures. It is one of the reasons the intelligent and cunning spider had a dual role in many mythologies—both helpful and deadly. The web of protection, for instance, under certain circumstances could be viewed as a spinning illusion, a web of entrapment, or a poisonous plot.

Twists and turns on one's path, especially those which are wholly unexpected, were also accepted as the work of the gods, and in many cultures it was believed to be a result of a spider god twisting the threads of fate. Indigenous people would not have missed the irony of an arachnophobe turning into an arachnologist and are likely to have attributed Uetz's transformation to the influence of this trickster figure. Only the Trickster, both a mythological creature and, according to depth psychologists, a universal element of our psyche, could have engineered Uetz's enrollment in that biology class. It was not a decision his conscious personality, understandably concerned with comfort and safety, would ever have made.

The Trickster, in its spider or other manifestations, personifies the energetic power of the total psyche to overthrow the personality's best ideas about how to proceed, tricking it into taking unexpected action. Since Uetz would not have opted to take a class on spiders, the Trickster would have arranged inner and outer events so that Uetz would unwittingly choose this particular class and "miss" hearing the class description until he was already enrolled and sitting in his seat.

Although feared as an upsetting, unpredictable influence, the Trickster was also considered a cultural hero and sacred creator of the world who brings to people the inspirations and energies of creativity. Consider that Uetz was tricked into discovering work that he is passionate about. His apparent misfortune was actually his good luck—one of the signs that the Trickster is afoot.

Both bad luck and good reside in the Trickster's domain. As a spoiler of plans, the Trickster often brings loss and what we perceive as bad luck, entering a situation to punish pride, arrogance, and insolence. The Trickster also chastises those who seek closure prematurely and, in doing so, cut off the creative possibilities of a situation.

When the Trickster presence is felt in our life, it helps to know that this energy is aligned with an authentic push in our psyche toward expansiveness. Although the Trickster's lack of concern for our fears, the culture's taboos, or social appropriateness is unnerving and can feel punitive, its demands for a change of direction or stillness is a call for a necessary alteration of some kind. Far from being unreasonable, its energies try to align us with deeper patterns of fulfillment, presenting opportunities for growth disguised as frustration, pain, and misfortune.

The Spider Trickster. In some Native American traditions, the Trickster was Coyote or Raven, and as mentioned in the chapter on biting insects, the Navajo had an Insect God and Trickster named Beotcidi, who was erotic and outrageous. But in many native traditions, also as noted earlier, the Trickster was personified as a spider. The Oglala Dakota tribes call their Spider Trickster Ikto (or Iktomi) and Unktomi. They say Ikto was the first being who attained maturity in this world. Known to be more cunning than human beings, Ikto as cultural hero named all people and animals and was the first to use human speech.

The West Indies Spider Man Anansi is also a Trickster figure, as is Ananse, a chief character in the folktales of the Ashanti people. In fact, in Twi, one of the chief languages of West Africa, the word for spider is "Ananse." Like Iktome, Ananse brings divine forces into human life by passing through the hidden, and largely unconscious, boundaries people erect and defend for protection. In myths, Ananse often takes human shape, walks with a limp, and speaks with a lisp. A little bald-headed man with a falsetto voice, Ananse lives by his wits, besting other animals and people by cunning and humor. Despite his greed and selfishness (the qualities he punishes in humans), Ananse is admired by the Ashanti people, who enjoy tales of how he outwits those stronger and more powerful than himself.

The Trickster Today. Although Western culture doesn't acknowledge this archetype known as the Trickster, except to call it bad luck, it still operates within us and our society. We tend to think we left behind this energetic pattern and its chaotic influence with the advent of our control-oriented technologies, but we didn't. We left behind only the context through which we might understand its emergence.

It is easy to see why creeping creatures like spiders have often been elected to carry the Trickster energies. Outwitting those who are stronger and more powerful is their lot. In today's world the Trickster has more opportunities than ever to wreak havoc, primarily because of our manipulations of nature and our assigning creeping creatures to the bottom rung of the hierarchy of value. Anytime we start believing that we can manage nature and control or eradicate certain species, our actions and attitude will seem to invoke the Trickster.

Perhaps the cotton boll weevil embodied the Trickster energy when it devastated enough cotton crops to force the farmers of Enterprise, Alabama, to diversify, a move that turned out to be more profitable than growing only cotton. And maybe it is this force at work making us look foolish when we import a predator species that turns out to have no appetite for its intended prey.

If we made room for the Trickster and the energy of our unconscious, we would spend less energy fighting to maintain the status quo and more energy looking for new avenues of growth. We would also laugh at ourselves more often. Personifying this energy, we could resurrect Anansi or Iktomi, or imbue Insect Man, an obscure comic hero from the sixties who could transform himself into any insect, with the power of the Trickster. And working with this energy, we could invite its gifts of grace and creativity, and gain from an outpouring of its benefits.

Good Luck and Protection

Good luck and profit also reside in the domain of the Trickster. The good fortune thought to be brought by the spider was often related to money matters, and the belief in a money-bringing or gift-giving spider is widespread. Also common is the idea that killing a spider will bring bad luck or monetary loss, and that a spider amulet will attract money to the bearer. The Romans even carried a precious stone engraved with a spider image for luck or little spiders of gold or silver to bring good luck in their trade dealings.

In the Near East a spider's presence at a wedding was thought to bring good luck and blessings to the marriage. In these cultures placing a spider in the bed of the newlyweds was the best thing well-wishers could do for the newly united couple.

North American Chippewa Indians of the Great Lakes region hung spiderwebs on the hoop of infants' cradleboards to bring good luck. They believed the web catches harm in the air and prevents it from reaching the baby. A modern researcher suggests that the only harm these webs caught was mosquitoes.

The spider has also been thought to bring good luck and protection to travelers. The myths of the Kiowa of the southern great plains of the United States tell of cultural heroine Spider Woman who is a protective spirit of travelers. Other stories tell of how spider webs protect travelers fleeing from harm. When Mohammed fled from his enemies at Mecca, for example, he hid in a cave. Suddenly, in front of the cave a tree grew and a spider wove a web between the tree and the cave. When his enemies saw the unbroken web, they didn't search the cave because they thought that no one could have entered it recently. They passed by and the prophet emerged unharmed. A similar legend is also told about Jesus when he was hiding from Herod's cruelty.

King Frederick the Great had his life saved by a spider when it dropped down from the ceiling and into his cup of chocolate. Dismayed, Frederick asked for another cup. The cook, who had poisoned the first cup of chocolate, interpreted the king's request for another cup as knowledge about the assassination plot and killed himself on the spot. Only then did Frederick realize that he had been saved by the spider's action. To show his gratitude, and in tribute, he had a majestic spider painted on the ceiling of this room, where it can still be seen today.

A contemporary example of how a spider may intervene in a situation and offer protection comes from native healer Bobby Lake-Thom. He recounts the time when he was washing dishes and a spider dropped down from its web and dangled in front of his face. After getting Lake-Thom's attention, the spider proceeded down into the sink and into the drain. He tried to communicate with the spider, even talking to her. "Oh, hello little sister. Why did you come to visit me, what are you trying to tell me?"[3] Then all of a sudden the spider crawled out of the drain and up on the sink and stayed there as though watching him. Curious, he reached down into the drain where she had been and found a shard of broken glass. He realized that if it wasn't for the spider, he probably would have turned on the disposal and sent shards of glass flying out of the drain. He might even have been injured. The spider had protected him, and he thanked it.

The Intelligence of Spiders

Following the native way, Lake-Thom wasn't apprehensive when this spider appeared to be watching him, but was open to its communication. It's this kind of openness that would firm our connection to spiders. In his memoirs, composer Grétry described a spider that would descend above his harpsichord when he sat down to play. He didn't mind the attention and took it as a sign that the spider had exceptional taste in music.

Whether a spider descends or not, the feeling of being watched by this creature is common. Not everyone is comfortable with the idea, however. It implies that there is a consciousness, a center of intelligent awareness in the spider doing the watching. This sensed intelligence of spiders was one reason that the Spider Trickster in his cunning could outwit larger and more powerful animals. More than just stealth and shrewdness, spider intelligence was thought to be comprised of many different aspects including an ordering function that helps the spider make essential connections in web construction.

The ordering aspect embodied in the spider is also manifested in the stringed figure game we know as cat's cradle. One of the oldest of art forms and found in cultures all over the world, in Zuni and Navajo mythology these configurations were taught by the helpful and sacred Spider gods. The different patterns, like native stories, were more than just amusing. They were a way for the people to order their psyches and keep their minds clear. Cat's cradle figures helped them relate their lives to other powers and other lives in the universe including the stars and sun. The complex patterns also taught that relationship was a necessity for health in the individual and the tribe. One elder explained:

> We need to have ways of thinking, of keeping things stable, healthy, and beautiful. We try for a long life, but lots of things happen to us. So we keep our thinking in order by these figures and we keep our lives in order with the stories.[4]

For native people a primary aspect of spider intelligence was a wise knowing of matters beyond human senses, and so they routinely sought the spider's favor and guidance. In the Cheyenne and Arapaho societies, for example, the word for spider actually conveys the idea of intelligence. Even today these tribes hold spiders in high esteem, and in ceremonies the women pray briefly to the spider to ask for its intelligent perspective. The spider's wise nature is also the reason that in Pueblo myths the creatrix Spider Woman is sometimes called "Thinking Woman," and the world is considered her brainchild.

Jumping Spiders. Anyone who has spent time observing spiders generally comes away feeling that spiders have some kind of intelligence, including contemporary spider enthusiasts like biologist Mark Moffett. Moffett is particularly enamored with jumping spiders. While acknowledging the role of instinct in spiders, Moffett also sees something else in the behavior of jumping spiders.

> When I watch them sneak about—leaving sight of their quarry to follow complex routes through the vegetation in search of a better angle for attack—I am struck by flexible behavior that seems to reflect intelligence.[5]

Interestingly, if we feel we are being watched by a spider, traditional scientists might even agree with us, if the spider was a jumping spider. This group of about four thousand species, comprising the largest of the 105 families of spiders, not only jumps, but it also has big front eyes that can detect motion in virtually all directions. Jumping spiders can also distinguish detail as far away as twenty times their own body length, unlike most spiders, which are virtually blind. Moffett says that if we look at a jumping spider, it will probably look back with its headlight eyes scanning our face, until they reject us as prey. In fact, their vision is so much like ours they even watch television, reacting to television images of other spiders or flies.

Many times the sense that spiders are aware and intelligent frightens us. If we haven't examined our culture's beliefs about creeping creatures, we may fear, however irrationally, that their intent is malevolent. It's a common reason cited to explain a dislike of spiders. In the recent science fiction movie *Starship Troopers*, enormous alien spider-like creatures with swordlike legs join forces with other giant extraterrestial insects to annihilate the human race. When the creatures are discovered to be under the direction of a single bug—a jellyfish-looking blob of a creature with a tarantula-like face—the fact that it is intelligent is an idea almost too horrifying for the humans to entertain. And when they finally capture it and subject it to sadistic and invasive procedures in a laboratory, the movie audience applauds.

Watching Spiders

Perhaps the fear of being watched is connected to an impulse to watch spiders and learn from them. There are correspondences between us and spiders that only observation and intuition can reveal.

When magician David Abram was in Indonesia, spiders and insects were his introduction to the spirits. From them he first learned of the intelligence within the nature of nonhumans, the correspondences between their form of awareness and his own, and their ability to "instill a reverberation in oneself that temporarily shatters habitual ways of seeing and feeling, leaving one open to a world all alive, awake, and aware."[6]

As he sat one day watching a spider climbing a thin thread that was stretched across the opening of a cave, he saw how it constructed each knot of silk in the web, and he marveled at its skill and surety. Then his vision caught another thread from another web and on a different plane than the first web. It was complete, with its own center and its own weaver. Although the two spiders spun independently of each other, to Abram's eyes they wove a single pattern with intersecting threads. Then he realized that there were many webs being constructed, all radiating out from different centers—patterns upon patterns. He felt as though he were watching the universe being born, one galaxy at a time.

Abram tapped into a way of perceiving that is common to indigenous people. The Hopi tribes of North America have a legend of a "Spider Man" whose web connected Heaven and Earth. And the revered grey spider of the Pima Indians of southern Arizona was said to have spun a web around the unconnected edges of both Earth and sky to help the Earth grow firm and solid.

The secret schools of ancient India used webs to symbolize how the gods constructed the universe by connecting the realms of light with those of darkness. And they referred to the builders of these threads of invisible force that held the newly born universe together as the Spider Gods and their ruler as The Great Spider.

Others of a philosophical or religious orientation who have watched spiders constructing their webs also report feeling they were witnessing the act of creation—the microcosm revealing the macrocosm. Spiritual teacher Omraam Mikhaël Aïvanhov, who observed the natural world closely and was open to its wisdom, taught that we can learn from the spider how God created the world, because its web is a mathematically perfect construction of an entire universe.

Spider Webs

Although Abram acknowledged the role of genetic instructions enfolded in the spider, he also acknowledged the intelligent awareness within the spider that made it receptive to its specific environment. Arachnologists recognize it too. Spiders are able to make the many creative adjustments needed to weave a particular web in a particular moment and environment.

Spider webs come in all sizes, shapes, and orientations. The largest and most developed of all webs are the aerial ones spun by tropical orb weavers that work at night, relying on touch alone. These webs are sometimes an astonishing eighteen feet in circumference. To begin the web the orb weaver spins a single "bridge thread." Perched on a twig or branch, she releases this silk thread from one of her spinnerets (small structures on the underside of her abdomen) and lets the wind carry it away. It's an act of faith. When the bridge thread connects to a surface, usually a branch or tree trunk, she tightens the connection and then travels along the thread reinforcing it with additional threads. Then the work of constructing the web begins in earnest.

In the late eighties, Yale researchers discovered that some spider silks reflect ultraviolet light, invisible to the human eye but not to insects. Since many flowers reflect ultraviolet light to attract pollinating insects, these ultraviolet-reflecting silk threads may be designed to mimic blossoms and fool unsuspecting insects. And in at least one tropical species, the spider itself reflects ultraviolet light.

Spiders whose webs do not reflect ultraviolet light often weave designs into their webs using a special thread that has reflective properties. Many ancient indigenous tribes believed that the geometric patterns and angles in spiderwebs were evidence that spiders created the first alphabet and oversaw the art of language and writing. Perhaps the design is an encoded message, an invitation to insects, for scientists have determined that these decorated webs capture fifty-eight percent more insects than do plain webs. Entomologist Thomas Eisner thinks web designs also serve as signals to birds to prevent them from flying into the webs and breaking them.

Ballooning. Although silkworms and butterflies also can make silk, spiders are without question the masters. Spiders use their silk-making ability from birth, even the ones that don't construct webs. If newborn spiders aren't immediately eaten by tiger beetles, ants, wasps, birds, and other spiders, they perch on a twig or blade of grass and release a thread of silk and let the wind carry them away.

The thread, called the parachute or balloon thread, is let out by the spider as fast as the air currents will take it. When it is several feet long, it is buoyant enough to lift the tiny spider up and into the air. In some parts of Europe, during "gossamer season," spider silk from newborn spiders "rains" from the sky. Like silk parachutes, sometimes these threads carry their small passengers astonishing distances, although they may get eaten along the way—easy prey for swallows, swifts, flycatchers, warblers, and dragonflies.

In 1832, in Charles Darwin's famous account of his voyages, he describes several thousand baby spiders landing on his ship when the vessel was sixty miles from the nearest land. And rockets, balloons, and airplanes have sampled the air at thirty thousand feet and found spiders.

The Art of Weaving

Web spinning is not a mechanical endeavor, but an art that requires the spider's attention and clearheadedness. When spiders were given mind-altering drugs, researchers found that they spun abnormal webs. Under marijuana's influence, for example, spiders left large spaces between the framework threads and inner spirals. Caffeine caused the spider to spin haphazard threads, and the webs of spiders on benzedrine were erratic and often remained unfinished.

Studies subjecting spiders to other unusual circumstances have demonstrated repeatedly that web-building requires considerable resourcefulness and ingenuity. In the early nineties, for example, NASA sent a spider into space to test the effect of zero gravity on web building. Since the spider astronaut couldn't use her body weight as a guide, at first she wove a misshapened web. In only three days, however, she spun a near-perfect web.

It is easy to see why indigenous people went to the spider to learn the art of weaving. The average garden spider constructs a web using three or four kinds of silk and up to 1,500 connecting points—in less than an hour. In Navajo myth, Spider Man taught Navajo women how to make a loom and Spider Woman taught them how to weave on it. And both deities required woven fabrics as offerings. Several Argentinian tribes also recount a similar myth about the origin of weaving.

The Symbolism of Weaving and Spinning. The symbolism of weaving and the woven cloth has complex and ancient roots; we have already noted its association with the feminine principle and creation. Clarissa Pinkola Estés, Jungian analyst and keeper of the old stories in the Latina traditions, says that female deities, like Spider Woman and Grandmother Spider, are the essence of the wild instinctual self. They are the Life/Death/Life mothers who control "what must die, what shall live, to what shall be carded out, to what shall be woven in."[7]

The most famous myth about spiders and weaving is that of Athene and Arachne. In pre-Hellenic myths, the goddess Athene, also a spinner of fate, could incarnate as a spider, and when she did she was called Arachne. But the myth that most of us are familiar with is a reinterpretation of the Athene/Arachne unity by later Hellenic mythographers. This later version portrays a mortal woman Arachne as Athene's rival in the weaving arts. In a contest, when an impertinent Arachne demonstrates a weaving skill that surpasses the skill of the goddess, Athene is enraged and starts hitting her and tearing her tapestry apart. Arachne flees to the woods where she puts a rope around her neck and tries to commit suicide. Athene takes pity on her and grants her a new life as a spider fated to spin and weave forever. This popular interpretation of the myth has undertones at odds with the helpful holy images of Grandmother Spider in indigenous tribes and may have been changed to undermine the spider's (and women's) creative power, for in the Greek myth the spider form is given as a sentence, and the act of weaving (creating) thus becomes the compulsory labor of an indentured creature.

Since all acts of creation are the spider's domain in ancient and indigenous symbology, spinning and weaving also have fertility and sexual implications. For a woman to create a child, for example, she must bring together from the whole of her being all the different elements—chemical, biological, and psychological—and weave them into a single unit. The umbilical cord that nourishes the growing child while the mother "weaves" it into the world is like the spider's "thread of life" that emanates from its body.

This strong connection to the feminine principal and to the power of creating life is one of the primary reasons that the spider was demonized in the male-dominated Judeo-Christian religious tradition. Regarded as an image of Satan, the spider as demon was thought to ensnare men's souls through the seductive wiles of prostitutes. The spider's webs were also despised in this tradition and regarded as vehicles to entrap the unsuspecting, or as works of vanity without value. To complete the condemnation, the spider's ambushing techniques were considered treacherous and Judas-like.

In contrast, Buddhists, who believed the spider was the weaver of the web of illusion and a creator, taught that just as a spider catches flies in its web, so must a seeker of enlightenment capture and destroy the lust of the senses. Also in contrast are the holy Spider figures in countless aboriginal cultures. For many of these groups the spider's web was thought to catch souls and hold them until funeral rituals were completed to send the dead on their way to the next world. And the Chibcha Indians, North Andean natives who hold the spider in great awe, believe that boats made of spider webs allow the dead to cross the lake of death.

The Medicine of Patience. Since weaving requires great industry and patience, Navajo people would often take a spider web that had been woven over a hole and rub it on a baby girl's hands and arms so that when she grew up she could weave without tiring.

In an Osage myth, a spider offers its medicine of patience to an Osage warrior who ventures into the wilderness to find an animal that would show itself to him and become his totem. Thinking he knows best what animal would be a good totem, he only tracks the deer and ignores signs of all other animals. One day, with his eyes focused downward on deer tracks, he stumbles into a large spider web. A spider, now at eye level, offers to be a totem for his clan, pointing out that it possesses the great virtue of patience. "All things come to me," the spider said proudly. "If your people learn this, they will be strong indeed."[8] The warrior, recognizing the wisdom in the spider's words, returned to his village to make the spider the totem of his clan.

A spider also taught patience and persistence to Robert Bruce, the fourteenth-century Scottish hero-king. While Bruce was hiding from the English, he watched a spider trying to weave its web across a portion of the ceiling. The spider tried six times but failed each time. Then as the creature began its seventh try, Robert watched more intensely than ever, knowing inexplicably that he was about to see whether or not another attempt to defeat the English army would be in vain. When the spider succeeded on that seventh try, an inspired Robert saw its success as a fortuitous sign. He renewed his campaign, and against all odds, finally freed Scotland from England in 1318.

The Nature of Silk

Today, in our hunger for new technologies and products, the silk of spider webs captures more attention than the art of weaving and the patience and industry it requires. Although early Christians equated the spider web with human frailty and a vulnerability to illness and death, spider silk is anything but frail. In fact, spider researchers are intrigued with its strength and elasticity. Like some synthetic fibers, spider silk is five times stronger than steel. But unlike synthetics, it can stretch up to 130 percent of its original length. It is also nonallergenic, waterproof, and stable at high temperatures.

Orb weavers typically produce a variety of silks, each secreted by a different set of glands and extruded from their spinnerets. One set of glands produces silk for draglines, which are the threads that spiders trail behind them and use to drop through the air to safety. Other glands make silks for capturing prey. These sticky threads, which form the bulk of the web, hold a struggling insect in place until the spider arrives to immobilize it with its venom. Still other glands produce tough fibers for wrapping prey, reinforcing the web, or spinning airtight cocoons with insulating properties comparable to goose down.

Most research has concentrated on dragline silk. In nature, adult females can spin five to six feet of silk a minute. With characteristic impatience, in laboratories we tease silk out of spiders mechanically at twenty feet a minute. Whether or not that procedure may shorten the spider's life is not a concern.

Spider research is primarily funded by the military, which began looking at dragline silk in the 1960s. The goal was to construct a synthetic fiber out of the silk for use as parachute cord and in flak jackets, tents, sleeping bags, and other military items. Medical equipment companies are also looking at the silk to make strong, nonallergenic sutures, artificial tendons, and implantable medical devices.

Tribal people have long used spider's silk. New Guinea tribes twist the silk into small gill nets and fishing lines. Natives in the South Pacific used the silk of large web-building spiders to make pouches for tobacco and arrowheads, but no one has succeeded in producing large commercial quantities of spider silk—although a Frenchman once tried.

Fashion designers frequently call spider researchers to ask about using the silk for fabrics. It's an impractical source of fiber, however, since spiders are difficult to raise in large numbers, and their silk is so fine it would take five thousand spiders spinning their entire lives to produce enough fabric for one dress.

The Mysteries of Spider Silk. Spider silk starts out as a soluble protein secreted by one of the spider's abdominal silk glands. It becomes a solid thread after passing through a narrow tubelike duct on its way to the spinnerets that forces all its molecules to align in the same direction. Research to find the gene that controls the production of the protein has led to other discoveries. Two interacting proteins, for example, are involved in producing the dragline silk. One gives it strength and the other elasticity.

Research in the last ten years has focused on inserting the silk-producing genes necessary for making both proteins into bacteria or yeast, thereby turning them into factories for making silk. The ethical questions related to such endeavors have already been posed in our look at transgenic crops and our investigation of bacteria.

The silk-producing gene was isolated and cloned by a Massachusetts military research team and is currently being used for products where price is not a concern, as for skin wrappings to treat burn victims, artificial tendons and ligaments, and coatings on implanted heart values and vein grafts to help promote cell attachment.

The Original Networkers

As research on spider silk continues in earnest, the image of the spider's web, a universal symbol of connection and relatedness, has returned to modern culture with the advent of the World Wide Web (WWW) on the Internet. This cyber-based "spider web" connects computers, their users, and information all over the world. In 1994 this vast, continually expanding, global network had more than two million host computers and an estimated forty million users—and its size was doubling every year.

As the original networker, creator of the first alphabet, and patron of language and writing, the spider and its medicine—creativity, intelligence, industry, and patience—is available to contemporary individuals. When environmental psychologist James Swan was asked to edit a book about networking, he was eager to start the project but developed a block when he started working on it. Following his own advice on consulting with nature for guidance when faced with a problem, he sought help at one of his favorite places in the San Francisco Golden Gate Recreational Area. After making an offering of cornmeal and singing a prayer, Swan began to wander through the wild landscape. Within a few minutes he felt drawn to a particular spot. He looked and looked for what might be pulling him. Then suddenly he saw a spider at work weaving its web. He watched the spider for a few moments, and it occurred to him that he was witnessing the original network builder at work. He watched how the spider carefully laid out the structure of the web and then wove it together, one knot at a time, and with the greatest of patience and attention to its design. He says:

> I went home, drew a picture of the spider, and put it over my typewriter. Almost immediately the ideas began to flow and the book took form, inspired by the original networkers.[9]

For the culture at large, Spider as networker and its web as a symbol of creativity and relatedness have interesting ties to Peter Russell's global brain and his vision of humanity's potential for evolving into a nervous system or brain for Gaia. Russell compares the near-instant interlinking of people all over the world through the communications technology of the WWW with the way the human brain grows. He believes that if the rate of data-processing capacity continues to grow at its present rate of increase, "the global telecommunications network could equal the brain in complexity by the year 2000."[10] When this happens, if there is enough cohesiveness and positive interaction, a new order could emerge that would revolutionize humanity in the same magnitude as when the Spider Gods first wove the webs of heaven and Earth together to create the universe.

A Spider City. We've already seen one biological version of the WWW in certain eusocial insects, in particular in species like Argentina ants, who are not territorial but visit and live in each other's nests, cooperating socially in all manner of activities. Surprisingly, there is also a similar version in the world of spiders. A species in the Mexican state of Veracruz lives and hunts in vast interconnected spider cities. Although less than one-tenth of one percent of the world's spiders are social and live in communities, this Mexican species creates a colony by constructing interlocking webs. The largest colony discovered so far has an estimated 160,000 spiders, with each individual spider anchoring its web to its neighbors' webs to create a single great web that covers the distance of a football field or longer. Because of their interlinking webs, each spider expends less energy to capture food. Prey that ricochets off one web inevitably becomes ensnared in a neighbor's web. Perhaps we too will one day learn the benefits of living in a socially cooperating planetary society and forsake our fiercely defended territoriality for greater synergy and an opportunity to develop more of our potential.

Helping Spiders

While spiders are reentering the culture as potent as ever in their new metaphoric garb, actual spiders are losing ground due to habitat loss and pesticides. Protecting the spider is a way of giving back to a creature who still weaves webs in our psyches. Protecting them is also a way to support a balanced ecosystem, since all spiders are vital predators in their specific habitat. The rare Tooth Cave spider, for example, is one species that needs protecting. This creature spends its entire life in Tooth Cave and three other limestone caverns near Austin, Texas. A mere tenth of an inch long and blind, it lives with six other endangered invertebrate species—a pseudoscorpion, two harvestmen, a ground beetle and two mold beetles. No one knows exactly how many of each of these six species are left, but the cave is not secure. Its greatest threat comes from real estate developers, although some threat may be offered by fire ants who have built nests near the cave and may enter it seeking these creatures as food.

The Great Raft Spider. Another spider in trouble is the Great Raft spider, Britain's rarest species. Luckily for this hand-sized spider, a company has intervened to help it survive. In 1997, a British water company, perhaps run by a spider enthusiast, set a great example by ignoring disgruntled customers and setting aside twenty million gallons of reserves to raise drought-reduced pond levels in the spider's 325-acre reserve. The water company plans to pump seventy-two thousand gallons a day from a nearby borehole into the reserve ponds for up to six months.

Although some customers and gardeners were angry that the company was elevating the spider's need for water over their own (the company had only appealed to them not to waste water because rationing might result), the company took the criticism in stride and remained firm on its position in regard to helping the spiders. A company spokesperson, replying to complaints with the compassion and wisdom of an elder, said, "How can one possibly equate the life or death situation facing the spiders with hosepipe bans? We know we are doing the right thing when one considers the risk to the survival of this rare spider if nothing is done."[11]

Relating to Spiders

Relating to spiders—beyond protecting them and gleaning wisdom from watching them and reflecting on their ways—is an avenue open for exploration, but kindness and compassion is always appropriate. In his 1950 book for laypeople called simply *The Spider*, naturalist John Compton gave the highest marks for loving spiders to a policeman. In 1936 this officer was on duty at Lambeth Bridge keeping traffic moving safely on that busy thoroughfare. Despite his preoccupation with traffic, he noticed a very large spider trying to cross the road. Knowing it would be killed, the policeman "immediately held up the traffic while the spider crossed with slow and dignified gait—to the great joy of all the onlookers." [12]

When people I meet learn about my interest in creeping creatures, many report proudly that they always help spiders out of the house. Some use a device advertised to remove spiders from the house without harming them. It looks like a pair of scissors except that one of the blades is a disc and the other is a dome that fits precisely over it. When the spider walks onto the flat part, closing the scissor handles closes the dome over it. The spider is then trapped safely and can be easily transported outside. Another device, designed to solve the problem of a spider trapped in the bathtub, is a spider ladder that can be suspended from one of the water taps to let the spider climb to safety on its own.

Helping spiders out of our buildings is a good step toward coexistence—unless of course, as Christie Cox discovered, the spiders belong in the house. When Cox was staying at the home of her friend Gail, a practicing interspecies communicator, Cox was left alone one morning when Gail had an appointment. As she sat down at the kitchen table to eat breakfast, she dropped a piece of food on the floor and bent down to retrieve it. That is when she saw two spiders sitting close to her foot. Their "spiderness, and their alarming hairiness"[13] sent her scrambling back in fear. She rummaged through the cupboards quickly and found two plastic cups that she dropped over the spiders, immobilizing them before her fear immobilized her. She didn't know if they were dangerous or not. One part of her wanted to just leave them trapped beneath plastic until her friend returned but another part of her didn't want to look so incapable. She decided to take them out of the house. She carefully slid a piece of paper under each cup and then, one at a time, took them out of the house and released them into the bushes. She reentered the house, pleased with herself.

When Gail returned several hours later, Cox told her that she had helped two spiders out of the house. Gail was not pleased and told her that the tarantulas had been born in the house and were an important part of its ecology, eating flies and keeping the insect population in balance. Later in the day she told Cox that she had been praying for their well-being, as a storm was approaching and they would be in danger without adequate shelter.

Three days later Cox saw Gail outside the house in the front sitting quietly under a tree in the midst of her yard of foot-high grass. Cox later found out Gail was calling silently to her spiders to come home. "Within fifteen minutes," said Cox, "those two spiders came crawling up out of that huge wilderness and onto her leg."[14] Later Gail told her that the spiders had found their adventure in the outside world interesting but had asked if it was all right to move back into the house. They also told her that they were aware of Cox's fear and that they appreciated the fact that she had been very gentle about taking them out of the kitchen. And Gail reported that they forgave her for removing them from their home and suggested that her fear of them was actually a fear of her own creativity.

Cox listened, her natural skepticism flaring up, but then decided, "Those spiders did respond to [Gail's] silent call. And when I saw, over the next few days, what else Gail was capable of, I was ready to believe almost anything of her."[15]

To acknowledge the intelligent awareness of spiders positions us to learn from them and perhaps to sense, as David Abram sensed, the intricate web of life with its interlocking points of connection that support physical existence. To acknowledge as well that the same awareness poised within the body of a spider may appreciate a thoughtful and considerate response to its presence is equivalent to dropping from our own dragline thread into a sea of possibility—with an act of faith equal to a spider's.

13

Insect Initiators

> On this day God is the name by which I designate all things which cross my willful path violently and recklessly, all things which upset subjective views, plans and intentions and change the course of my life for better or worse.
>
> —Carl Jung

For all our fear, we are fascinated with creatures that can inflict pain or cause death. It's a universal phenomenon present in every culture and in every age. While cultural beliefs dictate acceptable responses and influence the way a group expresses its fascination, the draw of these species, in otherwise diverse people, suggests that an archetype—a pattern in the human psyche—is at work.

Archetypes are not some invention of psychology, but are universal constants, beyond individual experience and culture. Archetypes arise from the collective unconscious—a sort of psychic storehouse for the whole human race. According to Carl Jung, they represent the primary structure or blueprint of the psychic world.

When an archetype is activated it releases a power comparable to the power released by splitting the atom. The resulting energy restructures external events and aligns our inner, subjective worlds with them. Like an electrical or magnetic field, we know an archetype only by its effect on us, by the way it patterns observable elements and organizes our experiences.

Inner and outer problems, emotions, and people activate archetypes and in those situations, these blueprints have enormous power as "organizers of meaning" to influence the way in which we consciously or unconsciously interpret the chain of events. And archetypes shape us—even in our responses to dangerous insects.

Fascination and Physical Survival

The psychological factors that underlie our fascination with creatures that could harm us is linked both to physical survival and psychological maturity. It is the influence of these primal forces, supporting our instinct to survive, that promotes a healthy fear of these creatures and alerts us to danger. Its opposite, irrational fear, as we have seen in our discussion of spiders, is by its nature disproportionate to the situation and usually involves ignorance about the creature and an assumption of malevolent intent.

The psychological and symbolic aspects of our fascination with dangerous creatures are complex but have a common denominator. They are all connected with a push toward psychological maturity and the cultivation of certain qualities—a process of inner growth, a turning toward our soul or true nature, that Carl Jung called individuation. Creatures that demand our attention and are impervious to our conscious desires serve this process faithfully. Their symbolic role in ancient cultures is intricately intertwined with the archetypes of renewal: birth, death, transformation, and rebirth. Encoded in our bodies, this archetypal process of initiation is activated in our psyches at pivotal times in our lives.

The Power of the Small. Wise individuals exercise caution when outside in the domain of wild and sometimes dangerous creatures. They are more alert, more likely not to charge blindly through the area. Such individuals know that an encounter with a species that can bite or sting us, unlike a carnival ride or electronic thrill, demands more than our money and physical presence. It calls forth in us certain qualities—traits of the native hunter or the meditator. We sense in the potentially dangerous creature real power, and, unless we panic, we are likely to quiet our own noisy thoughts and restlessness and try to match its depth of silence. Sometimes our well-being depends upon how well we match it.

Silence in the presence of such power is the only appropriate response because it anchors us to our own center where the power can be matched and used to transform and initiate us. Our sensitivity heightened, a brush with these kinds of creatures typically leaves us raw, as though our terror strips away the layers of comfort that protect us from life. Their ability to hurt us also binds us to the present moment like few things can. And after the moment, when we are returned to safety, we notice that we feel more alive.

Small creatures have power unaccountable for their size. Mosquitoes can make grown men run, as can spiders and bees. Tracker and teacher John Stokes, who studied for many years with the Australian aborigines and now teaches people how to survive in wilderness areas, thinks small creatures make us aware that we live in an automated world of false power. Because they have real power, they fascinate us and effectively counter our own disproportionate sense of importance. Stokes teaches that "power can't be grasped until you go out in the bush and gain the knowledge of the heart—get humbled by heat, cold, the sting of a bee, the power of the spider."[1]

Angered or confused by the unpredictable and painful encounter, we enter, if only for the moment, the gaps—transition places where we are not in control. If we don't lash out, if we forgo the heroic stance and allow ourselves to be humbled and temporarily subdued, all manner of insight and fortune await.

Confronting real power frees us from the bonds of greed, desire, numbness, and the concerns of the ego for comfort and control. It also awakens our intuitive and imaginative abilities and gives us rein to approach the big questions about who we are, where we come from, and where we are going.

Encounter with a Scorpion

Lashing out at what hurts us is often our first reaction, a culturally supported response that has its origin in the view of the world as a battlefield. For many people, it is the only response they know. Yet, there are other responses, like the one found in a Hindu teaching story, condensed here, called "The Saint and the Scorpion:"

> A saint bathing in a river rescues a scorpion, cradling it in his hand as he moves with it toward the riverbank. When the bedraggled scorpion realizes its new dilemma, it stings the saint, but the holy man keeps walking toward the bank. The scorpion stings him again, and the pain is so great that the saint staggers and almost collapses in the river. His agitated disciple, watching from the shore, tells him to put the scorpion down and leave it to its fate, saying that kindness is of no value to such a creature, for it is unable to learn from it.
>
> The saint ignores him and continues toward the bank, carrying the scorpion. The scorpion stings him a third time and the pain explodes into his head and chest. Smiling a blissful smile, he collapses into the river. The disciple rushes into the water and rescues the saint who is still smiling and still carrying the scorpion. When they reach dry land, the saint sets the creature down and it quickly crawls away.
>
> The disciple asks the saint how he can still smile after the scorpion nearly killed him. The saint acknowledges that the scorpion's sting almost killed him, but explains that it was only following its dharma or nature. "It is the dharma of a scorpion to sting, and it is the dharma of a saint to save its life.... Everything is in its proper place. That is why I am so happy."[2]

The saint who responds to the scorpion while staying centered in his own nature is able to match the power of the stinging creature—to match each sting with a deepening of his commitment to save the scorpion's life. Instead of dealing with such incidents of pain in a reactive manner and flinging the scorpion away, the saint demonstrates that we can move more deeply into our own center and from that place match the power of the sting with the power of compassion.

A Modern Encounter. To a population that has embraced the idea of the world as a battlefield, the saint's behavior is more than puzzling, and we are likely to dismiss it as an anomaly peculiar to sainthood. We are more comfortable with killing or running from what hurts us. A director of a county-wide gardening extension service, for example, wrote an article on how to kill scorpions after a visitor brought to the extension office a dead one preserved in a glass jar. She wrote that the sight of the scorpion triggered memories of a time when she stepped on a scorpion in her house and got stung. Incensed by the stabbing pain in the arch of her foot, she grabbed a book and started smashing the creature. When it was dead, she called Poison Control. They told her that unless she had a severe allergic reaction to the sting, it would only cause swelling and soreness in the affected area. "Soak it in ice," they advised.

Later she discovered that nearby house construction was the reason a few scorpions had moved to her home. Their habitat had been disturbed. She also read that they sting only when provoked. Understanding their reasons wasn't enough to quell her fear of having them in her house and possibly stinging her again. The rest of the article focuses on getting rid of them. She recommends crushing, stomping, smashing, or squashing them and using pesticide sprays. While admitting that the scorpions may be able to hide from the poisons, she says she wants to saturate her house with pesticides—her feelings overruling good sense. She concludes the article by recommending smashing as the preferred method of control and restating her feelings of disgust toward the dead scorpion that arrived at the extension office in a jar.

This type of article is popular in the culture. It holds up a mirror showing us what we already believe, so there is a kind of shared misery in the telling of such a tale—a version of the psychological game, "Ain't it awful." In these kinds of situations, more information about the creature doesn't help. Facts don't alter an emotional bias if the person is unaware that he or she has a bias. Identifying the bias is the first step. The second step is seeking a way to coexist harmoniously. With that intent a priority, more knowledge about the creature helps, because it feeds the imagination and balances the fear. And greater help is available still when we can step back and view an encounter symbolically.

Necessary Humbling. What is often humbled in a painful encounter with another creature, especially a small creature like a scorpion, is the self-important, inflated parts of ourselves. Those parts mask our general fear of the unknown and our resistance to the pain of being overcome and changed. "Forget about transformation and renewal," protests our personality who fights for order and predictability. It is this familiar aspect of self, playing at king or queen, that prefers safety to knowledge. And it is this aspect of self that builds an empire on false power and scrambles for position and visibility among other false leaders.

I suspect our task, and a monumental one at that, is not to withhold ourselves or defend ourselves from that which would help us grow strong and help us move closer to our true natures. "What we choose to fight is so tiny, what fights with us is so great," Rilke reminds us in "The Man Watching."[3] When we let go of our resistance to pain and change and can actually seek out these transitional places where the subjective and objective worlds intersect—in dread and expectation—we will have sufficiently altered our way of being in the world so that avenues of thought and action previously unavailable will open to us.

If we trust that the creatures of the natural world that move into our lives bidden by unseen powers are intent on arousing us and helping us grow, we can learn to submit to them. As Marlo Morgan learned to surrender to the swarms of Australian bush flies, perhaps we can also let go and refrain from erecting elaborate defenses, or engaging in righteous retaliation. Maybe we can enter the small and great initiations that our soul brings us without doing battle with forces and creatures that are ultimately allies of a fundamental natural self at home in the world. There is power in our defeats and in our surrender and blessings due for those messengers who disrupt our familiar world. As Rilke so eloquently explains in the last passage of the same poem:

> Whoever was beaten by this Angel...
> went away proud and strengthened and great from that harsh hand,
> that kneaded him as if to change his shape.
> Winning does not tempt that man.
> This is how he grows: by being defeated, decisively,
> by constantly greater beings.[4]

Allies of Growth. In shamanic traditions it was understood that other species are way-showers to the mysteries and messengers for the divine powers operating within us and within the universe. As initiators, insects and arachnids are impeccable, and, as we discussed in the spider chapter, aligned with and frequently activated by the Trickster archetype. We can neither appease nor bargain with them. Their task in dreaming and waking is to arouse us out of our complacency and push us past the edge of what is familiar and comfortable. And once beyond the edge, where vision is possible and energy is available, we are transformed and renewed—and we may even return with the powers of the creatures that initiated our transformation.

The arousal methods of insects and the tactics of other species can be very persuasive. A Zen story describes the anxiety that accompanies these passages.

> "Go to the edge," the voice said.
> "No!" they said. "We will fall."
> "Go to the edge," the voice said.
> "No!" they said. "We will be pushed over."
> "Go to the edge," the voice said.
> So they went...
> and they were pushed...
> and they flew....[5]

One of the roles animals play is to get the human over the cliff. Some push, others chase. Some use pain to nudge us toward the cliff and away from safety. Stinging and biting insects might torment us until we jump.

If we understand their intent, their alignment with our souls and the forces of growth, we could try to lift ourselves up in surrender. We could appreciate that our fear or anger is a process that includes pride, doubt, helplessness, and self-protection. We could admit that we are afraid of the pain of a biting or stinging creature and afraid of the unknown—despite its promise of renewal. Accepting our fear shifts it and allows us to come into a new relationship with the unknown, to see it as mystery and as an important step in initiation. Aligning our personal wills with a greater will, we could then wait with fear, dread, and hope for the visit that marks the beginning of a shift of consciousness and new life.

The Gifts of Pain. One of the reasons this process of growth looks so foreign to us and feels so frightening, despite its archetypal nature, is the fact that we have split our rational awareness from our natural, image-making self. We can't access the symbols that map the process and give it meaning through concepts and rational analysis. And as long as we are alienated from our wilderness self, we are largely cut off from experiencing the spiritual and psychic energy that emanates from its potent symbols and images.

Another reason we are unfamiliar with this process of growth is that we have been raised to avoid pain and discomfort. Pain-relief products and services permeate the culture. Yet, pain is a great teacher.

Stephen Levine has discovered in his work with the dying that our reactions to physical pain offer insight into our attitude toward life in general, and the more we push pain away, the less energy we have for living. Pain stirs our grief and brings up long-suppressed anxiety and unfinished business. Whereas simple awareness has a healing power. He suggests that we can use each moment of unpleasantness, each insect bite or sting, if you will, to learn how to meet all unpleasantness, all pain, and investigate how resistance turns pain into suffering.

Thus discomfort can teach us how to live and can take us to places otherwise inaccessible in our normal conscious state. Each incident provides an opportunity to stay in the present moment and bring awareness to the places inside us that recoil and harden in resistance. Softening around them, as Levine advises, lets us penetrate the armor that keeps life at bay and lets us live closer to the mystery. No one can live a pain-free life. Knowing there are gifts in painful experiences helps redeem them and helps to temper our responses if it is another species that brings the pain.

In her book *Pain: The Challenge and the Gift*, Marti Lynn Matthews considers pain as a guide, a biofeedback system that lets us know what is healthy and unhealthy for us:

> There is integrity in pain: it is not punishment but a force that pushes us into expansion. Without a push, we would never take the leap that would allow us to fly free.[6]

Transforming Weaknesses into Strengths

Some cultures advocate the use of meditation and ritual to transform unpleasantness and discomfort and to further self-understanding. As we saw in the "Mirrors of Identity" chapter, Gyelsay Togmay Sangpo deepened his wisdom and compassion by attending to the lice on his body. In *Path With A Heart*, Jack Kornfield tells the story of a poisonous tree. On first discovering it, most people only see its danger. Their immediate reaction is to cut it down before someone is hurt. This is like our initial response to dangerous creatures. It is also our first response to other difficulties that arise in our lives, such as when we encounter aggression, compulsion, greed, and fear, or when we are faced with stress, loss, conflict, depression, or sorrow in ourselves and others. We feel great aversion and want to avoid it, or get rid of it. In the case of the poison tree, we cut it down or uproot it. In the case of an insect, spider or scorpion, we poison or stomp on it.

Others who have journeyed further along the spiritual path discover this poisonous tree and realize that to be open to life requires compassion for everything. Knowing the poisonous tree is somehow a part of them, they don't want to cut it down. From kindness, they create boundaries around it. Perhaps they build a fence around the tree and post a warning sign so that others aren't poisoned and the tree may also live. This is a profound shift from judgment and fear to compassion. Applying it to the mosquito, we might search for a vaccine to protect the insect from the malarial parasite, while taking measures to prevent mosquitoes from feeding on people already afflicted with malaria.

Another type of person, who is extremely wise, comes upon the poisonous tree and is happy because the tree is just what he or she was looking for. This individual examines the poisonous fruit, analyzes its properties, and uses it as a medicine to heal the sick. By understanding and trusting that there is value in even the most difficult circumstance, the wise person's actions benefit a great many people. This may have been Wagner von Jauregg's attitude when he used the malarial parasite to save thousands of syphilitics from a slow and painful death. As mystical poet Jalaluddin Rumi said, "Every existence is poison to some and spirit-sweetness to others. Be the friend. Then you can eat from a poison jar and taste only clear discrimination."

Perhaps, on the deepest levels, we already know that within adversity is a gift. It is why we are so fascinated with the creatures that can arouse us and initiate our transformations. In dreaming, our psyche informs us in symbolic language about the role of fear, pain, and death as prerequisites for growth and renewal. And some deep aspect of self must activate the internal and external events, bringing us the creatures and experiences we need to initiate our own rite of passage and move toward healing and growth. The soul will use whatever it can to move us along.

Following shamanic wisdom, we could support our initiations by naming the creatures that we fear as our allies. It's a practical way to begin courting their power. What happens if, following Hildegard of Bingen's lead, we call dangerous creatures "glittering, glistening mirrors of divinity"? Are these not the passwords to open the door to our greater identity? Maybe all we are required to do is to meet adversity and pain with our respectful attention and be willing to learn from it. Instead of killing the creatures with the potential to harm us, seeing with the eyes of wisdom, we can dance around them and allow difficulties to become our good fortune.

The Scorpion

The scorpion is a venomous creature whose threat of a painful sting warns us to be careful in its presence. It also has the symbolic power to initiate us into psychological and transpersonal realms known to shamans and masters of esoteric mysteries. Learning something about both roles will position us correctly for coexisting with local scorpion species and benefiting from their symbolic instruction.

To the average person, scorpions look menacing, although they are not particularly aggressive toward humans, and they use their venom to capture food and defend themselves from natural enemies. They usually sting people after being disturbed suddenly—a startle response that anyone who has been abruptly awakened from a deep sleep can understand.

A newspaper in the Far East featured a man with an earache who went to a doctor. The doctor discovered a scorpion in the man's ear. The creature had probably entered the canal for warmth and shelter and stung him only after the doctor grabbed it with a metal instrument to forcibly evict it.

Predictably, the popular culture serves us a one-dimensional view of the scorpion. In a contemporary tale from a *Journey Into Mystery* comic book, a scorpion living in an atomic research lab gets hit accidentally with delta radiation. As its cells rapidly multiply, it grows larger and larger and breaks into the lab, where it attacks the helpless humans. A scientist yells for someone to call the airbase for help. Before they can reach the phone, the scorpion smashes through the wall and out of the building, and the humans realize that the creature has understood what they said. To their added horror, the scorpion sends them a message by means of thought waves:

> Hear me Humans! I have always had the seed of consciousness! But now your radiation has brought it to life. Now after years of utter helplessness, at last I have the strength and intelligence to pit myself against you! Now I shall gather my brother scorpions.... I shall enlarge them with the same radiation that I absorbed! We will be an invincible army of giants. Then we shall conquer mankind.[7]

The desperate humans finally defeat the scorpion using a combination of hypnosis and trickery before he can gather his brother scorpions together.

As we have seen in every chapter, the images and practices so prevalent in our culture revolve around power over others—power over a savage natural world bent on destroying us at the first opportunity. It is an uncomfortable way to live, looking over our shoulder. And we don't have to read comic books to absorb the ballistic context that prods us into battling these creatures.

Sometimes our fascination for creatures like scorpions is capitalized on by companies who may not understand it, but see in it a potential for a product that will sell. A company in the southwestern United States kills tens of thousands of scorpions annually to produce scorpion paperweights, bola ties, and refrigerator magnets. Living scorpions and black widows are encased in a lump of plastic resin and sold as souvenirs of the "Wild West." The company obtains its scorpions by paying people to enter the desert at night with an ultraviolet light. Under ultraviolet radiation, a scorpion glows (a feature that scientists believe may attract insects) so these hunters can find and capture them easily. The captured scorpions are then taken to "pourers who decant liquid resin into molds and then stuff living scorpions into the plastic after it is partly hardened."[8]

Our desire to own dangerous creatures, safely enclosed in plastic or otherwise rendered harmless, is a sad commentary on our relationship to wildness and real power. We can't reconcile our fascination with our irrational fear that these creatures will try to overpower and harm us.

Master of Extremes

Scorpions evoke a great deal of fear, considering that the vast majority won't harm people. They're not insects, but they belong to the class Arachnida and are relatives of spiders, daddy longlegs, ticks, and mites. Six families of about fifteen hundred species comprise the order Scorpionida. The largest family in the order has over three hundred species including those dangerous to people.

Scorpions have lived on the earth almost 440 million years. In Scotland, archeologists uncovered the fossil remains of a scorpion sixteen inches long. According to most experts, the largest ones alive today are the eight-inch black Emperor Scorpions that live on the west coast of Africa. The smallest scorpion, in contrast, measures only a half-inch in length.

One of the most venomous scorpions is Algeria's fat-tailed scorpion. Although not particularly large, it is responsible for eighty percent of all reported stings, a third of which prove fatal. Fat-tailed scorpions lived their lives for centuries under rocks or in shallow burrows on hillsides. Then, when people built houses on hillsides, scorpions used houses for shelter so they could hide during the day and feast on the insects living in the house at night. They especially like wet places, because humidity and water attracts insects for them to eat, and they often crawl into shower heads, baths, and toilets and congregate around outdoor water troughs and wells.

Scorpions also frequent shoes for warmth, and in Trinidad scorpions have been responsible for human deaths in this way. A necessary precaution in those areas is to always shake out one's shoes before putting them on.

Scorpions eat the adult and larval forms of cockroaches, beetles, crickets, and spiders, as well as centipedes, millipedes, and even small lizards and rodents. Some species feed on other scorpions, an appetite discovered to be critical to the population control of the prey species. In turn, centipedes, ground beetles, praying mantises, spiders, bats, and lizards all prey upon scorpions. So do adult elf owls and roadrunners in the southwestern United States.

Some scorpions live as high as 6,000 feet in the European Alps and 16,000 feet in the Andes. Other species live beneath the bark of palm trees and inside burrows in rainforests. Still others dig in dry, sandy deserts and even under rocks in areas where it snows. Most live solitary lives, meeting only to mate or kill. A couple of species, like the common stripe-back scorpions, are the exceptions. They fill the spaces in rotted logs, stacked on top of each other like sardines.

Another exception is the large black Emperor scorpion that looks ominous, but is gentle and, by most reports, reluctant to do any harm. Male and female Emperor scorpions live together peacefully and raise their young for two years or longer.

Scorpions look like a small lobster with two main body segments and four pairs of walking legs attached to its trunk. The abdomen has twelve segments with the last five making up the tail and stinger. The curved stinger of the scorpion receives venom from two sacs housed in the end of the tail. The scorpion's venom is a complex liquid containing various poisons, each of which is effective against different prey species, injuring their soft body parts or nervous system. Only a few species, about twenty out of fifteen hundred, have poison strong enough to kill an adult man or woman, and the venom of most of North America's forty or so scorpion species is not deadly to humans. Most scorpion stings feel like the sting from a bee or wasp—painful, but not life-threatening. The study of the venom of certain scorpion species has led to the development of a drug for treating strokes, and scientists expect that scorpion poisons will yield other useful drugs in the future.

Scorpion Mating Dance. Scorpions grow by shedding or molting their covering, usually an average of five times until reaching adult size, and then they live an amazing fifteen to twenty-five years. They mate for the first time when two to seven years old, depending on the species. Males seek out females, and before mating they perform an intricate and attractive dance. Grasping the female's claws with his own, the male leads her as the pair moves together. In some species, the pair seem to be kissing because the male grasps and nibbles the female's mouthparts with his own. Sometimes they dance for more than thirty minutes, moving together and covering a lot of ground.

Young develop in the female in a manner more similar to mammals than most other arachnids, the development or gestation taking from three to eighteen months. In fact, the longest scorpion gestation periods rival those of sperm whales (sixteen months) and African elephants (twenty-two months). Scorpions give birth to live young, and their average litter is twenty-five babies. The newborns crawl onto their mother's back for protection and transport, and because the scorpion mother has no breasts, they feed on her back by osmosis.

A scorpion's eyes are located on the hard shield-shaped piece that covers its front portion. One pair of eyes is located near the center, and two clusters of two to five small eyes are located near the front edges. It is surprising then to learn that scorpions don't see well. Their eyes, however, are more sensitive than any other animal to low levels of light. In fact, they can distinguish light from dark so well they can navigate using shadows cast by starlight. Shunning the daylight, they hunt for food at night.

Scorpions are also highly sensitive to vibrations. The hairs on their large clawed arms detect and magnify vibrations in the air, serving as miniature antenna or "ears." And although scorpions do not have voices, some species make a noise with their legs to scare away predators.

Scientists have long been fascinated by the ability of scorpions to survive extreme conditions. These remarkable creatures don't seem to need much oxygen and can be underwater for several hours and still revive. Masters of extremes, they can also go without water for three months and without eating for a year. They can survive in the hottest desert and live even after being frozen in ice. And when the French were testing nuclear weapons in the Sahara, scorpions withstood more radiation than any other creature.

Intellectual Appreciation without Love

These attributes and abilities have captured the attention of scientists and science writers alike. Although Eugene Marais, the author of a mystical and poetical treatise on termites, had a long friendship with a scorpion and even acted as her midwife, what is missing in most other tales about scorpions is an emotional connection to these creatures. And it is missing because most of those who study and write about scorpions don't believe it's possible.

As awareness about environmental issues heighten, there is a current trend to champion species people generally don't like—such as scorpions. But the majority champion them in such a way that the emotional bias is left intact. As we've noted repeatedly, without conscious examination few people, if any, are immune to the distortions in the lens through which they view other creatures. Despite the complex role of the scorpion in ancient cultures, for example, one science writer introduces the historical role of scorpions this way:

> Throughout history and across almost every cultural boundary, scorpions have had a rotten reputation. And if the truth be known, they deserve it. They're nasty and they're not afraid of anybody. They can kill you or throw you into a seizure. Even those species whose venom is relatively innocuous can deliver stings of incomparable pain, 'Like flaming bullets twisting inside you,' as a victim once put it.[9]

The emotional context set, she lists the scorpion's admirable abilities, concluding that nature is strange and wonderful in its myriad of life-forms and that we should try to respect scorpions for their abilities. Imagine how much more powerful her call to respect them would be if it were made within a context that invited kinship.

The Destructive Aspect of Scorpions. A recent book on animal symbology draws upon a range of traditions, including ancient forms of animal work and myth-making. The section on scorpions begins, "Scorpions are purely destructive in most mythologies."[10] It is a definitive statement that requires an explanation to remind the reader of the role of destruction in life.

Destruction was understood in ancient times as the flip side of growth. To destroy means to destructure, to break apart, and this breaking apart of the old is a prerequisite to new life and new structure. The hull must crack open before the seedling can sprout, and the caterpillar must relinquish its worm body and surrender to the harsh hand of transformation that would reshape its form and give it wings. Symbolically then, the scorpion is a destructuring agent, a constellation of energies in service of our potential and a critical force in the process of renewal.

If we look at the human psyche as the open evolving system it is, we can use system theory to further illuminate this destructuring aspect. System theory teaches that new structures or systems begin to organize themselves once there are enough new elements present to initiate the process. In the psyche, the building of this new aspect is undertaken behind the door of our awareness, for the epicenter of change or growth lies beyond our conscious control. What's more, we don't even sense what has been fashioned within us until the old structure is broken away—that's the destructive part symbolized by the scorpion and the part of the process that hurts. From the personality's perspective, the destructive events are only experienced as disorienting and painful and something to try to eliminate. Luckily, however, we can't (although we can distract ourselves or numb ourselves with drugs). Eventually enough pieces are broken so that we can glimpse the new structure underneath—the hidden gift of the soul.

Most ancient cultures recognized the scorpion's role in this renewal process. In a stamp seal from 3300 B.C.E., the rosette of the Great Goddess Inanna, goddess of the underworld, is protected by the pincers of two scorpions, a testament to its sacredness and role in the psycho-spiritual descent. The journey to the underworld discussed in the chapter on flies was accepted as a necessity of life. It was a time of disintegration in which the former identity was lost and the unshielded person had to descend into his or her own forgotten interior realms to emerge, finally, reordered and renewed. It would be akin to what we refer to as the midlife crisis, although not limited to that passage or our contemporary understanding of it, for psychological descents happen at all ages. The scorpion was associated with that internal call to grow because it held both the weapon of disintegration and the antidote necessary for healing, reintegration, and return.

Demonizing of the Scorpion. The scorpion's ancient association with the Great Goddess (in particular, with her destroying or destructuring aspect) led to its becoming a sign of evil in the Judeo-Christian tradition when they demonized the female deity it served. This pronouncement of evil stripped the scorpion of the context through which we once understood its power to destroy and clear the way for new creations. Remnants of the old context remain. In Buffie Johnson's *Lady of the Beasts*, for example, Johnson includes pictures of scorpions drawn on pottery made by the Mimbre. These ancient people of the American southwest associated the scorpion with fertility and depicted a scorpion figure on a dish surrounded with symbols of the goddess. The spiral of life and death on top of this figure illustrates the central theme.

Johnson also shows a Sumerian bowl from the Hassuna Sawarra period that is decorated with eight scorpions surrounding a naked divinity figure with windblown hair. The scorpions swirl about her in a double swastika pattern that symbolizes rebirth and infinity.

Ancient Egyptians also revered the scorpion. They had a scorpion goddess of writing called Selket (or Serqet), who wore a scorpion on her head. Selket was also portrayed as a scorpion with a woman's head. Despite her fearsome power over death, Selket appears beneficent when associated with Isis, the Egyptian mother goddess, probably because as myths tell it, scorpions loved Isis. Seven scorpions accompanied Isis when she searched for the remains of Osiris (another story of dismemberment and reintegration), and it was said that scorpions never stung her. Scorpions are even supposed to have saved her son when he was stung by a deadly scorpion—all because of their love for her.

Selket the Scorpion Goddess symbolized resurrection into new life beyond earthly existence, when she wore on her head a sun disk and the horns of Isis, who was also identified with the cow goddess Hathor. As the link between the living and the dead, Selket was the one who helped the dead accommodate themselves to their new state of being.

Guardians of Higher Zones

In many spiritual traditions, scorpions were thought to guard the threshold to new levels of consciousness. Initiates had to prove their readiness to assimilate a new level of being and they did this by confronting the guardians of the gate. Both the ancient Babylonians and the Assyrians, for example, revered scorpions, believing them to be guardians of the gateway of the sun or enlightenment. Gilgamesh, the great hero of the Babylonian epic poem of the same name, confronts scorpion guards in his search for the answer to everlasting life.

That epic poem and all myths, fairy tales, and dreams use symbolic language to show how the archetypal process works in the human psyche. The universal symbols they employ also have correspondences in the religious practices and beliefs of diverse tribal societies. All such symbols emerge from the collective unconscious, which contains all the archetypes or basic patterns of human experience and self-awareness.

Jungian analyst Marie Louise Von Franz calls myths master teachings that express the basic dynamics of the human psyche—the purest and simplest processes in human psychic development as it moves toward wholeness. The setting of a myth describes an inner condition. The characters represent integrative or disruptive energies operating within the psyche of a single individual. The animals of myth, like mosquitoes or scorpions, in contrast to fanciful creatures of legend and fable, are usually taken from nature or depart only slightly from their form in nature. As links between gods and people, they represent qualitatively differentiated manifestations of powers, and we can best understand them in a symbolic context.

A Borneo tale, condensed below, called "The Myth of The Greedy Brother,"[11] provides a lesson on the correct stance to take when challenged by the guardians of more expansive levels of consciousness and the consequences of approaching the gate without being psychologically ready.

> A young man who had lost his way in the jungle came upon a house. He entered, and soon what appeared to be people joined him. They offered him blood instead of water and maggots instead of rice. Being a gentle and sensitive soul, the young man was discreet, and, perhaps sensing that his reaction was being watched, he managed to choke down a little of both. When night fell, they brought him a banana leaf to sleep on instead of a sleeping mat. He thanked them and lay down on it. In the ensuing darkness, swarms of mosquitoes descended on his flesh. He held himself still, aware that everyone else in the house was quiet and apparently undisturbed. He gently brushed the mosquitoes away, careful not to injure any. They went away and did not return. In the morning, his hosts sent him on the right path to reach his home and gave him a hollowed out piece of bamboo. When he arrived home, he opened it and found gold and silk, which he distributed among his family.

The myth teaches that courtesy, caution, and surrender are rewarded. The young man who submits to his fearful hosts and does no harm to the swarm of mosquitoes that beset him at night receives gold and silk. His greedy brother attempts to secure the same reward, rushing into the same situation only to respond inappropriately. He is disgusted by the maggots and blood that his strange hosts offer him, complains about having to sleep on a banana leaf, and slaps at the mosquitoes, killing hundreds of them in the night. The next morning he too is given a box and sent home. He opens it on the way, wanting the treasure for himself, and scorpions run out and sting him to death.

The ability of the scorpions to wield their destructive power when faced with an individual's unwillingness to grow and change is part of their role as guardians to the gateways of enlightenment. As one of the ancient symbols of the process of spiritual enlightenment, the scorpion that stings itself to death symbolizes the vital energies within us that can illuminate or destroy. These energies, when transformed, become the treasure, the psyche's wholeness or Carl Jung's "Self." The fire of transformation which this creature controls through its sting can bring insight and growth to an individual as it moves up the spine through the chakras or energetic centers of development. And as we have noted, these same energies, if ignored and allowed to gather at the base of the spine, can also collapse in on the person. Since the scorpion venom is said to contain its own antidote, its gift of self-sacrifice and rebirth has, understandably, been both feared and honored.

Agents of Transformation. The symbolism of myths provides us with a context to redeem painful encounters with other species, firmly positioning them as agents of transformation. It is not a question of believing or not believing in this process of human growth—in the turning toward the soul or true nature and the breaking up of small and familiar faces of identity. It doesn't wait for our acceptance and understanding. In fact many times we don't recognize the initiations for what they are until years have passed and we have gained the perspective of time. But myths show us what is and what can be, depending on our attitude and actions.

The author who summed up the scorpion's mythological role as merely destructive is lacking a context that includes the role of destruction in growth. He supports his statement by citing the First Book of Kings in the Old Testament, which links desert-dwelling scorpions to drought, wilderness, desolation, and a dreadful scourge. But wilderness and desolation in the Christian tradition is also part of the dark night of the soul, the testing of the individual's resolve and commitment to seek enlightenment. If we place symbols in too narrow a context, we are left with a vague negative association that depends on unconscious bias to fill in the gaps.

The author also states as evidence of the destructiveness of scorpions the symbolism of some Amazon tribes who believed the scorpion had been sent by a jealous creator to punish men for impregnating women whom he himself desired. Other native tribes of Central and South America, however, had a Scorpion Goddess of the Amazon River called Ituana or Mother Scorpion, who ruled the afterworld, receiving the souls of the dead in her house at the end of the Milky Way. She also oversaw the reincarnation of souls to new life and nursed the Earth's children from her abundant breasts.

More Discrepancies. In still another contemporary introduction to scorpions, the author states unequivocally that the ancient Chinese believed that scorpions symbolized pure wickedness. Yet in the ancient Chinese belief system, it is scorpions and other poisonous and venomous creatures that are enlisted to *dispel* evil perpetuated by spirits or *gui*, which, according to the Chinese calendar, are most troublesome during the summer when the yin part of the year begins. In the Ming dynasty even imperial eunuchs wore badges of the five poisonous insects and creatures that combat sinister spirits—the snake, centipede, scorpion, lizard, and toad or spider. Today children still wear aprons with these images on the fifth day of the fifth month.

The discrepancies in interpretation and the slant of mythology and folklore are not only a function of the unconscious bias that operates when anyone sifts through information unwittingly looking for what they already believe, they are also a function of how we relate to power and the great fear of the disintegration process so vital to growth.

The Archetypal Influence of the Scorpion

The scorpion's archetypal status is suggested by the fact that civilizations around the world, from the ancient Greeks to the pre-Columbian Mayan culture, all saw a scorpion in the same grouping of stars, which includes the supergiant red star Antares, typically the "heart" of the scorpion. Scorpion star symbolism was the same in Babylonia, India, and Greece as it was for natives of the Americas. All the American versions, for example, like the Mayan culture of the Yucatan who called this constellation "scorpion stars," and the Pawnee and Cherokee Indians who believed Antares was the spirit-star of the Old Goddess, closely resemble Babylon's Goddess Ishara of the Sea, a scorpion-tailed mother who was believed to reside in the same constellation.

Also testifying to its universality, ancient astrological myths everywhere placed Scorpion at autumn equinox, tying it to the death of summer and the movement into the darkness of winter (with Aquarius the Water-drawer at winter solstice, Taurus the Bull at spring equinox, and Leo the Lion at summer solstice). Each represents one aspect of the renewal process operating in the natural world—and mirroring the process in our human psyches. Spirits of the four points of the year were sometimes called Sons of Horus, and it was the scorpions who killed Horus and sent him to his midwinter death (another story of disintegration, death, and rebirth). When Christianity transformed the twelve signs of the zodiac into the twelve apostles, they assigned the scorpion to Judas Iscariot—the betrayer. It is obviously not the favored role, but if we view it symbolically, it is a role necessary so that death and then resurrection can follow.

There is also a strong connection between the Scorpion and the Amazon warrior women of Amazon myth. It is depicted in the zodiac astrological chart, which has twelve "houses" ruled by various "signs." The eighth house is Scorpio, and it is ruled by the fixed water sign of the female warrior or healer.

In the zodiac, the Scorpion is linked as well to the planet Pluto, a planet of regeneration which was thought to wield the death force. Vickie Noble, author of *Motherpeace*, a Tarot deck and book that honors matriarchal consciousness and the wisdom known in Goddess-based religions, also links the sign of Scorpio to Pluto, whose influence changes us from the deepest place inside ourselves. According to Noble, in its relationship to Pluto the "Scorpion represents transformation, death and mysticism—the three central experiences of the shamanistic mysteries [and it] rules the sex organs, the deep unconscious, and the ability to channel healing energies."[12]

Today the scorpion is still potent, an archetypal initiator of the process through which we strive for psychological maturity and spiritual enlightenment. Those who are actively engaged in this process would do well to study the rich mythology of scorpions, taking special note of when these creatures make an appearance in dreams or in our houses, or when we encounter them outside.

If, following the native tradition of taking as a guide the first creature we see when we start our day, we see a scorpion, its presence might signal a time of descent leading, in turn, to an opportunity to deepen and grow. Being open to these forces will radically shift how we experience the descent, although a certain amount of fear and anxiety always accompanies the stripping away of what is outworn and to be discarded. Encountering a scorpion might also relate to our ability to survive a situation of extremes, whether emotional extremes or difficult outer events. Buddhist initiates are instructed to wield the sword of knowledge as the scorpion wields the sting in its tail.

If a scorpion has shown up in our life, we might also recognize in this creature a way of being in the world that is similar to our own way of hiding our vulnerabilities behind the threat of stinging words and actions. We might even discover that we have an affinity for scorpions as one teenage boy discovered when he met his first scorpion in a pet therapy program.

Scorpion Therapy. In a novel variation of the popular animal visitation therapy programs, volunteers brought scorpions and other creeping creatures to a home for troubled teenagers, along with the usual assortment of dogs, cats, and rabbits. What they discovered is that creatures like the tarantula and scorpion reached certain teenagers in a way that a kitten or rabbit could not. During one visit, when the knowledgeable volunteer talked about the scorpion and the manner in which its fearsome appearance hid a benign temperament, one boy in particular dropped his sullen and withdrawn stance and began eagerly asking questions as though in the scorpion he recognized his own way of being in the world.

Affinity for certain creatures surfaces in so-called chance meetings. In short, we know when we meet a creature for whom we have an affinity because of the intensity invoked in us by the creature. I suspect whether we name it fascination, fear, or love, it is still just one of the faces of a primordial interspecies bond that calls us home inside ourselves.

The Scorpion Eaters. There are stranger aspects to affinity than initial fear and loathing. In Morocco and Egypt there are people known as "scorpion eaters."[13] It's a profession that belongs to a family. By apparent heredity, the members of these families are immune to scorpion stings. They claim their immunity is due to eating scorpions generation after generation.

People outside the profession call on scorpion eaters (usually men) to clear out scorpions from vacant houses. The scorpion eater accomplishes this by going into the house and squatting on the floor in the middle of the room. Then the scorpion eater whistles softly. The scorpions appear from their hiding places and walk across the floor toward him. He picks them up and puts them in a bag. When the scorpion eater is finished and the house is clear, he takes the scorpions home and he and his family eat them. Eating the scorpions is not a hostile act, for these people revere scorpions—as kin and ancestor. It is rather an act of assimilation and an opportunity to become more scorpion-like, as well as an opportunity for the scorpions to live through them.

Laurens van der Post writes that "the task of every generation is to make what is first in us new and contemporary."[14] If we apply that to the scorpion, our challenge today is to look for new metaphors that combine ancient psychological and spiritual truths concerning creatures of power with the latest scientific observations about their unique form and mode of being in the world.

Mother Hathor, the English woman of the Foundation Faith who had found cockroaches to be friendly creatures, also learned about scorpions when she was living in Xtul, Mexico. She remembers lying in a hammock watching the scorpions walking up the wall, larger ones followed by smaller ones: "It was a real family feeling, a weird thing to feel about scorpions, but you could sense a relationship to them."[15] She said she lost a lot of fear living there. She realized that everything had its place and that being confident and trusting is all the protection that anyone needs.

Sarah the Sensitive Scorpion

A remarkable story shared by dog trainer Vicki Hearn even opens the door to the possibility of relationship with a scorpion. In her book *Animal Happiness*, Hearn tells the story of Warren Estes, a herpetologist who lived with snakes, kangaroo rats, all sorts of insects, and "Sarah" the scorpion. According to Warren, Sarah had trouble at holiday times when his relatives came for a visit. They would look at Sarah and express disgust or fear. He said it took some time after his family left for Sarah to feel good again. He also cautioned the visiting Hearn that if she was afraid of Sarah, she shouldn't try to hold her, because Sarah wouldn't like it.

The visit was memorable for Hearn. What first caught her attention was how Estes talked to Sarah as he held her:

> He would speak to her, and she would respond. He talked to me as well about the various evidences that she was nervous or relaxed. He stroked her poison sac. He admired her for her courage, and for her fine poison sac.[16]

Months after the visit, Hearn remembered the look of Warren's desert-rat brown hand, and the scorpion named Sarah glistening there. At first, the creature's glisten appeared menacing even to Hearn who loved animals. But as

Warren talked about Sarah, it became "the glisten of something very much like consciousness...He talked with her as some men talk with nervous thoroughbred fillies, cajoling, admiring, gently admonishing."[17]

> The knowledge in his hand as he gently held Sarah, just as she liked to be held,...there is no room in the animal rights controversies for that knowledge. It does not sponsor or finance any political point of view.[18]

Hearn didn't really know how to think about Warren and Sarah. She admits the possibility that he understood something at some level, for he never got bitten or stung by any creature. If Sarah had been a dog instead of a scorpion, perhaps Hearn would have more readily understood and accepted what she had seen. Our assumptions about certain species get in our way and make us forget that we all share a language beyond words, a more direct kind of knowing. Encoded in our bodies, this "language of the heart" travels the trajectory of our thoughts and perceptions, arriving before our words.

Warren related to Sarah as an intelligent, spiritual being capable of responding to love—and she was. He could see that she was comforted and calmed when he held her in a gentle loving way, and he had faith in her and didn't worry about being stung. Sensing his love, Sarah also demonstrated her faith in him, relaxing in his hand. In *The Souls of Animals*, minister Gary Kowalski says that when we relate to others in this way, "we become confidantes—literally, those who come together with faith. And it is through faith—not the faith of creeds or dogmas, but the simple 'animal faith' of resting in communion with each other...that we touch the divine."[19]

What Hearn saw when Warren held Sarah—just as she liked to be held—was a man and a scorpion resting in communion with each other, touching the divine. If we make room for that truth inside ourselves and ground it in our daily life, it can transform us and everyone with whom we come in contact. Such is the power and inherent healing potential of our connection to other species.

Part IV

Rejoicing in Insects

- Nation of Winged Peoples
- Finding Our Affinities
- Strange Angels
- Following Mantis

14

Nation of Winged Peoples

What the caterpillar calls the end
of life the Master calls a butterfly.
—Richard Bach

Butterflies are our favorite insects. Their beauty has elevated them above other insects, allowing them to escape most of our negative projections—but as we shall see, not all.

Counting on our adoration, a United States congressman recently asked his colleagues at the Capitol to designate the monarch butterfly as the national insect. He argued that this familiar black and orange butterfly, a favorite across the country, enhances the beauty of the environment and signals the need for protection and conservation of the natural world. Others have debated just as intensely that the honeybee deserves to be the national insect. The matter has not been settled.

Preferences aside, few would disagree that butterflies are charismatic. Ranked fifth in a study conducted by researchers at Yale University to determine our preference for different species, butterflies were topped only by dogs, horses, swans, and robins. Their exquisite colors and association with flowers charm us, and they have been called "flying orchids." Conservationists consider them the "wolves and whales of the invertebrates." They hope that butterflies can generate enough public interest and support so that we will rally to save their endangered habitats, which are also home to countless species that we don't find charming.

Some companies are already protecting threatened and endangered species. Standard Oil of California protects the endangered El Segundo blue butterfly, an insect that needs the flower of sea cliff buckwheat for its tiny caterpillar larvae. In the mid 1970s, when entomologists discovered a colony on a two-acre plot near a Standard Oil refinery, the company (perhaps trying to brighten its tarnished image for the growing environmentally aware sector of the public) agreed to create a sanctuary for them.

In Northern California, Waste Management, Inc., has adopted the endangered bay checkerspot butterfly as its mascot. This rare beauty lives on the company's property overlooking Kirby Canyon landfill. The company deposits $50,000 a year in a trust fund to protect and enhance its habitat and encourages visits by schoolchildren.

An international group concerned with species survival has chosen swallowtails as the first group of invertebrates on which to base a conservation action plan. One of their representatives is the largest butterfly in the world—the Queen Alexandra's bird wing of Papua, New Guinea. The health and abundance of swallowtail butterflies in the tropics reflect ecological trends in rain forest areas, providing a quick and reliable take on the health of entire ecosystems.

Finely tuned to their environment, butterflies are one of the first species to react when the climate changes. Recent studies, for instance, confirm that global warming is driving some species north, like the Edith's checkerspot butterfly. Chris Thomas, senior biologist at England's University of Leeds, thinks these biological changes are more than just a matter of curiosity and may have major implications for agriculture, medicine, and conservation.

The exacting ways of the butterfly have made it the emissary of ecologists who work to restore one small piece of a damaged ecosystem. Since butterflies, as both caterpillars and flying creatures, are finicky about their surroundings, their presence in a damaged area means that a good measure of balance is still operating. Sometimes restoring a butterfly to a habitat means saving another species that is in symbiotic relationship with it. If the metalmark butterfly were threatened, for instance, a plan to save it would have to include protecting the ants that guard its caterpillar. Luckily ecologists know that. An endangered blue butterfly in Britain was recently brought back by not only protecting the wild thyme the adult feeds on, but also by protecting a species of red ant that cares for the newborn caterpillar in its nest until it emerges as a butterfly—ten months later.

We generally think flowers are enough to attract and keep butterflies in an area, but they are not. Fog will deter many sensitive species from feeding on the flowers in an area. And butterflies often perceive simple divisions of their territory, such as those caused by roads and fences, as impassable.

To complicate things further, many caterpillars feed on just one species of plant—often one we consider a weed. This need for specific plants has pushed several species in the United States into extinction after suburban development destroyed their habitat. Entomologists estimate that eighty percent of all butterfly species are just barely surviving in degraded habitat. Ten to twenty percent of the species need the real habitat and will disappear in the near future without it. Those butterflies are not going to make it without our help. Fifteen North American species are now listed officially as endangered or threatened, and at least seventy are candidates for the Endangered Species List.

Pesticide Poisoning. It is not surprising, given butterflies' finely tuned biology, that their absence in an area typically signals that pesticides have been used. While many insects suffer from exposure to commonly used pesticides, butterflies fail noticeably and right away, almost always dying out. Even Bt, the natural insecticide discussed in our look at beetles and agriculture and considered by many to be a safe, organic pest control against gypsy moths and other insects, destroys butterfly caterpillars as well. And Demanol, another product used to kill gypsy moths, has also been implicated in killing many species of butterflies in West Virginia and the Carolinas.

Although space prevents discussing the war against the butterfly's poor relative the gypsy moth, aerial spraying to eradicate the moth continues even in areas where the moth has all but disappeared and despite the fact that many people are voicing their concern. Robert Spears is currently writing a book on the war against the gypsy moth and has published a Web site devoted to this insect. There he details the "dubious doings"[1] of those agencies and pesticide companies involved in perpetuating the war against this insect. For those in areas where the gypsy moth is advertised as a real threat to the trees, the reader is encouraged to visit Spears's informative site and draw their own conclusions about where the threat lies.

For the Love of Butterflies

As epiphanies of the Goddess, butterflies have been worshiped throughout the history of our species. The Cuna Indians of western Panama worship the Earth Mother in the form of the Morpho butterfly. They call this deity the Luminescent Giant Blue Butterfly Lady and depict her on their brightly colored patchwork molas in intimate communion with her sacred serpent.

The butterfly goddess Itzpapolotl, whose wings are tipped with obsidian (a knife-sharp, black volcanic glass), is depicted in the Aztec calendar, and the ancient Mexicans worshiped butterflies as their god of love and beauty, considering butterfly eggs as the seeds from which happiness grows.

This strong association between butterflies and love is also present in the East. Emperor Gensõ, or Ming Hwang, allegedly let butterflies choose his loves for him by freeing caged butterflies in his garden and taking note of which maidens attracted them. And a Chinese scholar, known in Japanese literature as Rõsan, was loved by two spirit-maidens who visited him daily to tell him stories about butterflies.

Christian myths tell of butterflies originating in the Garden of Eden and then following their beloved Eve when she was banished from this paradise. In a Korean story about the origin of butterflies, a girl is betrothed to a man she has never met. When he dies before they can marry, tradition forbids her from ever marrying anyone else. She mourns at his grave, pleading with the spirits to let her know if the affinity between them is true, and, if it is, to break the grave in two. The grave splits open and she leaps into it. A handmaiden grabs for her but only gets her skirt, which breaks into many pieces, each turning into a butterfly that flutters away.

Selling Butterflies. The association of butterflies with love and happiness persists today. A number of enterprises, capitalizing on the insect's symbolism and our love of these winged creatures, sell butterflies for use at weddings, store openings, and other happy occasions. One company sells the painted lady butterfly. At a wedding, instead of throwing rice, each guest releases one of these butterflies from a specially designed box, filling the sky with bright moving color. Other companies use monarchs or yellow-and-black eastern tiger swallowtails.

While some favor using butterflies in this way, others are alarmed. Releases of butterflies for our purposes don't take the butterfly's mating and migration cycles into account. And some are invariably crushed in the rush to free them and have them take to the air simultaneously. Another fear is that selling butterflies will tempt some people to take them from hibernation (overwintering) sites, an act that could eventually threaten the entire species.

Poaching butterflies is a violation of federal wildlife laws. In 1995, three men were convicted of poaching, trading, and stealing more than 2,000 protected butterflies, including 210 that were listed under the Endangered Species Act.

Cities with butterfly hibernation sites within their boundaries have passed their own laws to protect butterflies, like the thousand-dollar fine for molesting or harassing a monarch butterfly in Pacific Grove, California. Unfortunately, fees and laws aren't much of a deterrent to those who realize the value of butterflies. Thousands of overwintering butterflies like monarchs, which are vulnerable and accessible, mean easy money to those who wouldn't hesitate to steal them. The Monarch Program, an educational conservation group, recently received a report that monarchs were taken from a Santa Barbara overwintering site and sold for a "role" in an upcoming movie.

The Shadow of Butterflies

We are less eager to protect moths, especially the small drab ones that gather around outside lights. In fact, if we were responsible for classifying moths and butterflies, we might readily divide them by their beauty and whether or not they fly in the daytime or at night. The distinctions between moths and butterflies, however, are more subtle than that, although both belong to the order Lepidoptera, a group that contains approximately 175,000 species of moths and 45,000 species of butterflies.

Lepidopterists consider butterflies specialized moths that probably evolved from a mothlike ancestor fifty to one hundred million years ago. What formally distinguishes a moth from a butterfly are small external differences that for all but the experienced eye would probably require a close-up inspection of a pinned specimen, or at least netting the insect—a practice of butterfly enthusiasts that is, happily, dying out.

Most of us nonspecialists assume moths are dull-colored, night fliers with feather-like antennae and furry bodies and butterflies are brightly-colored, day fliers with club-shaped antennae. But they don't separate out so neatly if we use the criteria accepted by entomologists. Although many moths fly at night, some fly in the daytime, and while many moths have dull coloring and feathery antennae, others do not.

Another important distinction is that moths spin cocoons in which to make the change from caterpillar to adult. Butterflies complete their metamorphosis inside a chrysalis, a hard covering that is actually the caterpillar's final molt.

The moth and butterfly are linked together in the symbolism of most ancient cultures by the cocoon and chrysalis. Both insects were believed to know the secrets of transformation from life to death to new life. Their fluttering movements are also linked to flickering flames and early fire gods. The Cherokee Indians tell of a small yellowish moth that flies about the fire at night seeking light. Because of its affinity for the fire, this tribe invokes this spirit to heal fire diseases like sore eyes and frostbite. And the song of the legendary musical butterfly of British Guiana is said to sound like the crackling of dry grass in a fire.

Goddess of Death. In some cultures the butterfly was not adored but feared. The shadow aspects of butterfly symbology developed when the ancient Goddess was demonized and the religions that had formed around her were suppressed. Butterflies then became demoniacal creatures often linked with witches. In Irish, Breton, and Lithuanian lore, for instance, the butterfly was feared as a symbol of the witch, or Goddess of Death.

Other shadow-based symbolism is seen in some mystical groups. The Gnostics, for instance, looked upon the butterfly as a symbol of corrupt flesh, and in Gnostic art the Angel of Death was portrayed crushing a butterfly.

In a later historical period, butterflies were separated into good and evil by color. Dutch still-life paintings illustrate this kind of grouping, with the painted white butterfly symbolizing virtue, resurrection, and immortality, and the red admiral, a black and scarlet butterfly, symbolizing death or damnation.

In general though, the night-flying habits of most moths linked these insects to death and sentenced them to carrying the shadow projections for all but nocturnal butterflies. In Bolivia, for example, the Aymara people took the presence of a certain rare moth as an omen of death. And the ancient Mexicans regarded one of the large night-flying moths (the owlet moth, or "Black Witch") with great fear as a messenger of death.

In early Christian mystical writings, moths represented the temptations of the flesh. The death's head sphinx moth, a large moth that sometimes makes a strident sound when it flies and whose thorax shows a skull-shaped marking, represented Satan and was an emblem of death for early Christians.

Today it is rare for us to treat the butterfly as an object of fear, although several years ago a tabloid featured a story about a rampaging six-foot butterfly in the Midwest which was finally shot by a farmer. We do, however, consistently view the moth with suspicion and mistrust. In the popular culture one example of our misgivings about these creatures is seen in the 1962 movie *Mothra*. In this classic horror film, an immense moth undertakes the destruction of Tokyo until the city officials decide to give up twin girls mysteriously linked to the insect.

In a "Weird Mystery Tale" comic, the connection between moths, fire, and death is featured. A forest ranger, more savage than wild, lives the life of a recluse, entertaining erotic fantasies and offering his soul in exchange for a woman. The next frame shows him killing moths whom, for some unexplained reason, he hates passionately. He kills them all except for the queen, who escapes. That night he dreams that she returns to avenge the death of the other moths. The next day he pursues the queen moth once more, intent on killing her. Then he sees a beautiful woman in the forest and grabs her, not knowing she is a moth-woman. She turns back into her moth shape and takes him into the fire, where they are both consumed.

From Caterpillar to Butterfly

The moth's association with death is not always negative. When the moth is a caterpillar, its "death" inside the cocoon is the signal that begins its physical transformation into a winged creature. Few have witnessed this change in moth or butterfly without making a comparison to death, resurrection, and renewal. In ancient religions the butterfly or moth spirit was considered a divine womb. The pairing of the double axe (reminiscent of a butterfly or moth with wings spread) with these insects was a prominent religious image that reflected an understanding that both death and renewal were born from the spirit of these winged creatures.

Indigenous tribes, understanding that the intrinsic patterns of life were reflected in the smallest of creatures, associated the caterpillar's act of transformation and rebirth with the renewal of nature each spring. The hoop dance of the Plains Indians, for example, is part of the annual celebration of winter's transformation into spring and includes a dancer emulating a caterpillar that turns into a beautiful butterfly.

For the Butterfly clan of the Hopi Indians, the butterfly is a totem. Every year, young men and women of this tribe perform the *Bulitikibi*, or Butterfly Dance, a ceremonial dance of renewal believed to bring good crops. A northwestern Brazilian tribe also performs a butterfly dance, presumably symbolizing rebirth or eternal life, at a festival honoring the dead.

Native Mexicans not only saw the butterfly as a symbol of rebirth, regeneration, happiness, and joy, but they also have a legend in which the powerful plumed serpent god Quetzalcoatl first enters the world in the shape of a chrysalis and then painstakingly emerges into the full light of perfection, symbolized by the butterfly.

Getting Out. As models of growth and transformation, insects in general are without equal. All must molt or periodically shed their external skeleton. The process of molting is known as ecdysis and comes from the Greek root meaning "getting out." Molting happens in one of two ways. In gradual or simple metamorphosis, which cockroaches, grasshoppers, and praying mantises undergo, the young insects called "nymphs" have the same general shape and features as the adult (imago). At each developmental stage, they molt or break out of their exoskeletons and emerge bigger than before.

The second, more complex, solution to growth is called complete metamorphosis and is the one adopted by caterpillars, maggots, and most grubs. Complete metamorphosis has an incubation or pupal stage during which virtually all body tissues break down and are reorganized into a new form, so immature insects or larvae look radically different from their parents—maggots don't look like flies, grubs don't resemble beetles, and caterpillars do not resemble butterflies or moths. The advantage of this form of development, which the vast majority of insects use, is that it allows the species to take advantage of very different habitats.

Metaphors of Growth

As we've seen throughout this book, patterns of life are seeded in nature and, prior to this age, were often woven into a cultural or religious context. The stages of simple metamorphosis that a grasshopper undergoes, for instance, were recognized symbolically as being the power we have to break free from earthly concerns. This ability was the reason that the grasshopper was respected as a symbol for the soul in early Christian iconology.

The stages of complete metamorphosis evidenced in beetles, flies, moths, and butterflies mirror in some universal way any kind of growth that involves changing from one identity to the next. Even in plant morphology there is a correspondence. One of the first to recognize it was German biologist Hermann Poppelbaum. He observed that each stage of insect development is "a pictorial image of a corresponding botanical structure of a flowering herbaceous plant."[2] For example, butterfly eggs are like plant seeds, while the compressed, molded form of the butterfly pupa is like the contracted flower bud enclosing the developing flower.

All instances of such radical change include, of necessity, the darkness of an interim period where the process that permits the reorganizing can take place. People of widely diverse cultures and religious orientations, observing this process and its startling result, saw in it a model applicable to human life. In ancient Egypt, for instance, the butterfly was the emblem of Osiris, who was confined after his death in an oak coffin until he arose again.

Transforming Ourselves and Society. This image of transformation is as relevant today as it was in ancient times, for groups as well as people. On Earth Day 1990, author and educator Norie Huddle published a book called *Butterfly*.[3] In it she tells the story of the metamorphosis of humanity from a nonsustainable society to a "butterfly civilization" that fulfills humanity's potential, bringing peace, health, prosperity, and justice to all the Earth's inhabitants. In her now popular analogy, she compared where we are today to where the caterpillar is after having consumed enormous amounts of food. Like the insect, we are encased in our chrysalis and are approaching metamorphosis. For the caterpillar this is a critical time when certain cells, which biologists call imaginal cells, begin to develop and the process of building the various parts of what will be a butterfly begins in earnest—although from the outside it looks as though nothing is happening. Huddle says that the culture's imaginal cells are the individuals, groups, and activities that are building new patterns—clumping and clustering together and sharing information. She sees all attempts to hold to the status quo as merely the caterpillar's immune system rigidly and reactively clinging to its old worm ways and not recognizing the imaginal cells as its own—until such time when enough are in place for the winged form to emerge.

In her 1998 book *Conscious Evolution*, futurist Barbara Marx Hubbard also relies on the caterpillar-to-butterfly metaphor to explain the metamorphosis of humanity from its polluting, overpopulating phase through a whole system transition that eventually promises to fulfill our potential as a fully conscious society. And the late Willis Harman applies the insect analogy to explain how whole system change works in our culture and how business will be transformed from its economic and financial foundation to one that values all life-forms and the planet. Like Huddle, he likened where we are in the process of this changeover to the caterpillar within its chrysalis and approaching, through blind instinct (and perhaps grace), its transformation. The communities and organizations already formed and working toward radical change—and under the considerable influence of the feminist, ecological, and spiritual movements—are the imaginal cells in the business community. When the current structure comes down, those "imaginal cells" will be there in full operation, and business will have been transformed.

In keeping with the resurgence of butterfly images and metaphors, on June 26, 1997, at the United Nations in New York, when heads of state from around the world assembled to reassess the progress made since the first Earth Summit, Alan Moore, founder of the Butterfly Gardeners Association, organized a release of butterflies to symbolize humanity's new awareness of the fragile beauty of the Earth. Moore, who has since then devoted all of his time to raising awareness about the potency of this symbol for the millennium, believes that humanity is emerging out of its cocoon and beginning a new chapter of human evolution where peaceful coexistence and responsible stewardship will be the norm. He believes the butterfly, as a symbol for the Renaissance of the

Earth and the dawn of world peace, coupled with organized butterfly releases could inspire people to work harder for that Renaissance.

Models of Spiritual Growth

In the beetle chapter we saw the archetypal transformative processes at work in our lives in the story of the bug that emerges from the table to enjoy its perfect summer life—a hidden life that comes forth when conditions are right. In the butterfly we have another image for this universal process. The idea of a hidden life is attractive. It gives us hope that a beautiful and winged life is also within us, a life whose egg has been buried under layers of societal and familial beliefs. To assist our transformation, all varieties of the spiritual path tell us that our thoughts and actions will help bring about our metamorphosis into the species we were meant to be. Our thoughts and actions are our imaginal cells. Taking our lead from the insects and trusting that at times we too must be in darkness, maybe we can enter life's deep wellspring without erecting elaborate defenses.

In *A Mythic Life*, Jean Houston also sees the metamorphosis of society as a renaissance with implications for our personal growth as individuals. "The culture is being so newly reimagined that it necessitates a rebirth of the self."[4] She believes that when the door to the caterpillar self has been unlocked and the wall to the human soul breached, we will be filled and subsequently directed by a flood of new "butterfly questions" about who we really are and what life is about.

In spiritual traditions, the human being who works for his or her own transformation into a more perfected state by leaving the agitation, noise, and anxiety of ordinary life has frequently been compared to the caterpillar entering a pupa to become a butterfly. In fact the three parts of the process of metamorphosis closely resemble the three degrees of the Mystery School through which a person unfolds his or her divine nature—from a state of initial helplessness and ignorance, to a disciple seeking the truth and entering the tomb, and, finally, to an unfolded, enlightened being.

The pupa provides an ideal image of serene contemplation and a promise of new life from the perspective of certain Eastern religions. Even today in the Himalayan mountains, some Asian ascetics live for years in almost inaccessible caves existing on a minimum of food in order to expose their souls more directly to the light of divinity. Later they become great gurus, comparing themselves to the huge butterflies that inhabit the valleys of the Indus and the Ganges.

Christian Symbology. In Christianity the caterpillar symbolizes Christ. According to Louis Charbonneau-Lassay's *Bestiary of Christ*, in the fifth century the Pope declared that "Christ was a worm, not because he was humbled, or humble himself, but because he was resurrected."[5] In fact all worms that underwent a metamorphosis were emblems of Christ. The transformation of the caterpillar, more than that of any other larvae, represented the broken body of

Jesus transformed into the resurrected body which emerges from the darkness of the tomb. A bright yellow butterfly common in the provinces of western France has even become the symbol of the resurrected Christ (called the Easter Jesus) because it is the first butterfly to emerge in March or April.

In this same religious tradition the caterpillar also represents the Christian who must pass through two preparatory stages before becoming a butterfly. The first stage, symbolized by the chrysalis or cocoon, is death. After death the human soul reaches its goal of resurrection and eternal life, symbolized by the butterfly.

Esoteric traditions describe a transitional place where people go after dying. The purpose of the place is to allow a transformation to take place. In *Living On*, Paul Beard, former president of the College of Psychic Studies of London, describes "Summerland," the initial stop on the after-death journey, as a waiting place, the chrysalis where people are transformed from human beings bound to the laws of time and space into ephemeral spirits.

A Sign of Resurrection. In the United States, the butterfly is also a well-known sign of resurrection and life after death, even within the Roman Catholic tradition. In Bill and Judy Guggenheim's *Hello From Heaven*, a compilation of the results of an extensive research project on after-death communications, the following story was included:

> After the funeral service for a young woman, her family went to the cemetery. While the priest was saying the final prayers, a large white butterfly landed on the casket and stayed until the ceremony was finished. A nun hugged the mother of the deceased woman, asking her if she had seen the butterfly and telling her "a butterfly is a symbol for the Resurrection!" The meaning ascribed provided much peace for the grieving mother.[6]

In another true account from *Hello From Heaven*, a different butterfly pays an unexpected visit during a Catholic funeral.

> While the niece of the deceased man was praying during Mass and thinking of her uncle, an orange and brown butterfly came fluttering down the aisle and flew around her and the family. Then it went over by the casket and then to the altar before flying away. She considered it a miracle, a sign that reassured the family and set their minds at peace. It was the only time she had ever seen a butterfly inside the church.

As a universal symbol of hope and escape from sorrow and death, the butterfly is unequalled. Recently in the news a young woman, who had been abducted as a child and held by her captors for a year in a van where she was repeatedly sexually assaulted, said the butterfly is her symbol of hope and renewal as she struggles as an adult to heal the wounds from that terrifying experience and emerge as a whole and happy individual.

Elisabeth Kübler-Ross, a pioneer in the field of death and dying, often speaks of the numerous drawings of butterflies she saw in the barracks at concentration camps in Europe. They had been scratched into the wooden walls by children and adults during the Holocaust. Today pictures of butterflies can be found throughout almost every hospice. This symbol of hope is also used extensively by many grief counselors, spiritual centers, and support groups for the bereaved.

The Darkness of Transformation. For those of us seeking personal as well as planetary renewal, the goal of wings is often overshadowed by the dying that takes place in the darkness of the cocoon or chrysalis. Those who have witnessed or experienced the struggle that accompanies the release of what is old and outdated within ourselves and within our culture might take heart in the fact that the transformation process in other species also teaches us that struggle is an integral part of the renewal pattern. There is a story told of a man who found a cocoon of an emperor's moth and took it home to await the moth's emergence.

> One day he saw that the insect had made a tiny hole in the cocoon, and he watched for several hours as the insect struggled to force its body through the little hole. Then it seemed to stop making any progress. In fact, it appeared as though it could go no further. The man, being a kind person, decided to help the moth, so he snipped off the remaining bit of the cocoon with scissors. The moth emerged easily, but it had a swollen body and its wings were shriveled and small. The man watched in happy anticipation for its body to contract and its shriveled wings to expand and unfold. But nothing happened. The moth lived out its life without ever being able to fly.

By saving it from its long struggle, the man had stopped the process by which blood is pumped into the body and wings in preparation for its winged life.

The Butterfly Soul

Just as the concept of a soul has been recorded in every culture known to history, so has the idea of a butterfly soul symbolizing emerging life and life after death.

In many cultures the butterfly and moth were believed to be the souls of departed ones. The ancient Slavs opened a door or a window to permit the soul, often in the shape of a butterfly, to leave the body of a dead person. The Kwakiutl tribe of northwest North America depicted the soul as a butterfly, and the Goajiro of Columbia believed a particular white moth in the house was the spirit of an ancestor come to visit.

Butterflies and moths were also considered to be human souls in the Far East. Certain Chinese texts, for instance, officially recognize sundry butterflies as the spirits of an emperor and his attendants. The Japanese treat kindly any butterfly that enters a house because they believe souls customarily take butterfly shape in order to announce that they are leaving the body for good. And in the South Pacific Solomon islands a dying person tells family members in which shape he or she intends to transmigrate—usually as a bird, butterfly, or moth. From that moment on, the family treats that particular species as sacred.

In Finno-Ugric mythology, the soul is also thought to depart from a dead body in the shape of a butterfly or moth, but even the soul of someone sleeping can exit from his or her mouth, drink water from a pond, and return to the sleeper. And in Slavic countries where the Great Goddess was demonized, it was believed that butterfly souls issued from the mouths of witches to invade living bodies when the true soul was absent.

In ancient Greece human souls were thought to become butterflies while searching for a new reincarnation. Carvings on sarcophagi show a butterfly soul flying over a corpse, skeleton, or skull, and the ancient Greeks placed gold butterflies in the tombs of loved ones to symbolize their reawakening to a new life. Psyche was the Greek word for both soul and butterfly and in art was often represented as a maiden with butterfly wings, or simply as a butterfly. In myth, Psyche's search for Eros is a psychological tale of individuation complete with the struggle and challenges inherent in any true transformation and rebirth.

Butterflies as human souls are also featured in many tales of undying love. In the Hawaiian myth of Hiku and his beloved wife Kawelu, Hiku enters the underworld to catch the butterfly soul of Kawelu who has died. Returning with it to her corpse, he makes a hole in the toe of her left foot and forces the spirit to enter, thus bringing her back to life.

In a true story shared by aficionado of the Orient Lafcadio Hearn, a Japanese man who loves an eighteen-year-old woman named Akiko mourns deeply when she dies. He moves near the cemetery and tends her grave, vowing never to marry. Fifty years later as he is dying, a large white butterfly enters his room and perches on his pillow. The man's friends and family believe it to be the soul of Akiko.

In every one of these cultures, the human body was considered to be merely the container of the soul. When death comes to the body, the eternal soul escapes as the butterfly or moth breaks through the chrysalis or cocoon.

Today the connection between butterflies and moths and the human soul is still strong. In parts of England people still believe that the spirits of the dead take the form of white butterflies. A nineteenth-century grave in Massachusetts shows a monarch emerging from its chrysalis and taking flight. And in *Hello From Heaven*, the Guggenheims report that the butterfly is the most frequently mentioned sign of an after-death communication. The following three stories taken from their chapter on butterfly and rainbow signs illustrate the ways butterflies and moths contact people.

Love is Eternal. In the first account, five months after her teenage grandson had died, a woman was sitting at her kitchen table looking out of the glass storm door. Suddenly a large monarch butterfly flew to the center of the glass. As it stayed there fluttering, she felt a strange sensation.

She called to her husband, and together they went to the door. The butterfly turned and flew to a large flower box at the far end of their deck. It flew around the flowers as they stood and watched for a few minutes. She felt, inexplicably, that her grandson was around. Mentally she asked her grandson to send the butterfly to the door one more time if he were really present. The butterfly immediately flew to the center of the glass, right to her face. It fluttered there a few seconds. Then she received a very clear telepathic message from the boy, saying he was alive and all right. After the experience, she knew she would see her grandson again, that there is life after death, and that "love is eternal."[7]

In another account, a retired police officer who had lost his teenage daughter Diana ten months earlier in an automobile accident was at home with his wife and some visiting relatives. As they sat outside on lounge chairs, the man noticed a butterfly. Immediately the thought of Diana inserted itself in his head. He thought, "If it's you, Diana, come down and tell me."[8] Without hesitating, the butterfly landed on his finger and walked up and down it. Then it went onto his hand and continued to walk back and forth on it. Astonished, he remembers that he was so close to the insect he could see its antennae moving. His wife looked at him as though she knew what he was thinking.

He got up, and the butterfly stayed on his hand. He walked to the house and went into the kitchen, butterfly and all. He told the butterfly he had to take a shower and that it had to go outside now. He opened the door and pushed it gently off his hand, watching as it took to the sky. It was an unbelievable experience. He had never had a butterfly land on him before. He went in to take a shower and cried. Later when he attended a conference of The Compassionate Friends, a self-help organization for grieving families, he learned that their symbol was a butterfly.

Luna Moth. Moths also have a long history of bringing messages to people. In fact, in Southeast Asia there is a species of moth that lives on a diet of tears, a substance which contains various proteins. Practicality aside, could there be a better image of comfort? Maybe the souls of those who have passed away return on wings and in silence to console and drink away the tears of the ones who remain behind.

Other moths bring comfort and assurance by their timely appearances. In the following account, also from *Hello From Heaven*, a teacher and her husband shared how they lost their teenage son. Two weeks after he died from a heart attack, the mother was in the kitchen when her husband called to her. She went outside to join him and there, in the middle of the day, was a large moth. It was chartreuse in color and about five inches across. The husband picked it up and placed it on the branch of a nearby bush. They watched it for a long time and finally it fluttered away. Later she looked it up in a book and discovered it was a luna moth, and luna means moon in Latin. Her son's hobby was astronomy. He had wanted to be an astrophysicist. She also discovered that the luna moth belongs to the family Saturniidae, and above her late son's desk was a picture of Saturn. The parents believe that their son sent this sign to let them know he was in a new life.

Synchronistic Events

The personal associations that make such occurrences so healing are always present in a meaningful coincidence or synchronicity. These events connect our subjective thoughts and feelings with the outer physical world, often bringing reassurance or guidance. They return us to a connection with life temporarily overshadowed by doubts, grief, or estrangement of some kind.

There is a magic to the timing of these visits that no amount of factual information or dispersion statistics on a species can dispel. We find ourselves suddenly in the embrace of the natural world, lifted into a terrain crisscrossed with tracks of meaning and seeded with divine forces.

In Helen Fisher's book *From Erin with Love*, a moving account of her daughter's struggle with cancer and her communications to her family after her death, Fisher describes the frequent visits of butterflies. The visits began four days after Erin died. A yellow and black butterfly appeared in the area where they scattered her ashes. Two days later at a ceremony celebrating her life that was held in a natural amphitheater on the campus of the college Erin had attended, a yellow and black butterfly flew down the hill and into the amphitheater. There it flew back and forth over the seats of Erin's family and friends, capturing their attention.

During the three weeks following Erin's death, her sister, who lived in a large city, was "literally divebombed by a yellow and black butterfly on two different occasions. In each instance it flew straight at her and she could feel the flutter of its wings in her hair."[9] The butterfly continued to appear to Erin's family and friends at significant moments—over two dozen visits in all—and brought them the message that Erin was alive and well, a message subsequently reinforced and delivered in a variety of ways in the months that followed.

Some visits are not as highly charged as the ones that convey news of a loved one who has died. Yet all speak of a dimension of life, a pattern of deep intentionality, that we are prone to forget in the dailiness of living. The presence of a large moth lifted me into this awareness one day years ago.

It descended on cinnamon-colored wings in the early morning hours of a summer day and took shelter in the protected alcove of my front door. Careful not to disturb its meditative posture, I studied the markings, the colorful eyes etched delicately on the faintly powdered surface of the outspread wings, the graceful curve of the fringed antennae, and the sturdy hair-thin legs. Even then I knew it was a polyphemus moth because of a childhood interest in butterflies and moths, but I wondered about its presence, as I had never seen one in the area before. It was some hours later, with the moth still on my porch, that I realized an illustration of this kind of moth graced the cover of a book I was currently reading about a medicine woman.

Native cultures traditionally viewed synchronistic occurrences as a sign of health. Ecopsychologist Leslie Gray, who began a shamanic counseling practice after spending ten years with native shamans and folk healers, said synchronicity is a consistent theme in her counseling. In fact, when a person fails to notice synchronistic occurrences in his or her life, it is a sign that something is wrong. When Gray's clients report having synchronistic experiences, they share that they feel seen and affirmed by something larger than themselves. And afterward, they're more congruent with their life and restored to personal power.

Mitchell Hall writes in *Orion Nature Quarterly* that when he noticed a series of meaningful coincidences that involved other species, he began to wonder about our mysterious connection to nonhumans. He tells of the time when his six-year-old son Ezra was walking with his mom, and a dragonfly brushed his cheek. "That dragonfly kissed me,"[10] the boy said matter-of-factly. Later that afternoon, Ezra hit himself with the blunt end of an old Boy Scout hatchet and started to cry. Hall was holding him when a dragonfly flew out of the canopy of a nearby tree. The insect buzzed around in a circle just above Ezra's head. When Hall told Ezra, the boy stopped crying to look at the insect. Hall told him that the dragonfly had come to let Ezra know that he loved him. Ezra calmed right down, and the insect flew away to the tree.

If it is true, as the ancients believed, that the more simply constituted creatures respond more readily to the unseen energies that have proved so maddeningly elusive to scientists, it is likely that insects at one time or another—especially flying insects like butterflies, moths, and dragonflies—will be involved in a meaningful coincidence that happens to us, especially if we are open to them and pay attention.

Synchronicity and Grace

Synchronistic events accompany crucial phases in our growth and are often called grace. Spiritual pilgrim Scott Peck describes grace as more than gifts:

> In grace, something is overcome; grace occurs in spite of something; grace occurs in spite of separation and estrangement. Grace is the reunion of life with life, the reconciliation of the self with itself.[11]

Often what is overcome, if only temporarily, is our general estrangement from life, our grief, our loneliness, or our uncertainty about a particular action. In keeping with their role as emissary and as model for growth and transformation, it should not be surprising then that often a butterfly or moth makes an appearance at a critical time in our life when we are contemplating a particular course of action or embarking on a new inner or outer path.

John Lame Deer says guiding spirits often enter a butterfly and direct the insect to fly to a particular young woman in the tribe. Usually landing on her shoulder, the spirit in the insect form then advises her telepathically to become a medicine woman.

Flying Boy into a Man. Psychologist John Lee reports in his book *Flying Boy* that he had his first out-of-body experience with a butterfly, and that since that time butterflies have appeared to him at just the right moment to remind him of the things he learned that day. The initial experience happened when he was alone on a boat dock reading, and a butterfly appeared and landed on his big toe. He followed his intuition and attempted to exchange bodies with it. The butterfly inched its way up his body, finally resting on Lee's hand, where it stayed. He stroked it gently and soon the exchange occurred. After experiencing the interior world of a butterfly for about forty-five minutes, he gently slipped back into his body, having learned lessons that, although untranslatable, would inform his life.

In the months that followed, he had periodic experiences with butterflies. His most memorable encounter came when he was working on an early draft of a manuscript about his journey to find his authentic masculinity. It was winter. A butterfly fluttered through his living room and landed on his desk "as if to bless my work and remind me of my changes from caterpillar to butterfly—a flying boy into a man."[12]

Sometimes an encounter with one of the winged beauties is not unusual given the time of year and the nature of how females attract males by emitting a scent, but it is significant, and memorable, because of the feelings invoked in us. For example, Milwaukee, Wisconsin, resident, Jean Collins, noticed some cocoons that were hanging from the high branch of a tree near her apartment. She thought how much she would like to have one for her windowsill so that she could witness the insect's emergence. Since the cocoons were out of her reach, she decided that if one wanted to "give itself to her" it would make itself available. Sure enough, a moment later her eyes spotted a single cocoon hanging from a hedge and close to the ground. She snapped off the twig it was attached to and brought it inside, placing it carefully in her windowbox. She had almost forgotten the still cocoon when several months later she walked into the dining room, glanced at the windowbox, and saw a large cecropia moth. She called neighbors and friends to come over and see this spectacular inscct and stayed up late watching it fan its wings as though preparing itself for flight. In the morning when Collins looked to see if the moth was still there, she was stunned to discover that not only was it there, but it had called an even larger moth to its side, and they were mating. For Collins this sight was a gift of power, beauty, and mystery. These glorious creatures had deepened her knowing about the need to balance the energies of the masculine and feminine inside herself and about how transformation always has its own perfect timing. A year later in her reflection on the experience, she says, "They had given me a new metaphor within which to make my own life choices."[13]

Other Signs

Traditionally when prodigious numbers of butterflies, moths, or caterpillars appeared, they inspired fear and were considered signs of doom. In Japanese history, for example, when Taïra-no-Masakado was secretly preparing for his famous revolt, there appeared in the city of Kyoto a swarm of butterflies so vast that the people were frightened. They believed that the unusual number of insects was a portent of evil and that the butterflies were the spirits of the thousands who would die in the coming battle.

During the Vietnam War, caterpillars were believed to be one of a series of signs that pointed to the demise of South Vietnam. Thousands, perhaps millions, of caterpillars appeared along the coast near Phan Rang in January 1975. The people connected the occurrence to a South Vietnamese legend that foretold of the collapse of South Vietnam. Devout Vietnamese Buddhists said it was simply the country's fate or karma to pay for their slaughter of innocent citizens in the fifteenth century conquest of Champa, an ancient civilization. When many South Vietnamese later fled the North Vietnamese Army, they were transported to the decks of American carriers. On deck, they clung in little groups to ropes stretched across the decks as a safety precaution, and observers who saw them said they looked like brightly colored caterpillars.

Message from the Monarchs. Insects in large numbers don't always spell doom, as Sherry Ruth Anderson, counselor and coauthor of *The Feminine Face of God*, discovered. When her father was seventy-five, Anderson first learned of his fascination with monarch butterflies and the mystery of their migration. He would ask her, shaking his head in puzzlement, "How can they make the long journey?" Even when she researched the phenomenon and presented her findings on the mechanics of migration, he would still shake his head and wonder out loud how they could accomplish the journey when they are so small and frail. He had somehow sensed a correspondence between himself—weakened after just coming through his second heart bypass in a decade—and the delicate monarchs and their ability to make such a journey. Eventually, Anderson figured out that when he asked about the monarchs, he was really asking about his own journey.

Later that year, when her father lay dying in a coma, the family gathered. Her niece Catherine flew in from Boston and asked to talk privately to her grandfather. She came back after a long time in his room, eyes red from crying. Later Anderson and her niece headed for a walk on the beach. Catherine shared that she was angry with her grandfather and had wanted to let him know how upset she was that he never even tried to get to know her. And now it was too late.

As they strolled along, they began to notice that they were being enveloped by a mass of orange and black butterflies—monarchs. Hundreds of them were keeping pace with them as they walked. In all her years of growing up in that area, Anderson had never seen more than two or three butterflies at a time. Monarchs like the ones she and Catherine saw that day were only seen in cities with overwintering sites. She kept thinking about her father and his fascination with monarchs, but she didn't say anything to her niece. Then Catherine stopped and nodded toward the butterflies.

> "This is Pop-pop," she said calmly, using the name she'd called my father when she was a child…. "He's come to say good-bye to us. It's okay now. It's okay how he was, not understanding us and making mistakes. I forgive him." She paused, reflecting. "I'm glad he was my grandfather."[14]

When she finished speaking, the multitude of butterflies that had been accompanying them changed direction and flew directly into the wind where they disappeared swiftly. Anderson says she wanted to tell Catherine how these monarchs, these long-distance survivors, had been coming for a whole year to visit her grandfather, the survivor in the hospital bed preparing for his own long journey, but she was speechless. Two days later her father died, completing his journey "just fine."[15]

Migration Mysteries

To entomologists, large groups of butterflies are either an outbreak or a signal that a migration is underway. Migrating is a phenomenon that has long intrigued people and one which has been interpreted in different ways. In Ceylon, for instance, Buddhists believe that the butterflies proceed to the top of a mountain called Adam's Peak, where, after paying homage to the "footstep of Buddha,"[16] they return cleansed of their sins. And the Malays, who are Muslims, think of the butterflies' flight as a yearly pilgrimage to Mecca.

Entomologists believe migrations are triggered by shorter days and cooler nights. For these specialists, large numbers of butterflies or moths are just another unsolved mystery. In 1924, three billion orange and black painted lady butterflies flew at about six miles per hour from Northern California into Southern California and on into Mexico. What triggered this migration or the whereabouts of their final destination isn't known.

In recent years millions of painted lady butterflies again filled the sky in Northern California. Entomologists admitted happily to reporters that they were puzzled by the numbers. Unlike monarch butterflies, painted ladies don't have an annual migration. But every ten to twenty years they seem to flourish so well in their native habitat that perhaps they run out of room. In this particular instance, entomologists offered an unimaginative theory that heavy winter rains may have triggered a population explosion. Maybe they are right, but then again, maybe the butterflies appeared to tantalize and arouse those individuals needing a sign.

Migrating butterflies are always a sight to behold. When yellow cloudless sulphur butterflies migrate, people call them "traveling butteries." Thistle butterflies, whose larvae keep thistle plant populations in check, migrate across the same territory as the painted ladies. One intrigued observer counted an average of fifty to one hundred butterflies per minute passing over one given spot.

Moths also migrate. Night-flying moths can even migrate at night navigating by the moon and, if there is no visible moon, by the stars. One spectacular American tropical moth, a daytime flier with a wingspan of three to four inches, has puzzled entomologists for years. Large groups of these moths fly at a good clip out to the open sea with no land ahead as a stopping place. Scientists don't understand how inherited behavior can lead to so many deaths by drowning year after year. Nor do they know where the caterpillars of this species of butterfly live or what they eat. One ambitious entomologist tried to count these moths during their mysterious journey to the open sea and reached a total of 54,615 individuals between sunrise and sundown over five consecutive days.

Counting Butterflies. Counting butterflies is becoming as popular as counting birds. For over twenty years the National Butterfly Association (NBA) has orchestrated a butterfly count (like Audubon's annual bird count) on the fourth of July involving 2,000 people in more than 250 locations, and each year the Morristown, New Jersey, all-volunteer NBA issues a report on what species were seen and at what frequency. Besides being fun, the count provides important information on rare colonies and endangered species.

Many people are attracted to the group because of its emphasis on using binoculars to see butterflies—instead of nets to capture them—a policy adopted by bird watchers (called "birders") everywhere. Jeffrey Glassberg is a butterflier, which means that he not only watches butterflies through binoculars, he also keeps records of how many he sees, takes photographs of them, and cultivates butterfly gardens. Glassberg is zealous in his advocacy of non-netting. He says his field guide shows him how to identify butterflies by sight only, without netting or killing them. He also maintains that if we identify butterflies without catching them, we become better observers and learn more readily about their flight patterns and behavior.

The Nation of Winged Peoples

A butterflier and a birder are linked in their love of flying creatures. Flying creatures, in turn, are linked by their ability to take to the air, a feat much admired in native cultures. The Oglala Lakota grouped flying creatures together, calling them "Winged Peoples"[17] and associated them all with the powers of the wind. Included in this group were birds, butterflies, moths, and dragonflies—the latter venerated because of their ability to escape a blow.

Other traditions grouped flying insects and birds together because all were considered messengers from other realms. Not only did they bring messages from those who have died, but they also were believed to bring messages from other kingdoms. Interspecies communicator Penelope Smith believes that some forms of life, including butterflies, dragonflies, and hummingbirds, have the role of being "inter-kingdom or interdimensional messengers."[18] These envoys seem to adjust their vibrations to interact with the life rhythms of several dimensions and often bring messages from other animals or plants.

In *Animals: Our Return to Wholeness*, Smith shares the story of Marcia Ramsland, who received a message from a young white pine tree—via a butterfly. Ramsland had observed a yellow-green cast on the tree and considered chopping it down when a butterfly appeared in front of her and advised her not to. Realizing that the butterfly was acting in the tree's behalf, she took its advice. In time, the tree's color changed to a more normal appearance, showing that it was not deteriorating as the woman had thought, but had only been going through a growth phase.

A recent news item tells of another butterfly-tree partnership. A twenty-three-year-old woman who calls herself "Julia Butterfly" has been camping out in the branches of a 200-foot-high redwood tree for over five months to stop Pacific Lumber Company from cutting it down. "My spirit led me here, and I mean to stay with it,"[19] she explained to reporters. When asked about her name, she said she'd had an intense spiritual experience with a butterfly as a child. The insect had landed and remained on her hand during a long and trying walk, fascinating and fortifying her.

Butterflies and Nature Spirits. In folklore, there is a long association between butterflies and moths and the realm of nature spirits or fairies. Ted Andrews, author of *Animal Speak*, a book on the mystical powers of other species, says that prior to giving a workshop on fairies and elves, he was meditating at a nearby nature center in preparation. When he opened his eyes, he was surrounded by a dozen or more black and yellow butterflies. Several were even in his lap. The occurrence was significant to him. He knew that in traditional angelology, yellow and black are colors often associated with the archangel Auriel in her role of overseeing the activities of the nature spirits. He took the butterflies as a positive indication of the supportive energy that would accompany him at his workshop.

The clairvoyant Austrian scientist Rudolf Steiner also connected insects to nature spirits.[20] He maintained that insects were a vital link between the physical world and the wavelengths of these energies—especially the fire spirits. That insects operate using wavelengths undetectable to our human senses has already been established by traditional science. In fact philosopher-scientist Philip Callahan (who explained how plants tell insects that they are sick and how insects communicate with each other) thinks that a vast communication grid is operating, composed in part of billions of insect antennae that receive wavelengths and link Earth to the cosmos in a symphony of vibration. This image of continual communication between the universe and insects (and all life-forms right down to cells) echoes the perennial wisdom passed down in traditional cultures that all is sentient and capable of communication. It also makes the notion of insects acting as messengers of invisible realms very plausible, given their numbers and communication apparatus.

Helping the Messengers

The weaving of scientific and esoteric knowledge provides a backdrop of complexity and mystery in keeping with the enormous mystery of our existence, here and beyond. The winged messengers' communications underscore the interdependence of our inner life and outer experiences, all held within the web of some vast, if largely unknown, universal network. All support the idea of a consciousness-directed universe in which our intent is of utmost importance and our every thought, word, and action is witnessed.

The way into the numinous underlining of our existence begins with a step out of the familiar house of the self, past the end of the known road, and into the borderland places which the insects love. To plan for experiences that confirm our deepest longings for meaning and connection requires that we stay alert. Inattention is a major impediment to these kinds of experiences. Another impediment is the fact that many species of the winged tribe are endangered. If we allow them to disappear from the Earth, their habitats replaced by concrete and the clamor and neon lights of shopping malls and amusement parks, will they only be able to contact us in dreams? And if they are no longer a part of the physical world, who will inspire us with their beauty, give us hope by letting us witness their transformations, and bring us comfort and assurance that our loved ones, who have passed on, are close by and just fine?

Will those of us who are the recipients of the Winged People's messages respond to their plight? Will those groups who have chosen the butterfly as their symbol of hope give hope to the creatures on Earth who are propelled by the beneficent unseen energies of creation? Will they add their voice to the voices already protesting the demise of the plant-loving caterpillars and the eradication of the water-bound dragonfly larvae, innocent victims of our current agricultural practices and war against the gypsy moth and mosquito? Can we assure them that there will be a place for them to live and thrive and perpetuate their species? And will all of us who love their beauty speak on their behalf and stand up to those who, for whatever reason, are unable to value them over economic concerns?

What rejoicing there will be among all forms of consciousness when we can acknowledge the role of the Nation of Winged Peoples as messengers of those realms within and beyond our living Earth and follow them when they beckon us. And what celebrations we can initiate when our own imaginal cells mature and we can finally see in these creatures a reflection of our own winged life, ready to unfold and take flight.

15

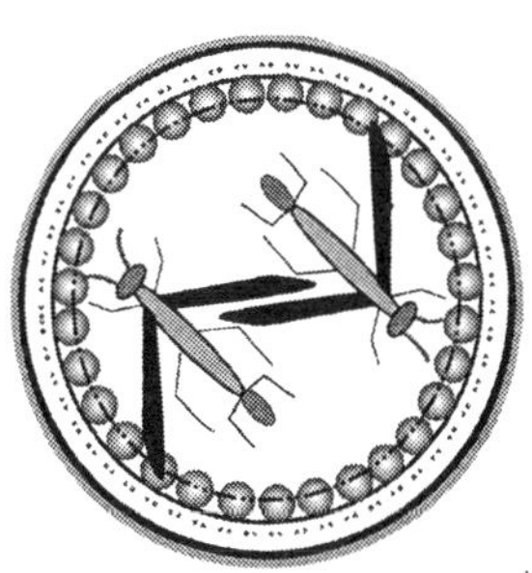

Finding Our Affinities

People protect what they love.
—Jacques Cousteau

Certain insect species will stir us, as though calling us into their sphere. I call this "affinity." It's a special feeling for, a fascination with, or a readiness to love a particular thing or being. It's Scott McVay's "Siamese connexion,"[1] and Edward Wilson's "biophilia" in its most specific, personal form. It is also an aspect of what James Hillman has called our innate "defining image,"[2]—our mark of character that attempts to guide us, by acts of providence or fate, to its fulfillment. Consider that Thoreau had an affinity for the pond and meadow of Walden. He had places inside him the right shape to respond to these aspects. When he wrote, "I am made to love the pond and meadow," he understood that his perception of them ignited something inside himself, connecting him with the outer natural world as firmly as the umbilical cord connects the unborn child with the mother.

From the age of six, after seeing his first swallowtail butterfly, Vladimir Nabokov wrote that everything he felt was dominated by a single passion—butterflies. Even the first glimpse of the morning sun would trigger in Nabokov a thought of how many of these winged creatures would venture out because of the warmth of the sun's rays. His singular passion for butterflies, awakened from that first sighting of one, also illustrates the intensity that is the earmark of an affinity. When explored without reservation, an affinity always leads the person to something beyond his- or herself, something divine and benevolent, as Nabokov discovered:

> ...the highest enjoyment of timelessness—in a landscape selected at random—is when I stand among rare butterflies and their food plants. This is ecstasy, and behind the ecstasy is something else, which is hard to explain. It is like a momentary vacuum into which rushes all that I love. A sense of oneness with sun and stone. A thrill of gratitude to whom it may concern...[3]

One reliable indication that we have discovered an affinity and entered a potent zone of connection and renewal is the sense of wonder invoked by the object of our affinity. Naturalists like David Brendan Hope have written about those holy moments when some creature pays us an unexpected call:

> The Visitation at the wood's edge is always fatal. It cannot be predicted or declined. You open your arms...A new world debuts..., a pattern hitherto unsuspected, a beauty so unprepared for, it must either be rejected or convulse the soul.[4]

Hope's description is reminiscent of a passage by English entomologist Alfred Wallace, the "high priest of entomology." Wallace, who died in 1913 at the age of ninety, made many contributions to the natural sciences. He is often remembered for his love of the foot-wide bird-wing butterflies. When he captured one during his exploration of the South American wilderness, he wrote:

> The beauty and brilliance of this insect are indescribable.... On taking it out of my net and opening the glorious wings, my heart began to beat violently, the blood rushed to my head, and I felt much more like fainting than I have done when in apprehension of immediate death. I had a headache the rest of the day, so great was the excitement.[5]

The intensity of Wallace's feeling suggests that his love of the bird-wing was an affinity. Meeting one with whom you have a special connection is never a commonplace occurrence. The wonder transports you out of ordinary time and space and the glowing feeling that results can last for hours or even days.

The Mystery of Affinity

Affinity is a mystery tied to our uniqueness. Because it arises out of our innate essence or soul, finding and expressing it is important for our emergence as full human beings. When we are in touch with our affinities, we are in touch with our essence, and we find latent skills and the energy to act. Our psychic connections with particular species or particular aspects of the natural world also encourage us to seek, in a spirit of celebration, their physical correlates in the natural world. Our unique daimon, angel, or soul guide who holds the blueprint of our character and calling arranges those appointments and helps us meet them. Affinities also underlie the impulses that make one person an entomologist and another a marine biologist. They explain why some people work passionately to save a butterfly from extinction and others direct their efforts toward clean air, saving the rain forests, or ending nuclear testing.

Yale entomologist Charles L. Remington loves cicadas, and finds them the most remarkable insects in the world. Patience and a love of mystery are two required characteristics for those with an affinity for cicadas. No one knows why, for instance, some cicadas spend seventeen years underground before emerging en masse for their final transformation into winged creatures. For Remington, the long waiting for their brief appearance is just the necessary price for witnessing their miraculous emergence and hearing their synchronized love chorus.

For the beetle hunter, the thrill of discovery is an integral part of his or her affinity for beetles, since only a small fraction of all species has been classified, and little is known about how most of them live. Arthur V. Evans and Charles L. Bellamy report in *An Inordinanate Fondness for Beetles* that "some coleopterists refer to their passion for beetles in terms of joy, excitement, wonder, delight, thrill, satisfaction, and fulfillment."[6]

In native cultures affinities were acknowledged as an important aspect of an individual's or a group's identity, and signs of its calling were looked for. Lakota medicine man John Lame Deer says that our bodies, senses, and dreams contain the secret knowledge of our affinities. If we listen for them, he advises, and follow our intuition, they will guide us to our unique path. He assigns the origin of an affinity to the work of the Great Spirit who wants people to be different and so makes each one love a particular animal, plant, or aspect of nature.

I suspect that there are impulses within humanity for each and every form of life and aspect of nature—impulses within people tied to all the areas that need our rejoicing, as well as our concern and hard work. Also encouraging is the idea that our individuality, reflected in the unique contours of our psyches and encoded in our bodies, is not a passive attribute but an active directing force. It primes us to respond to specific ideas and forms in the outside world. And when we find that outer counterpart, it resonates with something inside us and, with inner and outer pieces reunited, we feel empowered and made more whole. We then experience a direct knowing of the object of our affinity with a certainty, clarity, and precision beyond what reason can provide.

Bug Period. Knowing that affinities exist inside everyone could change our school curricula and alter the way in which we educate our children. Some researchers say that if children miss out on their "bug period,"[7] the years when children traditionally roam their local woods and creeks in search of what fascinates them, they lose a critical chance to bond with nature. They also lose an opportunity to find their insect affinity.

"To educate" comes from the Latin word "educare," which means to lead out. If we recognized that inside each of us are specific affinities, our task as parents and teachers would be to facilitate this discovery in ourselves and in our children. Then we could look for and trust the impulse that draws a child near a particular species and help "lead out" that affinity. It's a relief to know that we don't have to create the desire; we need only offer ways to explore it. Experiences that strike a deep resonant chord will then be worth a thousand lectures and textbooks filled with facts.

For the Love of Bugs. Entomologist Eric Grissell says that most of the entomologists he knows were born that way. Insects were their first love. The exceptions are those like naturalists or artists who come to the world of insects less directly. After twenty-five years of looking at this phenomenon, he believes that impulses or passions that lead one in a certain direction are outside of our personal control and beyond the influence of family and culture. James Hillman would agree. In his acorn theory, he proposes that we are born with our specific uniqueness and destiny packed into a tiny oak seed or acorn, and that its call is louder and its promptings stronger than the voices or demands of family or society.

Thomas Eisner, whose contributions to our understanding of insects' chemical communication are unsurpassed, calls his work with insects entertainment. "You can't become my kind of biologist as an adult—you have to have eaten a bug at age five."[8] One of his earliest memories is playing with pill bugs in his sandbox. During the war, when his family moved to Uruguay, he found himself in bug heaven, surrounded by the most abundant diversity of insects in the world. At seven, he found an injured swallowtail, and coaxed it to sip sugar water. He named the insect and kept it for days as a pet. From then on, it was insects for him.

Miriam Rothschild, a naturalist noted for both her scientific discoveries and her writing, has researched fleas, butterflies, birds, snails, intestinal worms, and wildflowers. She considers herself a born naturalist and notes that while her siblings were exposed to the same rich countryside as she was, they were as indifferent to insects, animals, and plants as she was enthralled. Finding a garden tiger moth and its woolly bear caterpillar in her garden captivated her, changing her forever from a gardener to an entomologist.

Affinity and Personal Abilities

The expression of an affinity is guided by personal abilities and environment. It is also shaped by opportunities available in the culture and by the culture's opinions and prejudices. In earlier chapters we met a number of individuals who found their affinities in the vast world of those who creep and crawl—like Geoff Alison, who has an affinity for insects in general and cockroaches in particular. His feeling for these small creatures and his uncommon sensitivity, although supported by his physical blindness, does not arise from it. If it did, we would expect that other blind individuals would also exhibit his same compassion and understanding of the creeping ones.

Blessings Disguised as Misfortune. Edward Wilson, also known as Dr. Ant, has been studying ants for over fifty years. An injury to one eye as a young boy left him with exceptional myopic vision in his other eye. Thus, even without a magnifying glass, Wilson is said to be able to count the tiny hairs on an ant's body. His vision is matched by a single-minded diligence that has been called positively antlike by those who know him.

In his autobiography, Wilson said that he just never grew out of his "bug period," and that his injured eye eliminated all chance of studying other creatures he was drawn to, steering him toward ants because they were easy to find and observe. At thirteen, he discovered that a new and better-adapted form of a fire ant species native to South America was well established in the Mobile,

Alabama, area. Later as a student at Harvard in 1959 he discovered the gland in ants that produces the chemical used to lay odor trails to food sources. In the following two decades he made major contributions to systematics, behavior, ecology, and biogeography. He attributes his originality to his handicaps. His poor vision in one eye gave him keen vision in the good one, and a learning disability gave him a tendency to "reconstruct much of what he learned, often with new images and ways of phrasing."[9]

Defining Affinity

The dictionary defines affinity as a relationship between biological groups involving resemblance in structural plan and indicating community of origin, kinship, or relation by marriage. If we examine our affinity with insects in terms of structural resemblance, our common excuse for a lack of empathy, we find a biological resemblance in a surprising number of areas. For example, an almost identical copy of the gene that helps an insect larva develop legs has been found in mammals and other vertebrates. What's more, a number of key genes that produce vertebrate and insect eyes are almost identical. In fact, the same basic genes guide the development of features such as the heart, head-to-tail body, and back-to-belly orientation, in a wide variety of insects and other animals.

New evidence also suggests a definite and unexpected structural link between insects and people. Scientists have long sought an anatomical unity in design that links insects and their kin with humans and other species with backbones. At last they may have found this structural homology*—not in the spine where most of their investigation has focused, but in the brain.

Although our adult brain and the brain of other vertebrates is not segmented, the posterior part of our brain develops in the embryonic stage from a prominent series of seven or eight initial segments called rhombomeres, which resemble insect body segments. Although not permanent, rhombomeres represent a second system of vertebrate segmentation and actually coordinate and organize some persisting features, including several of the prominent series of cranial nerves. Rhombomeres also correlate in a one-to-one fashion with the important segmentation of the gill arches that eventually form part of our tongue and throat.[10]

Supporting the idea of structural unity between insects and humans is evidence that genes map to rhombomeres in the same way that they map to insect segments, and that rhombomeres develop as compartments just like insect segments. It seems we have an elemental connection—an insect self—if you will, beneath the cortex that provides us with our reasoning self. William Jordan's speculation, triggered by observing cockroach communities, that all reason is

* Homology is a key evolutionary concept of similarity due to presence in a common ancestor despite later divergence.

supported by the characteristics of our primitive brain, takes on new ramifications if we consider the insect structure that gives rise to our oldest brain.

Affinity by Community of Origin. Affinity can also arise from common origins. In aboriginal cultures, for example, totemism is based on a mysterious affinity between a certain group of people and an animal or plant that they trace back to common ancestors. Certain nations and clans associated with particular species claimed animal descent—like the Greek Myrmidons, who reputedly came from ants.

Outside of totemism, new studies in microbiology have revealed that insects and humans have common ancestors in microorganisms. As the building blocks of all life-forms, microorganisms provide a contemporary foundation for affinity by community of origin. The current fear of this kingdom has unfortunately dampened our curiosity for this piece of knowledge, so we are apt to hold back from exploring this world with enthusiasm, like disgruntled children looking for their real parents. Given the right context in which to explore these life-forms, I'm convinced that good feelings will surface automatically, as they did for J. William Jean who treated bacteria as his friends and tiny business partners in the chemical research laboratory in which he worked.

Affinity by Attraction and Choice. Affinity or relation by marriage involves attraction, fascination, and choice. Again our daimons are responsible for our attraction and fascination and they structure our perception. Consider that the world is comprised of things we don't see, even when they are in front of us. An affinity gives us eyes for what we love. It activates some unknown ability so that one part of us is always looking for the object of our affinity even when we are thinking about something else. Thomas Eisner, who credits his many discoveries to his observation of insects in the wild, was primed to see insects and related creatures. He says, "I used to be able to drive along at night, stop, go backwards and pick up a millipede I had noticed out of the corner of my eye..."[11]

Art historian James Elkins, who wrote a book on the nature of seeing, says he was entranced by moths as a boy and hunted for them at night. He learned to see the particular shape of their bodies, the slight shadow they cast as they pressed against a tree and the faint difference between their patterned wing and a tree's bark texture. Twenty years later he still has the ability to spot them, even though he stopped consciously looking for them long ago. He says, "Something in my subconscious must still be scanning for that particular shape, and it breaks into my conscious thoughts to warn me when it thinks it has discovered a moth."[12]

Gary Lovell, an amateur entomologist with a deep interest in praying mantises, thinks he may have a "mantid predisposed mind."[13] One summer when he was walking through the parking garage in downtown Toledo, Ohio, where he worked, the thought of a praying mantis suddenly entered his head. He looked around and a few seconds later saw a female praying mantis on the chain link fence. He took her home and released her into his yard. He looked every day for the next two weeks and never saw another. Then another week passed and he was again leaving work. This time he was walking around the garage in the opposite direction when again "mantis" popped into his head. He started looking at the fence again and found a male praying mantis which he also took home and eventually released. He decided that he was scanning for the insect unconsciously or that each had put out some kind of a telepathic signal to get his attention, knowing he would search for them and escort them safely out of the concrete downtown.

The Underside of Fascination. Sometimes attraction and fascination are hidden in repulsion and fear, or, because of cultural beliefs, they are expressed in a way that distorts the original impulse and its intentional thrust toward wholeness. When the pull is strong and our reasoning ability is undeveloped, fascination for beings that look and behave so differently from ourselves might easily be felt as distaste, even horror. I suspect that both the insect enthusiast and the exterminator share an affinity toward insects, an impulse that initially draws them near.

Thus fear can be an indication of affinity. Recall George Uetz who was intensely afraid of spiders and enrolled in a class on spiders that led him to his career as a spider researcher. Entomologist May Berenbaum was afraid of insects for the first decade and a half of her life. It wasn't until she took a college course in entomology to learn which ones she should be afraid of that her affinity for insects was activated in a positive way. Fear gave way to admiration and then to passion. Now a leading entomologist, Berenbaum is zealous in her desire to change people's minds about insects. At her innovative Insect Fear Festival, Berenbaum educates the public about insects while entertaining them with insect horror movies.

Fear hints at a connection. The creature you fear has power and commands your attention when you are in proximity to it, and beneath the fear is fascination.

Affinities triggered by fear and expressed as hatred are also tied to fears of the unknown. If power or energy comes in an unexpected form outside normal classifications, we often think it evil. We have forgotten the shamanic context that helps us understand the raw intensity and sometimes strange packaging of transformational forces. If an insect is connected to an aspect of our essential self, we may fear it and the unknown territory within us that is its domain. Maybe we sense the creature's alchemic power and its intention toward us, and we run in terror from the dying that must precede transformation and new understanding.

Other facts also impact this equation. Direct experience and the context in which we learn about other species guide our impulses and often dictate their outer expression. In Florida, Texas, and other southern states with tropical climates, for example, an affinity for cockroaches or mosquitoes will have a difficult time emerging in a life-enhancing way. Such individuals have few choices.

They can become entomologists, exterminators, or science writers. Otherwise, it is likely that strong cultural beliefs will lead them into contests that shower negative attention on the insect through killing and ridicule.

Remember the artist who reinvented the flea circus. She had a curious mixture of fascination and a lack of regard for the actual creatures, typical in cases where affinity has been twisted by cultural and personal factors. In other people, a healthier expression of an affinity for fleas has found its way across culturally sanctioned prejudices. What marks it as healthy is its link to heartfulness, however rationalized. For instance, a friend of naturalist John Ray is reported to have given bed and board to a favorite flea, allowing it at certain times to suck the palm of his hand for nourishment. After sharing three months of friendship and easy compatibility, the flea died from the cold, and the man grieved.

Time and enough knowledge about the species also helps people develop an affinity for those creatures despised by the culture. Consider someone who has an affinity for lice. It is well known in certain circles that scientists who have devoted years to the study of lice sometimes come to regard them with a kind of dubious affection. Lice researcher Dr. W. Moore fed over seven hundred lice on his own body twice a day. Between meals he is reported to have tucked them away carefully in a nice warm box.

Dr. Hans Zinsser wrote a book on lice because he wanted to "present the case of the louse in the humane spirit which a long intimacy" had taught him.[14] During his experiments he had come to regard "these little creatures [with] what we may call, without exaggeration, an affectionate sympathy."[15]

Daniel Brooks, a zoologist at the University of Toronto, thinks parasites are "splendid creatures." Much of his career has been spent dispelling myths about them—like the misconception that they are primitive and degenerate beings. He teaches that "Parasites are successful, innovative creatures...[and that] if you compare them with related species that are free-living, parasites are often more complex. They give you a healthy respect for the power of little things."[16]

Beekeeper Sue Hubbell says that every entomologist she has known or has interviewed regards their selected insect of study to be beautiful. When she asked a woman mayfly expert, for instance, how she had chosen that particular insect to study, the woman told her it was because these miniature creatures were so beautiful. Another entomologist answered the question by inviting Hubbell to look at a tiny drab-colored moth under his microscope. She did and saw what he saw—a splendorous sight, a gleaming and shimmering beauty.

Facets of Affinity

The affinity of beekeepers for their bees even has a name: "bee fever." Hubbell says that when one falls in love with honeybees, takes the plunge, and acquires his or her first hive, they have bee fever. Those who are already afflicted with bee fever think it's a blessed malady.

The folklore of beekeeping contains many individuals with an affinity for bees. One of the most picturesque characters was the Englishman Daniel Wildman, known as The Bee Man, although he is also reported to have tamed wasps and hornets. Around the time of the American Revolution, Wildman traveled through Europe exhibiting his mastery over swarms of bees. Like "Chiquino of the Bees," Wildman could command his bees to settle on his head, his beard, or any other part of his body he designated. Once he was even carried through London in a chair almost completely covered with bees.

Allergies as Signs of Affinity. The German scientist Karl von Frisch, who discovered the dance language of bees, devoted his life to bees and minnows. He felt that a good biologist should regard the objects of his research as his personal friends, and so he did. When asked why he had chosen bees and was content to devote fifty years of his life to their study, he answered that being in right relationship with one species brings all creation into clear view and that "every single species of the animal kingdom challenges us with all, or nearly all, the mysteries of life."[17]

An interesting fact in light of von Frisch's devotion to bees is that he was allergic to bee stings. What's more, James Gould, who proved that bees use the information coded in the dance, is also allergic to bee stings. Perhaps these allergies reflect a sensitivity toward bees, a predisposition reflected in the body that contributes to the push of this affinity to seek the wisdom and secrets of the hive. In fact, Gould was going to be a molecular biologist until he tried a bee experiment one summer just for fun. From that time on, bees have been his speciality.

Vocation as an Expression of Affinity. Depth psychologist Stephen Aizenstat believes that inner impulses linked to the natural world direct us to work that we are passionate about, even when we are unaware of their influence. He encourages his clients to rediscover the aspect of the natural world that corresponds to an area of their current expertise. An aviator, for example, would look for what it is in nature that feeds this yearning to fly. Although Aizenstat didn't direct his insights specifically toward insects, he could have—his ideas are readily applicable to this kingdom. If insects were included in the aviator's search, he or she might discover inspirational models in the masterful flying and navigational skills of the dragonfly or housefly. Similarly, the architect, engineer, and carpenter might find inspiration in creatures like the termite or honeybee that engage daily in the work of planning and building structures. Musicians would be inspired by Grasshopper and Cicada. Poets and devotees of many religious traditions would see in Mayfly an archetype of life's swift passage—an exalted view of the day. Computer networkers and weavers would observe and learn from Spider and Silkmoth, and theologians would reflect on Mantis. Aizenstat believes that any time someone explores this connection between the natural world and his or her vocation, the result is a heartfelt empathy between the person and the corresponding creature. And once these connections are rediscovered, each would know, in a deep and essential way, what part of the restoration of the natural world he or she has a link to and perhaps what part he or she is responsible for preserving.

Butterflies and Computer Chips. Affinities help us to access the archetypal patterns of the natural world that underlie and inform all human creativity and invention. It is these universal patterns operating behind our specific affinities that provide the impetus and enthusiasms for our explorations. The Odyssey of the Mind, an international competition that challenges students to find creative solutions to unusual problems, acknowledged this link when it named its highest honor, the Renata Fusca Award, after a water insect.

Engineer Ioannis Miaoulis, one of many "biomimics"[18] who looks directly to the natural world for solutions to problems, believes that what he has learned about butterfly wings can be applied to semiconductor design. Miaoulis examined the ultrathin layered construction of a butterfly's wings, which makes some butterflies iridescent, and learned that it also lets the insect control how much heat energy is absorbed by its wings and how much is reflected away. He knew that its construction was the key to building a better microelectronic chip. Materials scientists have struggled for years with the problem of how to evenly heat microelectronic chips during production to avoid melting parts of the chip's thin-film surfaces. Using what he has learned about butterfly wings, Miaoulis hopes to make a wafer of silicon, mother of dozens of computer chips, that isn't perturbed by heat. If he succeeds, the invention could open the door to much faster processing techniques.

Our affinities for specific aspects of the natural world give us eyes to see connections across specialized areas of learning. It was mathematician Barbara Shipman's knowledge and fascination with bees, for instance, that allowed her to make the connection between the flag manifold, a geometric structure, and the waggle dance of honeybees. Without knowing about the dance, she wouldn't have been able to recognize it in the mathematical curves or link it further to the mathematics of quarks in a quantum field.

An Affinity for Beauty

Natural form is the foundation of our ideas of beauty and accounts for the aesthetic appeal of nature. We all have some affinity for beauty, impulses that provide us with still another natural connection to insects. Aside from their enormous diversity, complexity, and essential activities in maintaining the earth's ecosystem, insects are beautiful. French anthropologist Claude Levi-Strauss thought they were equal in beauty to the art we so carefully preserve in our museums.

Beauty is usually what attracts people to butterflies. A Native American myth teaches that butterflies were created because the Great Spirit wanted people to love nature and seek out its beauty. At first the Great Spirit put the colors of the rainbow in pebbles, hoping that their presence in the beds of all streams would be an incentive for the people to explore nature. But the pebbles were too beautiful to hide under water, so the Great Spirit called in the South Wind and commanded it to breathe life into the stones and lift them in the air. When the South Wind did, the pebbles rose slowly and flew away on beautiful wings—the first moths and butterflies.

Affinity as a Calling. Sometimes the pull of beauty is so great that it finds expression outside one's primary occupation. Jalaluddin Rumi could have been referring to affinities and the power of beauty when he said, "Let the beauty you love, be what you do." Many have honored their affinities, not by changing vocations, but by pursuing it alongside other work. Population expert Paul Ehrlich, for example, has an affinity for butterflies. Every summer since the early sixties, he and his wife Anne have traveled to the meadows of western Colorado to observe butterflies.

It is not unusual for a butterfly authority to be self-taught. The late Paul Grey, who was America's leading expert on a butterfly genus that includes many of the fritillary butterflies, was a carpenter by trade. Jeffrey Glassberg, president of the National Association of Butterflies, which sponsors the annual butterfly count, is a molecular biologist. Richard Heitzman, an internationally recognized authority on butterflies and moths, earns his living as a United States Postal Service mail carrier.

Nature photographer Kjell B. Sandved of Washington, D.C., embarked on a twenty-year journey to thirty countries in order to find and photograph every letter and the numbers 0 through 9 on butterfly wings. Because of his passion and perseverance, almost every elementary school has a beautiful poster of his discoveries.

Ronald Boender was an electrical engineer and entrepreneur who sold his Fort Lauderdale communications firm and puttered around the garden, retired and bored. He had always enjoyed watching butterflies, so he followed that interest and started cultivating plants to attract them in his backyard in Broward County, Florida. It was a life-shaping decision. In 1988, he founded Butterfly World in Coconut Cree, Florida, a three-acre haven of plants for rare, tropical, and North American species and today one of the largest butterfly research and education centers in the United States.

Legal secretary Ro Vaccaro was stirred to action when visiting Pacific Grove, California, one of the few wintering grounds for monarch butterflies in the United States. One glimpse of the thousands of monarchs from all over the West was enough. She sent her resignation back to Washington, D.C., and started Friends of the Monarchs. Recently, her 180-member group helped convince the seventeen hundred people of Pacific Grove to pass a tax hike to save from development a 2.7-acre butterfly resting site.

The Beauty of Moths. Beauty is what attracted John Cody to moths. He has an affinity for huge colorful silk moths that after sixty years he still doesn't really understand. In the 1930s when he was five years old, Cody saw a cecropia moth and was literally stunned by its beauty. He wanted to capture that beauty on canvas so he started painting them. Even after he became a distinguished psychiatrist in Kansas, he never lost his feeling for these moths. During his years of training and in the practice of psychiatry, he learned a lot about himself, but never about why, all his life, he has painted moths. He has found the impulse embarrassing at times and inconvenient, but it endures.

It has been decades now since that impulse first directed Cody to take up paintbrush and canvas and try to capture the beauty of these moths in watercolor, and today his paintings appear in museums around the country as well as in his own gallery. Since he prefers to work only with living creatures, his models are posed in natural positions, and his paintings seem alive.

Entomologists and dealers who know of him send him cocoons from around the world, which he hatches in his house. Once they emerge from their cocoons, silk moths have about five days to find a mate and lay eggs before dying. They do not feed. The digestive system they used as caterpillars has atrophied, and as winged creatures their energy is stored like a battery from their former life. In order to capture their exquisite color, Cody must paint them during their short lives. Using his hand for a perch, the great creatures will sit still as long as there is daylight

It is easy to see the beauty in butterflies and other iridescent or metallic-colored insects, but it is harder to see the beauty in a plain creature like the moth whose image was damaged by Judeo-Christian designations as a symbol of loss and decay. Cody has in some ways reversed the damage caused by those judgments. His work is a testimony to the power of affinity and could soon be the only visual record of these vanishing species. Through Cody's eyes and talented hands, the world has been allowed to glimpse the stunningly beautiful giant saturnid moths with their soft furry bodies and velvety wings.

Unexpected Beauty. There is beauty and inspiration in every species. A contemporary sculptor uses themes from scanned electron microscope photos of honeybees. At a quilt and textile exhibit a designer who creates "wearable art" featured a hand-painted, chiffon silk cape crafted to resemble a butterfly. New York artist Knox Martin keeps a species of the European honeybee on the roof of his building and says bees and other insects inspire his work. Inspiring too for those with the eyes to see is the beautiful polished wood-like appearance of the shell of a Madagascar hissing cockroach with its shiny black head that resembles a Japanese river stone. And consider the yellow-striped wasp with its tiny waist, the bright summer green of the katydid, and the astonishing gold eye of the lacewing.

Photographer Robin Kittrell Laughlin has seen and captured the beauty in familiar insects. She likes the way common insects became uncommon, actually extraordinary, when time is taken to examine them closely. "Their delicacy, their color, an odd face, or perhaps an antenna thrilled me,"[19] she said in the introduction to *Backyard Bugs*, a collection of her photographs.

Insect Artist Gwynn Popovac. Good artists teach us how to see. Visionary artists lead us to connections above and beyond cultural norms, and Gwynn Popovac is such an artist. Describing herself as an amateur entomologist, Popovac has always loved the design of insects, their texture, iridescence, and color. Her exquisitely detailed drawings celebrate insects in all their variety and intricate beauty. Enshrined in decorative borders, suspended in time, each insect is given the respect it deserves, and the viewer is left with a feeling of having glimpsed the essence of the creature, its mode of divinity, and the unblemished treasure that it is. Rudolph Steiner once said, "The artist does not bring the divine onto the Earth by letting it flow into the world; he [or she] raises the world into the sphere of the divine."[20] And that is an accurate assessment of what Popovac has done for the insects.

She recognized her affinity for insects some thirty years ago, as a result of an unusual meeting with an insect while sitting on a boulder in the middle of a creek on a midsummer day—a meeting sure to have been arranged by her daimon. A drab, many-jointed bug crawled onto the back of her hand. She almost obeyed a startle reflex and flung it away, but for some reason she checked herself and allowed this bug (which she later learned to call a nymph or aquatic larva) to crawl to the tip of her finger. Once there its back split open, and it extracted itself from its own body. This new creature was transparent and started pulsating. Soon what looked like wads of cellophane began to uncrinkle and—"vibration by vibration"—became flat glistening wings. "A few quivers more, and this drab creature of the creek bed had changed into an iridescent creature of the air," recalls Popovac. She watched the dragonfly resting on her fingertip a moment more before it soared off downstream with a rustling of wings. In a personal correspondence Popovac shared:

> The pleasing vibration that ran from my fingertip up my arm and down my spine persisted, and when I returned home, I painted an oil painting of the dragonfly: a Green Darner hovering over a stream. It was the first time I had ever given center-stage to an insect in a painting and it was the best I could do to repay the experience.[21]

Even without an artist's talent to capture the object of his or her affinity, most people generally know when they find an outer counterpart. It resonates to something within them that I am calling their essence; they feel empowered and, somehow, made more whole. It also can free up the necessary energy to work in behalf of that species. In a real and practical sense, we were made to love other species. Once in touch with our inner blueprint and the impulses that arise from its unique contours, our actions gain the transformative power to make a difference.

Finding Our Affinities

Although each person has a general leaning toward the natural world, most of us are drawn to a few particular species whose shape fits a corresponding place inside us. Considering the number of species that share the Earth with us, our search for these counterparts seems impossible. But affinity and synchronistic experiences go hand in hand. Experience indicates that the forces of affinity are magnetized or charged in some way and perhaps, with the intervention of our daimons, set up potentialities for fulfillment. Taking it further, the creatures are seeking us as well.

Poet and naturalist David Hope suggests that "certain people see things other people don't because those things desire to be seen by them."[22] There is much precedence for this idea of an inspirited, intentional world, proposed recently by ecopsychologists and depth psychologists and found in the works of great poets and other visionaries. Seeing involves in some mysterious way the will of the creatures that are seen.

Perhaps the endangered Palos Verdes blue butterfly willed Arthur Bonner, a former Los Angeles street gang member and drug dealer, to see its fragility and its struggle for survival in an asphalt-covered world, for this street-smart young man was touched deeply by the plight of this insect. He says he learned of its uncertain fate while he was participating in a conservation program aimed at giving at-risk youths a new start, and he identified with the butterfly in a profound way. He followed his feelings and started restoring the habitats it needed to survive. Now an employee of the same conservation program, Bonner says these butterflies helped his own metamorphosis into the person he was meant to be, keeping some vital aspect of himself from extinction. That is why he is now returning the favor and helping these butterflies recover their presence in the rugged hills of California's Palos Verdes peninsula, and leading inner-city kids there to see them.

Consider also that the discovery of a new species may be due as much to the will of that creature to be seen as it is to the efforts of the discoverer. Edward Wilson (and I suspect many other successful explorers) assumes a particular mind-set on a trip through a forest or meadow that allows him to sense something moving toward him before an actual discovery is made. Wilson calls it "the naturalist's trance, [or] the hunter's trance—by which biologists locate the elusive organisms...."[23] When Wilson enters this inner place, he is aware of the natural world and its impersonal rhythms in a way outside ordinary attention. Each step intensifies his concentration, and he can feel when something extraordinary is very close to him, moving toward him and toward discovery.

In Jean Houston's description of "Essence," this renowned teacher of transformation and human potential says that nonordinary states of consciousness (like the hunter's trance) are ways to access the power of our true nature and actualize what we were born to actualize. "Traveling to these places (of consciousness) with adequate maps, recognizing them as our soul's domain, and returning with their treasures is the active pursuit of Essence."[24]

The Will of the Green Ants. William Mann, entomologist and former director of the National Zoological Park in Washington, took a trip as a young man into the Sierra de Trinidad of central Cuba. When he lifted a rock to see what animals were hiding underneath, it split down the middle to expose a half teaspoonful of metallic green ants living in a small cavity deep inside. Mann named these remarkable creatures in honor of William Morton Wheeler, his major professor at Harvard and the reigning authority on ants.

Thirty-seven years later, at the start of his career in entomology, Edward Wilson started his own journey into the world of ants with a trip around the world, remarkably similar to Mann's journey. He remembered Mann's discovery as he was climbing a steep slope in the same mountains. He grabbed a rock for support and it split in his hand. Inside was a half teaspoonful of the same glittering green ant species.

Wilson acknowledged the event as one of his rites of passage. Consider the proposition that this green species of ants was seeking Wilson, hoping to affirm and bless his path. Perhaps those with an affinity for ants stir an archetypal ant self, a self-organizing ant field that, like the overlighting spirit of native teachings, orchestrates these kinds of synchronistic occurrences. The "discovery" linked Wilson to a line of others whose affinity for ants led to distinguished careers and in turn brought more ant knowledge to us all.

The Power of Affinity

We carry within ourselves a seed form of the living world. No matter how different or insignificant a life-form might appear, every presence has its own dignity, its own beauty, and its own ability to reflect divinity. There are correspondences within us for every life-form, and each plant, animal, or aspect of nature is for someone an affinity with the power to reveal his or her connection to the Divine. If we can start to acknowledge and respect that fact, we position our spirit correctly, and real change is possible.

The Butterfly Principle. Small changes made by even one individual who acts in accordance with the beneficent powers of creation have great power to produce systemic change in our society. It's the "butterfly principle" discovered in 1961 by Edward Lorenz, a meteorologist at the Massachusetts Institute of Technology. While working on a model to forecast the weather, Lorenz needed to extend his forecast, so he rounded off one number by .02 of a percent. Then he took a coffee break. When he returned, he found that his minute rounding off of a number produced a significantly different pattern for the weather ahead. From this event was born the understanding that minor changes in initial conditions produce major changes in dynamic systems.

In 1995, the butterfly metaphor for this understanding of how major changes occur in a dynamic system was popularized in the movie *Jurassic Park.* Since the wings of a butterfly generate wind by affecting tiny air currents, and because tiny inputs into dynamic systems can create major changes, the idea that a butterfly stirring the air in Hong Kong could influence a storm system over Boston captured our imaginations. We can see the butterfly principle at work when fractional temperature changes on the ocean surface turn tropical storms into hurricanes.

The butterfly principle implies that we can work for transformation in ourselves and our culture by deliberately influencing the right "butterfly wings."[25] The wings to stir are those that are most generative and positive in their effects—like affinities. Affinities bear fruit when we tend to them and create channels for their fulfillment. By discovering our own "Siamese connexions" and encouraging others to do the same, we embrace our greater identities and align ourselves with the Earth community. A line in the Talmud says, "It is not up to us to complete the task, only to do our part." Affinities point us in the right direction and show us what part of the natural world has claimed our heart and requires our dedication.

Although at times we may feel isolated, we are, in fact, all connected. If any of us follows our heart and intuition and responds compassionately to another species, attempts communication or communion with another species, or moves in any fashion in accord with the beneficent powers of the universe, all ultimately will move with us.

We are the butterfly wings, the strong and beautiful bug emerging from the table after a long period of unconsciousness, ready, finally, to see insects and praise them. Matthew Fox says praise is important for itself and to get the dark work with the shadow done. It is also practical. By praising insects, we regain a capacity for joy and a bigness of soul that attracts grace.

Insects come to us in our homes and gardens and in our dreams, zealous in their attempts to get our attention and stay with us. David Hope tells us they "will not let us go until we bless them."[26] How then can we bless them? "Point and cry out."[27]

Losing Our Affinities

Biophilia is the ground from which all affinities arise. When Edward Wilson proposed that biophilia is a need and a requirement for health and well-being, he hoped we would examine our motivations and pinpoint under what circumstances we cherish and protect life—and under what circumstances we not only let it die, but hasten its demise.

If we let the experts hammer out biophilia's particular characteristics and expression and can accept the strong evidence that biophilia exists, the environmental crisis is revealed as a ruptured emotional and spiritual relationship between humans and the natural world. We are members of a community whose welfare we have discarded. We are tied to a vast kingdom of creeping creatures that we have misunderstood, exploited, dismissed as unimportant, or tried to eradicate.

Quentin Wheeler, a Cornell University entomologist, estimates that at least 1.5 million and possibly as many as 50 million undiscovered insect species will become extinct worldwide as forests are burned to clear the way for farmland. Approximately nineteen insect species become extinct each hour. That's 456 species each day or 167,000 each year. What part of our collective consciousness and individual affinities vanish with each extinction, and how can we grieve adequately for such an immense loss?

The late Laurens van der Post believed that if we are to survive, not a single species must be allowed to go extinct except through some real necessity of life: "Life will never understand or accept any action less than necessity and everywhere it has eyes to see whether the law of necessity is obeyed."[28] And whenever a species does leave, he suggests that all of us join together with every other species on Earth to praise the one leaving and mourn its passing, knowing it was precious and knowing that it served the planet as it was created to do. Let our mourning also be for the part of the collective consciousness of humanity that was dimmed with its extinction and for that someone, somewhere, whose personal affinity for the one lost will never be stirred to wonder by its presence.

16

Strange Angels

> They come to us in dreams which is what angels are supposed
> to do. Startling, terrifying, sudden: is this the only way angels
> can now enter our world which has no openings for their welcome?
>
> —James Hillman

Ancient people looked to dreams, visions, and the creatures who appeared in them for guidance, warning, wisdom, and creativity, and they used various means to decipher their meanings. In tribal cultures, interpretation was often a group effort involving the tribe's council and medicine person. Since other species were seen as messengers from the Great Spirit, it was a priority for these people to understand the message in whatever form it came.

The shaman or healer who knew how to "read" an animal did not learn it from a dream sourcebook or bestiary but by observing the creature and reflecting on its ways. Encoded in its appearance and unique strategies for living on the Earth was its metaphor or medicine. Siya'ka, a Lakota native, says of a dream he had: "All the birds and insects which I have seen in my dream were things on which I know I should keep my mind and learn their ways."[1]

Blackfoot Indians, and perhaps all prairie tribes, not only paid attention to insects and other animals in dreams, but they also believed that butterflies, as emissaries of the Great Spirit, brought dreams and news to the dreamer. When they painted the butterfly symbol (an eight-pointed design that roughly resembles a Maltese cross) on their lodge, it meant that the other designs and colors on the lodge were not chosen by the one who painted them, but by the Great Spirit who then instructed the painter in a dream.

Although our specialists and a variety of sophisticated techniques have provided us with an abundance of facts about insects, we are not adept at translating those facts into meaning, especially in dreams. We are likely to accept a dream interpretation in line with the biases of the culture, until we realize that there is a bias. Even Carl Jung changed his beliefs about insects as merely "reflex automata,"[2] when he learned late in his life that bees are capable of symbolic communication and are, therefore, in all probability, conscious and capable of thought.

In a popular dream glossary, the assumption of insect as adversary and as an object of fear or distaste is clearly evident: Termites are reduced to meaning "slow destruction." Cockroaches are equated with "dirt,...neglect, carelessness, [and] lack of pride or self-worth." Scorpions imply "vengeance," beetles "the destruction forces at work in our lives, especially in our seed-thoughts, and affirmations," and flies "a real pest."[3]

What is evident in these descriptions, besides our culture's bias, is a lack of knowledge about insects that might honor their visit as "Other," expand our symbolic vocabularies, and lend our dreams more creative possibilities. Imagine a glossary that shows termites as altruistic and vital to the growth of the trees of our minds, that is, vital to the healthy formation and structure of our thought, abilities, and characteristics. What is eaten by them in a dream then might be our outgrown ideas, the deadwood of our minds.

Even indigenous people who have been educated in Western schools fall prey to the bias of our culture toward certain species. A recent book on American Indian nature symbols by a native healer who holds a master's degree in both sociology and psychology contains a mixture of native lore and Western prejudice. For those species which our culture has branded pests, he reiterates the cultural bias as though it were a general native view, calling flies a pestilence that brings disease and an indicator of inclement weather. He also says that the kinds of flies that feed on dead things and garbage are bad signs representing evil and dark forces.

In his statements we find little of the native view of flies as navigational wonders and guardians of fish, or of flies' ancient connections to valor. Also absent is the acknowledgment of the destructuring (recycling) powers of flies so vital to life's cycles of renewal. Instead, the Western myths that limit the fly to being only a carrier of disease and filth and the Judeo-Christian condemnation of the fly as evil are restated.

The same Westernized native also condemns cockroaches. He considers them dirty, a "bad" insect, and a "bad sign" that warns of forthcoming sickness, disease, and undesirable visitors. He also says that sorcerers work with these kinds of insects, sometimes sending swarms of them against other people.

Again we see the Western view of cockroaches as dirty, disease-spreading creatures. His statement that cockroaches can be sent on evil missions is reminiscent of the heroine Marcia in the story "The Roaches," who sent cockroaches to kill her neighbors. It is also like the heroine in the film *Creepers* who sent hordes of insects to punish a psychopathic killer. It is unfortunate that this perspective is being passed off as a native view, albeit unwittingly.

Brooke Medicine Eagle learned in her apprenticeship as a medicine woman of the Dawn Star shaman lineage that insects are messengers. She writes in *Buffalo Woman Comes Singing* of the time when the women of this line of seers began to signal her by sending shiny black insects with copper wings, and of when she had a numinous dream wherein she met an insect of great power and called her "Insect Grandmother." In the dream Brooke followed her into a cave that appeared to be an immense crystal geode. In her words:

> Insect Grandmother stood on her back legs and indicated other beetles scurrying around the room. Making each word clear, distinct, and emphatic, Grandmother said, "This is a communications room. My people use these crystals to communicate with each other over very long distances."[4]

In dreams crystals are symbols of Self and the union of matter and spirit, and they have long been believed to have energy-augmenting abilities. When Brooke Medicine Eagle awoke from her dream, she understood that the insect elder was letting her know that crystals can be used for communication in more direct ways than our technologies have currently employed.

Dreams and Affinity

An openness to insects, even without the broader context provided by indigenous teachings, can take an individual outside common opinion and beyond the pull of the collective shadow. Those who study insects are likely to dream about them. Imagine how insects might appear in the dreams of butterfly enthusiast and writer Vladimir Nabokov whose swallowtail butterfly set him on his life course. And British zoologist Miriam Rothschild in her anthology on butterflies and doves, *Butterfly Cooing Like a Dove*, says she knows of no knowledgeable entomologist who has not dreamed of a rare, desirable, or fantastic species of butterfly. Although to her these kinds of dreams appear to be straightforward wish-fulfillment dreams, she wonders what butterflies really symbolize and what hidden symbolism links an entomologist's daylight pursuit of insects to their dreaming. She says, "What are we *really* after when we set out with our green net, our pill boxes...?"[5]

Imagine too the dreams of entomologist J. Henri Fabre, who studied only live insects and in his long career completed ten large volumes on them. At the end of the third volume he wrote, "Dear Insects, my study of you has sustained me in my heaviest trials."[6]

Although entomologists are not a group likely to record and work with their dreams, at least one published example exists of a distinguished entomologist who says he becomes so attuned to the insects he studies that he often dreams he is an insect. The dreamer, Thomas Eisner, who is considered by many to be the father of chemical ecology, said in a *National Wildlife* magazine article that when he studied the ornate moth, he became so attuned to it, that he dreamed he was an ornate moth. In another dream, he was an ant talking to other ants in Spanish. Eisner also shares a dream where he is an insect talking to other insects and telling them about a dream in which he was a human being!

Interpenetrating Worlds

Since waking and dreaming worlds interpenetrate and influence each other, our feelings about insects naturally impact our dreams and our ability to understand them. The boundaries between waking and dreaming experiences are never fixed, but are fluid and permeable. It is not uncommon for a dream creature to make an appearance as a physical creature in the daylight hours, or for one we encounter in the day to visit us in a dream. As we saw in our discussion of butterflies and other flying insects as messengers, these kinds of improbable occurrences or synchronicities provide evidence for the existence of a unifying pattern that weaves together inner and outer events. If we are attentive to both waking and dreaming encounters, we have the chance to see the way they work together, building on each other and nudging and informing us as we grow and move into greater expression.

In *Spirits of the Earth*, Bobby Lake-Thom shares the story of how on a cold winter night, when his oldest daughter was meditating and focusing on a ritual for menstruation, a butterfly appeared in her room. In the dead of night Lake-Thom heard his daughter talking and laughing loudly. When he checked on her, she was sitting on the edge of the bed, but still in a quasi-dream state. She pointed to a butterfly, invisible to Lake-Thom, and called it "Grandmother." She told her father that the butterfly was now human size and kept changing from a pretty young girl to an old woman—and it wanted her to come over and give her a hug. He reassured her that Grandmother Butterfly was a good spirit and to give her flowers, cookies, and a hug. In the morning, as his daughter was relaying her beautiful dream to her parents—convinced that it was just a dream—a butterfly flew out from her bedroom. Lake-Thom and his wife told her it was an ancestor spirit wanting to communicate with her during this spiritual time of menses.

Delivering the Message. By letting go of any prejudice we may harbor against insects we allow them their role as the bearers of blessings and affirmation. But insects also bear the task, not always pleasant for the dreamer, of revealing unconscious material.

In a recent book on the healing power of dreams, the author writes that insect dreams are almost always unpleasant. What may be unpleasant, however, is not the insects, but having our attention drawn by them to neglected or out-of-balance areas in ourselves. Pushing past our defenses, they alert us to issues that we have ignored or denied.

A woman, who for weeks disliked both her job and her apartment, but had not taken any steps to change either, dreamed:

> I go into a bookstore. The manager directs me to a room in the center of the store with glass walls. Inside the room, hanging from the ceiling are several human-size cocoons. Some are withered as though the occupant had long since died. Others are not. Suddenly an enormous caterpillar

> emerges from one of the cocoons. It's very angry, and I know it wants to sting me. I back away as it moves toward me.

The threatening behavior of the caterpillar appears to be a call to action and an expression of the pent-up energies within the dreamer that demand transformation and a more expressive life. The dreamer is stuck, as is the caterpillar stuck with its worm body. Perhaps it is angry at having been deprived of its transformation into a moth or butterfly by what we can assume is the dreamer's lack of action.

A different dreamer has a similar dream. The dream setting is a small, dimly lit and poorly ventilated room. The dreamer opens up a cupboard and out flies an insect. Heading straight for the dreamer, it stings her, and she cries out. The setting of this dream describes where the dreamer reports she is in her life—closed in and barely able to see or breathe. A flying insect is trapped with her, perhaps a winged self, who released from its confinement in the cupboard promptly stings her. Perhaps the insect is trying to arouse her so she can make some necessary changes. Or, maybe by stinging her and making her cry out, the insect helps her express her frustration and despair.

When a dream insect has something to tell to us, it usually pursues us, gets on us, or bites or stings us. In these dream situations no thoughts of goodwill can turn the insect away from its mission. When they need to get our attention and steer us in a new direction or reveal emotions that have been suppressed, they are simply unstoppable.

In a northwest contemplative community, a woman who has a very positive attitude toward insects dreamed:

> I am walking by a path leading into the forest. On the path, coming out or going in, a man is struggling to escape from a swarm of bees. Then the bees start to move toward me, and I begin to send good thoughts to them. But they attack and sting anyway. I put my head down in a protective gesture. The back of my neck gets stung, and angry welts appear.

What appears negative and painful here suggests that some verity is operating that can't be arrested by the good intentions of the dreamer. As she tries to walk by the path leading into the forest, sending the bees good thoughts, they attack and sting anyway. What does their insistence mean, and who was the man who also struggled to escape them?

As we learned in the chapter on bees, a swarm is homeless, having left a hive or nest to seek a new one. Why are they responding as though they were being threatened? Do the bees want her to leave the road and follow the path through the forest? What energies swarm inside the dreamer, homeless energies not bridled to the road but to a new path? What is angry and aroused in the dreamer that she cannot eliminate by thinking good thoughts or putting her head down and shielding herself? Is she refusing to see or deal with something directly? Do the stings cause the angry welts or just allow them to appear?

Confronting the Shadow. Sometimes insects appear in dreams as manifestations of the shadow. The shadow as noted earlier is the archetype that resides at the threshold of self-conscious awareness. It defines what is "self" and what is "not-self." All the traits we don't like in ourselves reside in the shadow, where they are then projected on others. When we are awake, we identify the projections of the shadow by noticing whom we most dislike or fear. The shadow is also evident in whom we might choose to ridicule.

Whether we are awake or dreaming the shadow usually appears as the enemy—threatening or repulsive. It is easy to see, then, how the psyche might borrow an insect form to reveal an aspect of self the dreamer needs to work on. Emerging from the shadow and revealed in our dreams are all the forms of tyranny and exploitation that alienate one part of the population from another and permit us to control, manipulate, and treat as alien the nonhuman world. In those kinds of dreams, we harm ourselves by despising and rejecting parts we have been taught are "not us."

Dreamworker Jeremy Taylor says working on our dreams is a critical part of learning to "love the enemy" both within and without. "Dreams reveal that the despised and feared *other* is actually a challenge to our own narrowness,"[7] and when we work with our dreams, he assures us that compassion and growth will replace the fear of others and our feelings of alienation.

As an added incentive to work with shadow material, it helps to know that the shadow also contains our undeveloped talents and gifts. In fact, paradoxically, within the shadow is the very thing we need most for our healthy development. Myths from cultures the world over teach that renewal always comes out of what we consciously despise and reject. In order to receive the shadow's gift, we must overcome the fear or revulsion evoked by its dark (unconscious) aspect.

It helps also to know that confronting the shadow is not a one-time affair. It is a task that must be undertaken at every step of growth and development. During each cycle, the shadow takes on new forms, and each time it requires that we find our courage and engage in honest self-reflection to facilitate a reconstellation of the energies within the psyche and reach a new balance point.

More Messages from Insects

As we saw in our look at insects in the garden and in the fields, insects act as messengers in daily life, routinely conveying upsets in the natural balance of things. We have noted too how the message is often obscured because of a tendency to perceive them as the problem in situations where they trouble us or in areas where they appear in large numbers. In both the natural world and in dreams, the specific way in which the insect interfaces with humans or plants holds critical information to understanding the situation and acting appropriately.

A recovering alcoholic remembers one of her childhood dreams:

> I am in a house with my parents. We discover insects all over the place, covering the floor and walls. I feel repulsed and tell my parents that they

> have to call an exterminator. They tell me not to worry; the bugs will go away sooner or later. They want me to fix them a cocktail.

The dream has unpleasant aspects. The bugs are inside, covering the interior surfaces of the house. The dreamer is repulsed and her reaction may indicate that shadow material is involved. We might wonder what the insects, by their sheer numbers and presence inside the house, are covering up. Who or what has summoned them? Why do the parents ignore or deny the situation and focus their intention on getting a drink? Did the dreamer turn to alcohol to avoid dealing with what the bugs were hiding in her childhood house?

Dream Lice. Looking at dream insects as messengers and allies aligned with beneficial forces will always move the interpretation of the dream down a very different path than if their presence is viewed as hostile. A woman in her late twenties dreamed:

> I am at a desk working with others in a big, old-fashioned house. I had a little girl and I had to give her away. She comes back and we have a joyful reunion. There are lice-type black bugs swarming in her braided hair. "If only you knew what they were before, you wouldn't have sent me away," says the little girl. I take the bugs out of her hair and swear to her that I am going to get her back. We are both crying. Her head is all red with open sores from where the bugs were fastened to her scalp. I assure her that it is going to be all right.

What aspect of the dreamer is this little girl, given away in error, and why did the dreamer give her away? Did she think the little girl was infested with something unacceptable? We can assume the lice-like bugs are helpers since the little girl says to her mother, "If only you knew what they were before, you wouldn't have sent me away." But then why are they in her hair and on her scalp? Do they cause the open sores to open her to something, or do they cover them to protect her? And if to protect her, what caused the head sores, and are they sore as in angry or sore as in hurt?

This line of questioning opens the door to an exploration of the dream and its message that is precluded by viewing the bugs as the problem. And remembering the related Navajo myth of Louse and Monsterslayer, maybe the lice came to keep her company because they knew the little girl was neglected and lonely.

Understanding the Message

When a dream insect directs our attention to something we would prefer not to see, we defend ourselves against it and its message in a specific manner. James Hillman explains in his illuminating essay "Going Bugs" that the source of our suffering is indicated by the way we defend ourselves or rid ourselves of the insect. For example, a woman who dreams she cuts a bee in two—into better and worse or perhaps rational and instinctual—performs the dissection to feel safe. She creates safety by dissecting problems with sharp practical distinctions.

Another dream in Hillman's essay begins with the dreamer looking up and seeing insects crawling on the ceiling. He crushes them with a broom, sweeping them out of his up-looking attitude and restoring his view of a blank ceiling. Hillman calls this response taking the heroic stance. Exercising his personal will, the dreamer obliterates from his view what is new, unknown, and thereby threatening.

In another dream, a dreamer sets fire to a caterpillar in the garden, torturing it. Hillman suggests we ask what the fire is burning or transforming in the psyche and what the result of the transformation will be. Fire is an alchemical symbol of transformation and the caterpillar is also a symbol of transformation in its own right. So to focus on what is being transformed is an appropriate way to investigate the meaning of the dream. To side with the dreamer against the invasive insect or to side with the poor caterpillar against the cruel dreamer is to miss the complexity of the alchemical process and the encoded dream message.

Insects and Images of Illness. The same dream authority who described most insect dreams as unpleasant says diseases sometimes take the form of insects in our dreams. I would add for clarification that insects point out the disease or malfunction by their presence. Since most insects are destructuring agents whose job it is to recycle what is dead or dying, they are attracted infallibly and automatically by the very existence of the malfunction. Their presence is a warning, a message that something is out of balance and has called them to disassemble it if it can't be brought back into balance.

Images, so central to the language of dreams, are also the language of our bodies and underlie our symptoms and illnesses. In his book *Care of the Soul*, theologian Thomas Moore links the body and its organs to expressions of consciousness. Moore maintains that each part of the body is not just a mechanical apparatus, but also has its own imagery that can be tapped and harnessed by a person who wants to explore an illness to determine its message. The patient's images of the illness, whether from dream or imagination, have power because they arise from the roots of the malfunction and are linked to the individual's emotions, thoughts, and life patterns. Thus dream images can reflect the roots of illness in the same way that healing images often surface in imaginal therapies. Visualization exercises for cancer patients, for example, harness the healing energies of the body to the will and desire of the patient to get well.

Following Jung's lead, James Hillman suggests that diseases and afflictions are divine processes through which the gods reach us. In this view, dream insects that reveal physical diseases become divine instruments helping to provide the pathological experience that leads us to a sense of soul. It's not the awareness-heightening peak experience we typically strive for, but it does rivet our attention and sensitize us to the movement of the psyche and its needs.

We can follow the insects in our dreams and let them take us into the diseased or afflicted body part. When something is not functioning right, we can bless the creatures who make us aware of it, sensitizing us to it so it can "speak." Without the affliction, we would likely remain unconscious of how that part of the body is functioning, and without the imagery, we won't know what it needs to communicate.

Healing Dreams

In some ways, all dreams are healing dreams, even when they upset us or when we don't understand their message. Healing happens on many different levels, with and without conscious awareness, and in dreams it takes many different forms. Often insects allow us to merge our consciousness with them (as John Lee did with the butterfly), perhaps to renew our connection to life.

Before falling asleep, a woman, who gave herself a suggestion to have a healing dream, awoke exhilarated from a dream she titled "Moth Thought":

> I am working in the library and holding a plate with some crumbs on it from something I have just eaten. Suddenly I notice that two huge moths are perched on my plate side by side, eating the crumbs. They have enormous furry antennae and a fur ruff that encircles their heads. As they feed, I can feel the vibration of their antennae and their bodies through the bottom of the plate. I am very excited. The experience seems astonishing to me, and I run over to a coworker to share it. My coworker is alarmed, and I am exasperated with her fear of these lovely moths, who continue to feed quietly. I walk away thinking that I will try to share the experience with other people in the library. Suddenly, I have an impulse to communicate with the moths by trying to reach their vibrational level, which in a weird way I know is slower and lower than mine. I close my eyes and try to feel/think like a moth. Immediately, I feel a deep sense of calmness and slowness. A deep relaxation floods through me. It is "moth thought." For a moment I have a clear sense of the aerial perspective of a moth in flight and feel myself skimming over the library's stacks of books. Just as I have this "moth thought," with my eyes still closed in the dream, I feel the plate lighten. My eyes fly open as first one moth and then the other lifts off the plate. I am upset, not wanting them to leave. I yell to everyone, "Quick! Look! They're flying around! They're going!"

The dreamer woke up trembling with excitement. As in Daniel Quinn's powerful beetle dream that I shared in the preface, this dreamer found her beetle equivalent in these moths, moving away from conventional human thought (or stepping off the sidewalk as Quinn might say) to experience "moth thought." She felt as though she had touched some deep ground of her being where the mystery of our interconnectedness with other species originates. She identified the dream creatures as cecropia moths, a species that had seemed magical to her as a child and which as an adult symbolized the beauty and mystery of life.

Healing dreams are also the ones that help us in our life's work, affirming and clarifying a path already taken. Insect artist Gwynn Popovac, introduced in the last chapter, also steps off the sidewalk in a dream and receives a gift from the insects. The dream, which follows, not only reaffirmed Popovac's desire to create insect portraits, but it also presented her with the essence of her style—quick, short strokes in pen, pencil, and paint—as a satisfying and appropriate way to join with and reflect the vibration of insects. She dreamed:

> I was walking on a sidewalk with a group of friends. When the concrete ended, everyone stopped walking but myself. I continued off into an uncultivated area, a sort of marsh, and looped around. When I returned to the sidewalk and rejoined my companions, my head was filled with a high-frequency whirring of insects. I felt as though I were hearing something only I could hear, something I would have to try to translate to them in the best way I knew how—through my insect portraits.[8]

Transpersonal Healing Dreams. Dreams with archetypal themes have an unmistakable intensity and a message that encompasses, yet transcends the personal. Some reflect conditions in the culture of which the dreamer is a member. In the following dream, ants direct the dreamer's attention to what is wounded and needs healing in herself and in the culture. The ants also signify that energy is available for healing the wound. At the time the dream occurred, the dreamer was actively searching for a path of service while attending a graduate school. She had no obvious connections to insects or ants when she dreamed:

> I am driving home from school in my mother's white car. Ahead, coming toward me is a black boy riding a bike and steering it in a haphazard fashion. I don't know which way to swerve to avoid hitting him, so I stop the car. He continues toward me and stops just short of hitting the front bumper. I get out and walk to the front of the car. The boy on the bike has disappeared, and in his place is a man's head torn off at the neck. The man appears to be from India. Thin and impoverished looking, he reminds me of one of Mother Teresa's "poorest of the poor." He has ants crawling all over him. They are in every orifice—his eyes, ears, mouth, and the ragged neck area, which looks like raw meat. He moans softly, moving his head in obvious pain and distress. I am repulsed. I don't want to touch him, but I know I must. I say, "Can I help?" He doesn't answer, but turns toward me, and I repeat, more firmly now, "Can I help?" He says weakly, "You want to help me?" and I say with resounding clarity, "Yes!"

Since a full interpretation of this dream lies outside our intention to highlight the role of ants, we must risk flattening the dream's multileveled message and confine ourselves to a few ideas. A shadow figure comes directly at the dreamer, turning into a living head severed from the body and in obvious distress. The head or rational thinking faculties have been disembodied, ripped away from the body and its instinctual wisdom and heart. The image suggests that the masculine has been forcibly separated from the feminine, a simplistic but telling image for modern society. Ants are swarming all over the open wound of the neck as well in the eyes, ears, nose, and mouth. They are aroused as though their nest has been violently disturbed.

Remembering that the ant colony has long been thought of as a single living organism and compared in its functioning to the human body, we can see how on one level this severed head with its agitated ants could depict a rupture in the human community. As a sister-based society and symbol of the feminine, ants could also represent the altruistic propensity within an individual to voluntarily sacrifice himself or herself for the benefit of the whole.

When an ant nest is disturbed or any part of it is destroyed, every ant in the vicinity of the broken structure comes to help repair it. In the dream the ants are swarming all over the openings (orifices) in the severed head where new things might be heard, sensed, seen, or digested to invoke healing. Perhaps they bring the energy of the colony, the consciousness of community and wholeness, to heal the split, motivating the dreamer to affirm her own altruistic inclinations, that is, her willingness and capacity to serve society as a healing agent.

A Good Death. In another dream with transpersonal themes, a schoolteacher with no outer connections to bees enacts a sacred ritual where she deliberately kills the queen bee, who comes happily to her death.

> I am walking in a big open space. It is a cool morning, and the air is calm with only the hint of a gentle breeze. I am purposefully collecting the queen bee from each field. I feel a strong sense of my own femininity as I continue with my task.
>
> As I approach each field, I walk with a delicate white cloth in my hand. The queen bee comes to me happily as though it is her duty to sacrifice herself for "the purpose," although what that purpose might be is still unclear to me. We communicate with each other, our hearts joyful, understanding that the time has come. I feel only pure love toward this insect and become one with her as I suffocate her under the cloth. She is my sister.
>
> As I walk to the next field, I feel a deep sense of honor and respect for my sisters, the bees, and I celebrate with them in this collection process. We become one.

In this multilayered dream is another community of insects comprised of socially cooperating individuals who work toward the benefit of the whole. The fact that most of the bees in the hive are female and are ruled by a queen makes the beehive a sacred feminine symbol.

An image of a willing death is also in the dream—what tribal societies call a "give-a-way." This death is a "good death," because the one killed is willingly sacrificing her life for some great purpose that supersedes her personal survival instinct, uniting human and bee in the sacred act.

Interestingly, in nature a queen bee is occasionally killed by the entire colony in a formalized death ritual known as "balling the queen."[9] All members of the colony participate, forming a solid ball around their mother and pressing harder and harder. As the ball of living bees becomes progressively smaller, squeezing in on the queen, she is suffocated or crushed. Observers have always assumed that the death was initiated for some offense by the queen such as not being able to lay enough eggs anymore, which says more about the observer than the bees. And following this assumption, the ritual is typically considered a collective act of homicide or a mob lynching triggered by some unknown judicial agreement, panic, or spontaneous madness. This dream suggests a deeper truth. There is a cyclical nature to the queen's life, an appropriate end and closing of a circle of life. The act of taking her life is not done against her or without her permission. In fact, she enables it in some inexplicable manner, and it is through her willing sacrifice that some greater, if unknown, purpose is fulfilled.

The Black Butterfly Vision. Visions are like waking dreams. The person is awake but his or her consciousness is altered and normal time and space parameters do not apply. Psychologist Richard Moss, in his book *The Black Butterfly: An Invitation to Radical Aliveness*, shares his waking dream of a black butterfly which occurred at a critical time in his spiritual journey.

After experiencing a state of unease and anxiety, coupled with a period where ordinary boundaries were blurred, Moss was in a meditative state releasing all thoughts, both positive and negative, when he observed two butterflies dancing in the air. One was mostly black and the other white. They landed on a nearby branch, and to his amazement and delight, they mated, wings opening and closing as one. After some minutes they once again resumed their dance in the air. Suddenly, the black one flew to him, landing between his eyebrows. He says:

> At that moment life changed forever. The descriptive words that came in the following days all involve the imagery of Marriage and Union. I am at once the Lover and the Beloved. All of Existence confirms me and is none other than Myself.[10]
>
> In that moment, all of creation became a single consciousness, a state of indescribable glory and unspeakable peace. The fear that existed when I stood rooted in egoic consciousness was now the most exquisite nectar. I was suffused with a current of aliveness so transcendently blissful that there is no analogy within ordinary experience that even approximates it. It was a living bliss, but it was also the most profound intelligence. There was a flood of knowing, of understanding as though all of existence stood before me in its totality with its secrets uncovered and revealed.[11]

Moss's experience of unity, delivered by the black butterfly, forms the core of all mystical and religious traditions—underneath the myriad of techniques and cultural trappings. The vast self behind the thoughts and experiences of the personality or local self is accessible in dreams and visions. It is our expanded identity, the self that knows all things are a part of itself and that our souls touch the souls of every creature.

Although visionary experiences like the one Moss describes are unusual, an increasing number of people report experiences of oneness in which ordinary boundaries of perception that separate us from others blur, give way to a direct bodily knowing of the unity behind form.

Destroyer Turned Benefactor

In tribal societies, healing dreams and visions were the familiar domain of the shaman, but life-shaping dreams with other species could also occur to ordinary members of the tribe. Anyone who had a vision or dream visit by a creature had a responsibility to act on it in some way. In some tribes, members organized themselves into clans or societies with special rituals and responsibilities based on the appearance of another species in a dream, vision, or physical encounter. Among the tribes of the Comanche Plains Indians, for instance, was a Wasp Tribe that may have been organized around such a dream or vision. A vision or dream alone, however, was not enough to guarantee someone membership in the society. The person would have to display behavior indicating that he or she had received the creature's blessing or favor.

As we noted in our discussion of spiders and scorpions, creatures with the capacity to harm us are particularly potent symbols, constellations of energy that seek individuals who are psychologically ready to move to the next level of development. In dreams these kinds of creatures usually appear human-size to indicate their archetypal status. They take us, willing or not, on an inner journey that always involves a form of self-annihilation and a movement from the known identity into the unknown, where a new identity awaits us.

In the Blackfoot tribe, for example, a warrior formed the Society of Mosquitoes after receiving information in a dream. He had been hunting in a place thick with mosquitoes. Swarms of mosquitoes descended on him and bit him so relentlessly that he wondered if they were going to kill him. He took off his clothes and lay on the ground in an act of surrender. The mosquitoes quickly covered him, biting him until he lost all feeling and then consciousness.

Suddenly, he heard strange voices singing and calling all the mosquitoes together. The voice said, "Our friend is nearly dead!" Although his physical eyes were shut, he saw the mosquitoes sitting in a circle singing and jumping up and down. Dancing in the direction of the sun, they sprang this way and that way. Some were red and others were yellow. All had claws attached to their wrists and long plumes hanging from their heads. Then the voice addressed him, "Brother, because you were so generous and let us drink freely from your body, we give you the Society of Mosquitoes and make you the leader."[12]

The Blackfoot warrior opened his eyes and sat up, but the mosquitoes had gone, and he was alone in the woods. He returned home and shared his lucid dream with the tribe. The elders and shaman agreed that he should create the Society of Mosquitoes. The members of this society wore buffalo robes with the hair side outward. Some painted themselves red and others yellow, with stripes across nose and eyes. They wore plumes in their hair and eagle claws attached to their wrists to represent the "bills" of mosquitoes.

Note that in the dream, the man had to surrender and approach his death before the gift was given—once again the destroyer-turned-ally theme. The wise individual, that is, one who is psychologically ready, will surrender to the forces that come for him or her. And once emerged on the other side of these passages, they also receive the gifts of their animal-destroyer.

Since we can't change or control this process, imagine how much easier it would be if we embraced it. What if we didn't use the word "nightmare" to designate sobering experiences that humble us or leave us in pain and mindful of death? What if we welcomed them as opportunities to move into a more satisfying life? When we insist on renewal without fear or pain, and on resurrection without death, we run from what appears to be a nightmare and subvert the intention of our deepest identity.

Granted, we might always long to be imbued with knowledge of soulful matters without being bitten or stung, but if we resist the impulse to fight with the forces who come for us, we'll discover that in the midst of the fear and darkness is the gift we need the most in order to infuse our lives with new purpose and will.

The Power Animals in Dreams

As we discussed in the chapter on scorpions, the way we meet and work with adversity determines the outcome. Again, in all shamanic traditions the one who attacks and destroys the initiate is the one who becomes the ally and teacher after the trials of initiation have been endured. These confrontations with power resemble descriptions by yogis of the awakening of the kundalini and accounts of the path to enlightenment. So surrendering to the creature—losing the fight—changes misfortune into triumphant victory and adversary into ally and protector in what Joan Halifax in *Fruits of Darkness* calls a kind of "psychological homeopathy."[13]

Sometimes confrontations with power involve a particular animal with whom we have a special connection. In *Shakti Woman* Vickie Noble writes of the "power animals" that appeared in her dreams. Her first animal as an adult was a human-size Mexican orange-kneed tarantula. As a child she had been terrified of spiders. She couldn't even be in the same room with one, and, if she was, she would scream until someone came and killed it or took it out of the room. She had nightmares of spiders climbing all over her, lurking under the bed and covering the walls. At nine, she had a secret compulsion to trap yellow and black garden spiders in a jar. Then, she confessed, she would make a small fire and drop them into the flames. She refers to it as a "terrible fascination" and was relieved that this obsessive activity only lasted about three weeks before disappearing. As an adult she later understood that her behavior was not so much a psychological aberration as it was an intuitive and compelling recognition of the spider's alchemical power of transformation for her—and perhaps sent by what James Hillman would call her daimon.

She remained hysterically afraid of spiders until the time after her first marriage had ended. While she was learning how to be self-sufficient, she opened the garage door and met a large spider at eye level. She knew it was a black widow and worried that it might bite her children. She decided to kill it and performed the act calmly, without hysterics. She was surprised that she didn't feel any anger or fear, and she was able to deal with the spider without calling for help. She showed its lifeless body to her children and warned them to not touch that kind of spider.

That was the turning point. From then on, her hysteria about spiders disappeared as though it had never existed. Five years later, when her shamanic path opened and spiders began to visit her, she was ready. In the first dream an enormous Mexican tarantula visited her. "Dark and hairy with knees of orange crystal, gemlike and fantastical...she danced and jumped in the dream, taunting Noble."[14] Noble was properly awed, dwarfed, and scared. She awoke hysterical and in a cold sweat.

In the years to come, Noble entered periodic dreaming confrontations with this same spider in a gradual process of healing and rebirth. She always dreaded the spider's return, yet paradoxically looked forward to each opportunity as an initiation in which she had another opportunity to meet the spider in a more courageous and creative way. When a student brought her a cast-off skin from her live Mexican orange-kneed tarantula, Noble knew that the tarantula, in shedding its skin, was also teaching her to leave her old tight self behind and allow a new, more expansive self to emerge—a self filled with new health and abilities. Other tarantulas also visited her over the years in dreams. Some were black with red interlocking rings on their backs, and there was one that was her pet and drank milk from a cup she put on the floor for her. Finally, she danced with them.

Noble believed that the spider was Shakti for her—a yoga teacher. She imagined the spider dancing, creatively and endlessly active, going about her business with power and authority, as she needed to learn to do herself.

> I witnessed in her a dark jewellike sexual power, a raw potency I recognized as Tantric and that I would later come to understand as shamanic, female, and arising from contact with the Dark Goddess.[15]

Consider that in this modern age of threatened wildlife, dreams may be the only vehicle left through which insects and other animals can reach us and help us contact the vital forces of our own threatened nature. It would seem that in dreams, aspects of our wilderness self pursue us, waiting in the gaps to pounce, for we are its only prey. And when the dream swarm of bees stings us, or the tarantula grabs us by the neck and tears away our flesh, breaking the bones that contain our understanding, we have unwittingly entered those gaps, the spaces governed by the personal and transpersonal jaws of transformation and growth. Do we surrender, or do we harden in fear and reach for poison?

Can we live in the spacious heart of life and learn to dance with them, or do we cling to the hard edges of the uncertain and frightened mind?

Today, when Noble discovers that she is sharing the house with black widow spiders, she either lives in harmony with them or escorts them politely outside to a more appropriate place for them to live. She feels blessed by the presence of spiders and understands them to be messengers of the Great Goddess in her life. They no longer come as opponents in her dreams; they are allies now.

Insect as Angel

One of the most interesting views of insects in dreams is James Hillman's angelic interpretation that says that underneath our culturally driven aversion, insects may simply remind us of the supernatural and the miraculous. Ecopsychologist Paul Shepard reminds us that angels, like insects, focus our attention on the relationship between zones or fields of experience, widening our sensibility and awareness of an autonomous order. After all, a fear of insects, which is often without a rational basis, has more in common with awe than ordinary fear. Consider that religious visions, as disquieting as they are fascinating, always invoke fear and dread at the onset. And according to philosopher Rudolf Otto a sense of the uncanny and the eerie is the primal spiritual experience.

One of the dictionary definitions for angel is "a messenger, especially of God." After looking at how insects, aligned with our most fundamental identity, continually assist us, conveying all manner of information to our benefit, the interpretation of insect as angel is not as unreasonable as it might appear at first glance.

Insects, like angels, have remarkable powers. They can live everywhere, partaking of all the elements of the natural world, and as models of transformation and growth, they can multiply, transform, and camouflage themselves endlessly. As bearers of information outside the boundaries of what we know, they bring new worlds to our attention—which is what angels are supposed to do. By resolutely drawing us in dreams out of our comfortable human perspective, insects also bring us the potential of new consciousness. Sometimes it's the consciousness of a hive or colony and a degree of cooperation and commitment that few human groups ever obtain. Other times it's the consciousness of transition that lives on the edge of chaos, where new energy and vision are available.

Hillman explains his angelic interpretation further:

> This angelic view calls us to look again, respecting who they are, what they are, why they are in the dream, and further, how to meet them, even care for them—these miraculous shapes and behaviors, each intricate appearance, a superb archaic ability, faultless, pious, comical, grave, intense, seeking us out while we sleep.[16]

Perhaps insects are D. H. Lawrence's angels knocking at our door in the night: "What is the knocking? What is the knocking at the door in the night?

It is somebody wants to do us harm. No, no, it is the three strange angels. Admit them, admit them."

The Golden Cockroach

To close, I want to share one last dream from a woman that tribal people would call a "Universal Dreamer."[17] Marcia Lauck's visionary dreams, some of which are published in *At The Pool of Wonder: Dreams and Visions of An Awakening Humanity*, which she coauthored with artist Deborah Koff-Chapin, are not personal dreams, although they have touchstones to her personal life. Rather these dreams emerge from a sacred, transpersonal realm where all the archetypal patterns of humanity reside—a realm outside the familiar boundaries that most of us travel at night. They speak of the evolution of humanity's consciousness, its initiations, and the new ideas and cultural developments present at each stage. The following is an unpublished dream she had while this book was being written.

> Dreaming opens in the moisture-laden night air of the South Pacific. In the humid dark, all is silent but for the rhythmic chirping of the nocturnal insects. I am part of a ceremonial circle of women elders. We are the Dreamers who spin the great medicine wheels into the field of time and space, and we come from every continent on Earth. Our work this night is to bring down a new field of energy, a new pattern to initiate an evolutionary shift in planetary consciousness. As we move into the innermost reaches, the atmosphere thrums. There is a gathering acceleration of consciousness. Images of wombs, passageways, and deep ruby throats of flowers flash by in lightning succession and then we are spun beyond reckoning, travelers on a mission into frequencies unnamed by human beings.
>
> The dream folds into itself, and when our circle reemerges, I am in the center, held in the arms of the others as I labor to bring forth the fruits of this night's work. Contractions ripple through my body and the great waves push the new life into our midst. We look, wondering what the imago of this new cycle will be. At our feet, in the center of this sacred circle is a shimmering golden roach. Small sounds of satisfaction emerge from the others. This is a good omen for the future. The insect, seemingly lit from within, grows in luminosity and size until it becomes the dream. As I come back to waking, I feel it at the shoreline of human consciousness, embedded in the archetypal ground.[18]

The pairing of gold, that enduring essence that symbolizes the soul, with the ancient insect that has both preceded and accompanied us on our evolutionary journey signifies that some great work is underway—a deep reconciliation and healing called forth from the primordial wisdom that is held in the heart of creation. Lauck explains that "just as there is a unique task for each individual to unfold through the process of individuation, there is also a transpecies, planetary task that must engage all of us desiring to remain part of this world's story."[19] It is the "Great Work" that calls us to remember. Myths the world over teach that the seed of renewal is always to be found in something humble, in what has been despised and rejected. What quality, what energies does the golden cockroach bring to a world that no longer knows its true name? And how can we do our part? Admit them, admit them.

17

Following Mantis

The word came with Mantis and the word was with Mantis.
—Sir Laurens van der Post

Taking liberty with Denise Levertov's poem "Come into Animal Presence," a fitting tribute to the praying mantis might read:

Come into Mantis presence,...
Those who were sacred have remained so,
holiness does not dissolve,
it is a presence
of bronze,...[1]

When people speak of the praying mantis or mantid, they speak of presence. Part of this insect's appeal is its huge eyes and attentive demeanor. A flexible neck, independent of its thorax, lets the praying mantis move its head freely and direct its gaze to its surroundings. Many people report that its stare seems quite human, as though it were coolly observing them.

The confident demeanor of the praying mantis isn't just show. It's known for holding its own in almost any situation. Insect artist Gwynn Popovac has a photograph of a praying mantis that had just caught a hummingbird. And the first praying mantis that biologist Ronald Rood remembers seeing was in the middle of a busy street in Connecticut, standing its ground defiantly. Another time Rood saw a praying mantis confront a robin and assume such a threatening display that the bird flew away, apparently confused. And a friend of his filmed a battle between a praying mantis and a frog—a contest that ended in a draw.

The praying mantis has a reputation for courage that is legendary. A story included in *100 Ancient Chinese Fables* tells of the time when Duke Zhuang of the state of Qi was out on a hunting expedition and saw a large insect in the road. It lifted its front legs and was about to fight with the wheel of his carriage. The duke asked his driver what kind of insect it was, and the driver replied that it was a praying mantis—the one insect that only knows how to advance and never retreats and is so confident about its abilities that it routinely overestimates it own strength and underestimates its foe.

Impressed, the duke remarked that if the praying mantis were a man, it would be considered the bravest man in the world. He told his driver to turn the carriage to avoid the mantis. When men of courage heard about the duke's response to the insect, it is said that they thought so well of the duke that they pledged loyalty to him unto death.

The Art of Hunting

Most of us have seen a praying mantis. Two thousand species of praying mantis are scattered throughout the world, ranging in size from less than half an inch to more than five inches. In tropical regions, up to 350 species can inhabit an area. Although most of us place praying mantises in a class of their own, entomologists have classified them in the order Orthoptera that includes six families—including cockroaches and walking sticks. In the United States, all species of the praying mantis are known as the gardener's friend because of their appetite for other insects. One of the most common in this country is the Chinese praying mantis, which was imported in 1896 for general release, an attempt by entomologists to augment the insect control services rendered by the smaller native species of North America.

The praying mantis is also prey for other animals, especially bats. But even for bats, the praying mantis is no easy meal. A unique auditory system often helps the praying mantis escape these skilled night predators. Interestingly, for years the praying mantis was thought to be deaf because entomologists could not find its ears, although they had found ears on its relatives the crickets and grasshoppers. They were looking in the wrong place. Mantises have a single ear located in the middle of their body on the underside. The ear looks like a deep slit, about a millimeter in length, and two teardrop-shaped eardrums, which function as a unit, face each other from opposite walls inside the slit. This single ear gives the praying mantis hearing in the ultrasonic range between 25 and 100 kilohertz. Although a single ear can't locate the source of the sound (an ability that requires two ears, separated), nondirectional hearing in this range is particularly useful to the mantis when bats, using sonar to navigate, are in the area. When it is flying and hears a bat approaching, for example, it begins twisting left and right and then does a spiral dive to the ground—a tactic which often saves its life.

Emulating Mantis. The praying mantis is an insect-eater and uses an effective sit-and-wait approach to hunting. An exceptional hunter, its superior vision and reflexes let it seize its prey suddenly and forcibly with lightening-fast precision. In fact the time required to strike and capture has been recorded on high-speed film at fifty to seventy milliseconds. Its hunting abilities have even inspired a style of martial art. Four hundred years ago a man named Wang Lang in the province of Shantung based a style of Kung Fu (which mimics animals and aspects of nature) on the postures and movements of the praying mantis. This already renowned martial artist got the idea while watching a praying mantis capture and kill a cicada. Wang Lang refined a series of movements—hand and arm positions and strikes and grabs—based on his observing the mantis, and he successfully demonstrated his new style in several bouts with Shaolin monks, who were devoted students of the martial arts. Subsequently, the style, with modifications, was widely adopted.

Part of this style's success is due to the fact that it employs the art of stillness. Stillness is a necessary tactic for the mantis to escape detection by potential prey. And by holding still, the mantis can grab other insects that, unaware of the danger, venture close enough. The great Zen Buddhist teacher Daisetzu Suzuki found the stillness of the praying mantis a striking example of one of the great Zen articles of faith—that of action through nonaction. This attribute also inspired a form of meditation often taught in conjunction with traditional kung fu called chi kung. An individual practicing chi kung goes within and directs his or her body's life force (or chi) through the body's organs and systems and along specific avenues to strengthen and heal it.

One Who Sees the Future. In *Animal Speak*, an animal symbolism book by Ted Andrews, the praying mantis is also associated with the "power of stillness,"[2] and Andrews suggests it's a directive for people with this totem to go into the stillness and open up to prophecy. This connection between the praying mantis and prophecy dates back to the ancient Greeks. They called this insect Prophetess, seer, or the one who sees the future. Named after the priestess daughter of the Theban sage Teireias, Mante (whose name means Prophetess or One Inspired by the Mood), the Greeks took its huge eyes and attentive air as evidence of its soothsayer abilities.

They believed that when the praying mantis saw a traveler it could discern the traveler's destination and what dangers awaited him or her on the road. Better yet, it could show them with a clear gesture in which direction to proceed to avoid the trouble.

In early Christian symbology, which adopted the popular belief in the praying mantis's power of divination and its willingness to use this power on behalf of travelers, this insect was an image of the "Good Guide."[3] As the guide of souls, the praying mantis was a symbol of Christ, who provides advice and helps those who have strayed from the path of righteousness. Still another ancient idea, this one originating from the observation of the praying mantis' ability to blend into its surroundings, sees in the mantis an example of how our thinking can adapt itself to and reflect the ideas of those around us.

It may have been its reverent appearance in conjunction with its killing power that led to the ancient Egyptians' regard for this insect, which is found throughout the Nile Valley. Egyptologists report that Mantis acted as a psychopomp in place of the wolf- or jackal-god Ap-Uat, conducting the dead through all the gates that must be passed. A drawing of a praying mantis is associated with a funerary rite of revival called the Opening of the Mouth Ceremony, and the carefully wrapped mummy of a praying mantis was found in a small mud coffin at Deir el Medina and in the Tomb of Seti I in the Valley of the Kings.

The Judgement Against the Mantis. As noted in previous chapters, many insects suffered when authorities of the Judeo-Christian tradition began to view their habits under the microscopic lens of a language and an analysis removed from nature's cycles and the regenerative female power of life. Whereas the Egyptians, Greeks, and early Christians honored the praying mantis, the Christian tradition from which our contemporary version stems did not. In Charbonneau-Lassay's *Bestiary of Christ*, for instance, the habits of this insect were described and judged as cruel.

What was deemed cruel was not just the fact that the praying mantis feeds only on living things, but also the fact that the female frequently devours the male, starting with his head, during their sexual union. Unable to reconcile this behavior with the respect other cultures showed for the praying mantis, Charbonneau-Lassay concludes, "This cruelty, which is common to a number of insects, was probably unknown to the Greeks."[4]

Those who believe that the praying mantis is cruel often point to the Asmat tribe of Irian Jaya, New Guinea, for whom the praying mantis was a totem. The Asmats practiced head hunting, a practice related to their fertility and initiation rites. Taking heads was intimately connected to the tribe's belief that a person's vital energy was concentrated in the head, particularly in the brain. When they killed their enemies, it was important then to confiscate their heads, extract their brains, and eat them. It was the only way to render their enemies harmless and assume their strength. In keeping with their beliefs in the concentrated power within brains and skulls, they also saved the skulls of friends and used them as headrests. Calling them "father," they believed that these headrests protected them when they slept at night.[5]

Cannibalism. Although we have already touched upon cannibalism in the arthropod world, more needs to be said to shift its power over our imagination to a new track. The fact that the female praying mantis often begins to eat the male while they are still engaged in intercourse is an act that in the past, according to various accounts, has horrified many unsuspecting entomologists new to the study of these insects. And it still causes nervous laughter today. For all our lack of knowledge about insects, this is one fact that, once we've heard it, we never forget because it reinforces our misgivings about insects.

In the mind and art of the Surrealists, the praying mantis was a central symbol—primarily because of its cannibalistic mating ritual and humanoid form. These artists, whose thinking was shaped by Freud's theory on repressed sexuality, saw the female's behavior as an appropriate image for the potential of erotic violence. Although other insects and arachnids also engage in cannibalism, the Surrealists were greatly influenced by the description of the mantis' habits by entomologist J. H. Fabre, who said that the mantid's "sanctimonious airs are a mask for Satanic habits.... "[6]

Salvador Dali, one of the most well-known Surrealists, associated the praying mantis with the most frightening aspects of sex and eating—two activities that, as noted in biographical writings about him, were said to be a reflection of his fear of castration by women. Dali also loathed the grasshopper, a relative of the praying mantis, and associated it with his fear of being eaten by his father.

As mentioned when we looked at the black widow spider, entomologists suspect cannibalism may be a propagation strategy for some males in some species. It also boosts energy reserves. The cannibalism of siblings is a common behavior among insects and is practiced not only by newly hatched praying mantises, but also by our beloved ladybugs. In a ladybug clutch of fifteen-plus eggs, for instance, most eggs hatch over a period of two or three hours. Any eggs not hatched within this time period are eaten by siblings who emerged earlier. Five to ten percent of all ladybugs die this way—or we could think of it as living through their siblings. The ladybug larva that eats its egg-bound sibling greatly increases its chance of living long enough to catch its first aphid—a challenging endeavor at this vulnerable stage when aphids are larger than they are.

Theories About Cannibalism. Praying mantis specialists are still debating why cannibalism occurs during copulation. Some biologists have proposed that the male mantis needs to lose his head to copulate successfully—that is, its brain, concerned with surviving, inhibits the mating instinct in some unknown but important way. Others disagree. A few evolutionary biologists have proposed that the male mantis, like the male redback spider, goes willingly to his death to provide his partner with enough protein, calories, and nutrients to help her produce more offspring. His reward? Babies that carry his genes.

Still others have proposed that it is a "nonphenomenon," an artifact of laboratory conditions. Two proponents of this view, W. Jack David and Eckhardt Liske, claimed that in their studies males almost never died from attacks by females. But E. S. Ross, entomologist and insect photographer extraordinaire, disproved that hypothesis when he captured on film a female praying mantis in the field devouring her copulatory partner. More recent studies also confirm that the phenomenon exists in nature, but the extent of cannibalism (occurring about one-third of the time) has been exaggerated.

A Symbolic Interpretation. Switching to a symbolic lens, it is easy to see, for one thing, why the praying mantis was a symbol for ancient female power. The female insects are larger and stronger than the males and wield, like all females, the power of creating life. And this power of size and creative capacity may be one of the reasons our male-dominated society is so troubled by the behavior. Yet, if we only draw those correspondences that reflect our confusion and pet theories about sexuality, we diminish the praying mantis and reduce its behavior to an abnormal and bizarre act.

Taking an imaginative approach, perhaps the exuberance of the redback spider as he flips himself back into his mate's jaws holds the key to explaining the behavior in all other species in which the male is devoured while mating. Goethe says in "The Holy Longing," "I praise what is truly alive, what longs to be burned to death." Maybe these males are "truly alive" and on the deepest levels of being accept the necessity of dying in order to be transformed and reborn. By taking the act out of the narrow framework that reduces it to something foolhardy, the willing surrender of the males can be viewed as a great teaching about how everything is renewed. To be eaten means to be assimilated, to undergo transformation by death. As Goethe concludes in the same poem,

> And so long as you haven't experienced
> this: to die and so to grow,
> you are only a troubled guest on the dark earth.[7]

So while completing his final task, it is the fortunate male who returns to the feminine ground from which all things come. It is the ending of one phase and the beginning of another occurring in one great moment of union. By rejoining the female, both male and female become whole, and from this point of unity they are reborn and live again in their young, and so the cycle begins again.

And perhaps Nature, in her wisdom, has wired in certain circuitry that makes the act not only a good propagation strategy, but a peak experience as well. Some studies suggest that the brain releases chemicals under life-threatening conditions that block pain and produce a euphoric feeling. Esoteric sources say that all so-called violent acts of dying are not what they appear to be. There is an understanding, what Barry Lopez calls a "conversation of death," and once it occurs, as the strike to disable and kill is made, the consciousness of the one dying leaves the body immediately. What we observe then in the struggle of the body is not conscious agony but reflex.

By moving past our adolescent fears of sexuality and our fear of being overcome, we stop being "troubled guests on the dark earth" and open ourselves to the imaginative power of the mantis—a power illuminated in the mythology of the Bushmen. Through these people, who lived in intimate communion with Earth and the heavens, we may rediscover that in the praying mantis "those who were sacred have remained so. Holiness has not dissolved, it is a presence of bronze."

The Creative Pattern

For the Bushmen, the praying mantis is the "Spirit of Creation"[8] and a manifestation of their God come to Earth. In Bushman myths and legends, Mantis is the great hero-God and Trickster, both self-destroyer and self-renewer, complete unto himself. Although Mantis has supernatural powers, he is also very human and often gets himself into trouble by the tricks he plays on others. But the trickery of Mantis is not malicious. He is continually trying out different ways of behaving, learning from his mistakes, and trying again. And most of all, he is trying to learn about himself.

One creation story says Mantis was present at the very beginning of the world, and that the sacred honeymaker Bee carried him over the dark windy waters that covered the new Earth. Since honey was an image of wisdom for the Bushmen, it was fitting that its maker carry their God to his new home. After hours of flying, though, Bee became tired, and its wings were stiff with the cold of the endless night wind. As Mantis became heavier, Bee looked for a bit of solid ground where he could set Mantis down, but all was covered by water. Bee flew more and more slowly and gradually descended. Finally skimming over the water and fighting an overwhelming fatigue, he saw a great white flower floating on the surface of the water. The flower was half opened as though waiting for the morning sun. Bee laid Mantis to rest in the heart of the white flower and planted within him the seed of the first human being. Then Bee died, and Mantis came alive in the morning sun and created the first Bushman at the same spot.

The exploits of Mantis (also called Xhui, Cagn, Kaggen, and Kaang) form the body of Bushman mythology. Other African tribes who revered the praying mantis borrowed from the Bushman culture and from each other's cultures. For instance, the Hottentots' praying mantis god was really the Bushmen's spirit of creation, and in Swahili this insect is called *kukuwazuka* or "fowl of the ghosts."[9] In the mythology of the Thonga, the praying mantis was an emissary of ancestor gods, and it also appears in the religious doctrine of the Zulus. But it is in Bushman mythology that Mantis expresses his full creative power.

The Bushmen believe that Mantis created all things, giving them their names and colors, including the all-important eland, an African antelope which was their primary prey animal. Mantis also protects people from illness and danger, sends precious rain, and determines whether or not a hunt will be successful. Prior to setting out on a hunting expedition, the Bushmen consult Mantis as an oracle, and his response determines their action in the hunt.

The stories about Mantis refer to an ancient time when many of the animals, birds, and insects were people before they became animals. Mantis accomplished many miraculous feats through these animal-people, although he also had his own kind of magic. For example, Mantis could bring dead people and animals back to life again, and when confronted by danger, he often grew wings and flew away—usually to the water, which symbolizes a place of renewal and new beginnings. Mantis was also a great dreamer. Just before disaster struck, Mantis would have a dream, and the dream would tell him what to do to avert the disaster. Other stories tell of how he transforms small and unobtrusive things into something radiant—a significant aspect of the creative pattern he embodied. For example, he used an ostrich feather to create the moon and turned ashes into the Milky Way.

Teachings of the First People

Much of what we know about the myths and stories of the Bushmen is a result of the dedication and work of the late Laurens van der Post, who was born in the heart of their ancient land. As a child, he was taught by his family's half-caste Bushman servants to pray to Mantis. Little did he know that this exposure would work its way into him so deeply that as an adult, when he saw the Bushman tribes vanishing, persecuted by other African tribes and those who had colonized the area, he would go back to live with them and record their stories and myths.

Laurens van der Post thought it significant that even though the Bushmen resided in a land that contains the greatest diversity of animal and insect life, these people did not choose a bigger, more powerful animal like the elephant, lion, or water buffalo to be their God. They chose an insect. Van der Post believed they selected an insect because their instinct was not deceived by size and appearance. They chose Mantis as the image of their greatest value because no other animal could represent the creative pattern inherent in all life as well as Mantis. In Mantis is a blending of the full range of human and divine characteristics. Mantis is a creator come to Earth to experience all that he can, just as many spiritual paths teach that we are creators, here on Earth to experience all we can and learn how to manage our energy wisely.

The Bushmen are as close to being the first people or race on Earth as we will ever see. As root people, they epitomize what it means to be "primitive"—living without the sophisticated cultures that other indigenous people developed. The Bushmen's imaginative powers and intuition, as well as their physical senses, emerged solely from their mystical participation in a sentient universe. Why their images are important to us today has everything to do with how they mirror the archetypal wilderness or primitive self that we have banished to the unconscious in our own technologically sophisticated, but image-starved consciousness.

Bushman Myth and Our Unconscious Selves. It was van der Post's belief that the long procession of animals, birds, reptiles, and insects in the Bushman's imagination and society has contemporary psychological equivalents. In other words, every creature or element that plays a role in the Bushman's myths and stories has a specific symbolic meaning in our collective unconscious—that archetypal realm whose patterns organize our human experience and inform our dreams. Furthermore, as van der Post worked to decipher the "ancient, hieroglyphic code"[10] of these creations of the Bushman imagination, he uncovered a transcendent creative pattern, an archetypal foundation in spirit whose activation he believed was absolutely critical to our ability to make the right decisions at this time of crisis on the Earth.

The creative pattern of renewal that van der Post wrote about is essentially a religious pattern. Calling it a largely undefinable pattern, he risks defining it as a desire and capacity to create a new and greater expression of life. Contemporary healer Rachel Naomi Remen calls it life's turning toward its true nature or soul. As a universal pattern that makes us aware of our continually expanding inner range of impulses and potential, it also links us to our beginning as a conscious being on Earth and highlights the inadequacy of any static state of being. Laurens van der Post was firmly convinced that if we attune to this creative pattern inside ourselves, an expression of our wilderness self, we will have "the power to achieve a greater and more authoritative statement of life and personality."[11]

So attuning to the Mantis within, the oldest, most natural image of God, we tap the power of this underlying pattern and begin to restore our intuitive self to its rightful place in our consciousness. By doing so, we also embrace what is despised and rejected in ourselves and in our society and, as Mantis teaches, allow this underpinning of spirit to be transferred into an inner source of great creativity and radiance.

Visits by Mantis. Those new to the idea of "following" Mantis will be interested in knowing that Mantis still appears in our dreams and in synchronistic visits. Interspecies communicator Sharon Callahan, whom we met in the bee chapter, remembers feeling the presence of God when she first encountered a praying mantis. She was six years old at the time and playing outside when she saw her cat corner something on the patio. Drawn to investigate, she bent down until she was suddenly face to face with a large insect. Although at the time she didn't know that the insect was a praying mantis, its presence captivated her. While waving its "arms" as though it were trying to box with the cat, the insect kept its large eyes fixed on her with an expression, recalls Callahan, that was so much like a person, so conscious, that she was astonished. She followed an impulse to rescue it from a confrontation that even the insect seemed to realize would result in its death. She picked it up gently and lifted it to the highest willow branch that she could reach. It didn't move off her hand right away, however, but instead kept its eyes on her face. It was a revelatory moment, although Callahan would not be able to put it into words until many years later. As an adult she describes the encounter as an epiphany: "The I of me, and the I of the creature became one and we rested on the breath of God." And throughout the years at pivotal times in her life, she notes that a praying mantis appears, sometimes in person, other times in a dream or even in an object of art, but always with the "shiny conscious eye of that first encounter—God looking at me through the eye of the Mantis."[12]

Besides special visitations like the one Callahan experienced Mantis is also known to visit those engaged in learning about the Bushman people and their insect God. Many of these visitations occur while the individual is engaged in reading the book *A Mantis Carol.* This quiet, yet startling, story of a series of synchronistic events begins with a dream of a praying mantis and ends with van der Post's journey from England to the United States, where he meets a woman who had known an exiled Bushman, Hans Taaibosch, living out his life in New York City.

I first read *A Mantis Carol* in 1984. I remember the time not only because the book is a remarkable and moving tale, but also because a praying mantis showed up on my office door when I was halfway through the book. A vision of miniature perfection, its elegant posture and calm presence surprised and delighted me. I hadn't seen a praying mantis since I was a child growing up in Michigan. Seeing one at that moment, after I had just learned that the praying mantis was the Bushman's God, spun me out of my normal preoccupations and filled me with the wonder that accompanies improbable events.

A Quest for Meaning

A Mantis Carol begins with a letter written by Martha Jaeger to van der Post. Jaeger was a practicing psychologist in New York City who had had recurring dreams of a praying mantis over many years. When the insect dreams started occurring with greater frequency, she renewed her search to discover what the praying mantis signified. She read about the insect and consulted with her colleagues, but no one knew what it meant. It seemed as though its presence in her dreams was haphazard and meaningless, an intrusion in an otherwise orderly and understandable array of dreams, and so it disturbed her greatly.

As a psychologist, Jaeger believed that everything that emerges from the darkness of the human psyche has a meaning that wants acknowledgment and expression from the dreamer. Yet, nothing in her past or present experience seemed connected to this insect dream image. It was as though it had emerged out of an unknown part of her psyche, a dimension hidden from her open and clear desire to explore this inner realm in herself and others. Since she had no connection to the physical insect, its unfamiliar and reoccurring presence in her dreaming seemed subversive—an image that undermined her understanding of dreams and her work as a dream analyst.

Then someone sent her one of van der Post's books about the Bushmen and their god. She read it eagerly and then flew to England where van der Post was taking part in a symposium, so she could ask for his help.

Van der Post shared with her what he knew about the role of the praying mantis in the life and imagination of the Bushmen and how he saw the role of the praying mantis in a healthy society. Her dream confirmed for him what he had already suspected: This particular pattern of the imagination was critical to understanding the nature of human imagination everywhere.

He shared with Jaeger his conviction that the Bushman's conscious mind corresponds in some important way to our dreaming self and that unfamiliar images—like her dream image of the praying mantis—surface unbidden from our unconscious, trying to inform us of an "unknown and profoundly rejected self."[13] It was this instinctive wilderness self that van der Post saw despised and relegated to the bottom rung of the contemporary mind's hierarchy of value.

And it was the resurrection of this "first" self that he believed was absolutely vital if we were to heal our fractured spirits. Restoring it to its rightful place then was the only way to make ourselves whole again and return to our native place in the community of life. In turn, returning home would end the profound sense of isolation and lack of purpose and meaning that so many in our society report feeling.

Directed by Mantis. All this he shared with Jaeger, and it made sense to her. What's more, it transformed her dream pattern, which had seemed so meaningless, into a source of revelation—her natural self was calling her, demanding restoration. On her return home to North America, she had another dream of a praying mantis that confirmed the change she felt inside. Van der Post shares the dream in *A Mantis Carol:*

> She was walking barefoot through the green-gold grass of the evening of the last day of a great summer. She looked down in the yellow light of a setting sun and there was Mantis, sitting firmly and happily in position in the middle of her right foot where her toes joined it.[14]

She awoke filled with "an indescribable sense of well-being and a re-belonging to life, happier than she had ever felt before."[15]

She could decipher this dream because van der Post had told her that the powerful and gentle eland, the Bushmen's primary prey animal, was dearest to Mantis's heart and that Mantis was thought to sit between the eland's toes. She also knew why it was an appropriate seat of command because van der Post had explained something about the way the feet of the eland were designed to walk and run on desert sand. The toes on the eland's feet separate as each foot comes down on the sand to prevent them from sinking too deeply. Then the toes snap together with a sharp click that sounds like the snap of an electrical current when the antelope lifts its feet. It is in this position—at the source of this electricity—that Mantis sat, as if to say, "The way of this animal is my way."[16] Van der Post thought it was like a first commandment to the Bushman spirit, directing them to follow all that the eland evoked in their minds and hearts in order to follow their god.

For Jaeger the dream meant that she had restored her natural self to its rightful position of command. She thus decided that nothing could be better than to be directed in your life by one of the purest, most natural images of what people called God, or as van der Post had once described Mantis to her, "the voice of the infinite in the small." She knew that a part of herself that had been arrested for many years was on its way again, expanding and informing her and her dreams of praying mantises ceased.

When she returned home, she wrote van der Post and asked him to come to the United States and talk to her and her friends about the Bushmen and all the manifestations of their natural imagination that had made such a great difference to her life and spirit. He agreed, arriving in New York in the early fall with a full schedule of lectures to give throughout North America.

One of his first stops was Houston, Texas. His hostess picked him up at the airport and drove him to her home. As van der Post walked up the front steps, on the door "in an attitude of profound contemplation, as if waiting for a temple door to open, sat a large praying mantis."[17] His hostess said that although she and her husband had lived there for many years, she had never seen a praying mantis before. Van der Post said, "If I had had doubts, they were gone. I was traveling under Mantis' auspices."[18]

In New York City van der Post talked with many people, but it was in conversation with one particular woman who attended his lectures that he discovered the existence of Hans Taaibosch, an exiled Bushman who retained his remarkable vitality and enthusiasm for life while living in New York City—strong evidence of the resilience of the natural spirit.

Under the Auspices of Mantis

More common than any of us realize are the mantis-related synchronicities that occur to persons who are learning about the Bushmen and the role of Mantis in their mythology. It is as though the insect comes to accentuate the importance of the teaching or to inform us that we are under its guardianship. For example, while the late mythologist Joseph Campbell was at his New York City high-rise apartment reading about the role of Mantis as the hero/god in Bushman mythology, he felt a sudden impulse to open a nearby window that faced Sixth Avenue. When he did, he looked out to the right, and there was a praying mantis walking up the building and onto the rim of his window. Campbell said it was large, even for a mantis, and as Campbell studied the insect's face he could see its resemblance to the face of a Bushman.

Several years ago *The Sun* magazine invited its readers to share their favorite insect stories. A woman from Northern California wrote that while working on a difficult master's thesis outside at a table beneath an orange tree, she felt that she was being watched. She looked up and saw a praying mantis on one of the tree's low-hanging branches. The next day another one came to the table. One by one they came, until six different species surrounded her. She and her husband searched the backyard, wondering if they had been overrun by the insects, but they only appeared around the table where she was working. Over the next five weeks as she worked on her thesis, the praying mantises kept her company, two or three on the table, five or six in the tree or adjacent greenery.

She had recently given her husband *A Mantis Carol* as a birthday gift. While she was outside surrounded by them, he was inside reading about the Bushmen and their god. The day after she finished her thesis, all the praying mantises disappeared.

A friend of mine from Wisconsin, who was raised as a Catholic but found herself interested in other spiritual paths, went to Pendle Hill, a Quaker training center just outside Philadelphia, for a retreat. This was the same Pendle Hill that van der Post spoke at during his visit to the United States, because Martha Jaeger, who was chairman of the Conference on Religion and Psychology of the Quaker Society of Friends, had arranged the event. My friend had recently heard about the book *A Mantis Carol*, but she hadn't read it before she arrived. Guests of this center share the chores of meal preparation and cleanup, and she was wiping down the table when she noticed on one of the chairs what she thought was a green bean. Closer inspection revealed the green bean to be a large praying mantis. She called to another woman in the dining area, "There's a praying mantis here!" The other woman came over with a woven cloth in her hand, and with the greatest reverence urged the insect onto it. Then she took it ceremoniously outside. When my friend returned home, she read *A Mantis Carol* and felt it was a confirmation of all she had experienced during her retreat. Her experience at Pendle Hill was a turning point in her life.

Following Mantis

The pattern of renewal so evident in primitive tribes and in the dreams of every individual in every culture shows over and over again that the renewal of life and our source of renewal comes out of what is small and humble. Psychologists agree, telling us to look for renewal in what we despise, reject, and ridicule. Both ancient wisdom and contemporary paradigms of self and the world testify to the power and hidden radiance inherent in the vast kingdom of the insects and imply that revisioning our relationship to them will bring about a renewal of spirit beyond our wildest imaginings. Like the gift hidden in the midst of the Shadow, it seems that what we need most, individually and culturally, is this aspect of self and of the living Earth that our contemporary minds have rejected.

Living our lives on automatic (ironically, in the same mechanical way we accuse the insects of living), many of us report having lost the animating aspect, the soul if you will, that gives our lives their continuity and meaning. Gone too is the feeling of being known by the sentient world—as the first people, and the indigenous tribes that followed them, felt known by the plants, the animals, and the stars.

Feeling known by the Earth community would soon dispel our feelings of isolation. It would even make us healthier, for the sense of being isolated has been demonstrated in clinical tests to be a mortality risk factor. Until we can actually experience it, we must have faith that we are not alone. As the beetle in Daniel Quinn's dream said (a dream shared in the preface), the other nations of consciousness, the other members of our community, live just off the sidewalk. They hear our cries for help and answer in myriad ways. We are the ones who must remember how to listen.

Daniel Quinn's beetle told him that he was needed in the forest off the main path and that the other species had a secret to reveal to him. Quinn thinks the secret the beetle wanted to share was that they are part of us and we of them.

We are not strangers on Earth, but rather have grown out of life like the mosquito and the butterfly. In fact, it is because we belong to the community of life that we are so desperately needed. Without our participation, the community cannot be whole anymore than we can. By removing ourselves, we have fractured the spirit of the Earth community as well as our own spirits.

The process of retrieving our wilderness self and restoring it to its rightful place in our psyche is more than a first step toward healing the loss of soul that we have come to believe is normal. This personal and cultural reclamation project has power beyond what we understand. Recognizing and undertaking the task means we have positioned our will in full accord with the pattern of renewal inherent in all life. When that pattern is invoked by our intention, it will call in all manner of help to assist in our transformation and movement into greater wholeness. And when we inhabit our true natures, our outer lives will also become more true to what is most genuine and unique in each of us.

An important part of the restoration work involves regaining the capacity to see the divinity in all modes of being—whether in a distressing guise or not—and crying out in recognition.

A Rumi poem says,

> You should try to hear the name the Holy One has for things.
> We name everything by the number of legs it has;
> The other one names it according to what they are inside.[19]

To find out who they are inside and hear the name the "other one" has given them takes time and perhaps even regular conversations with bugs. Francis Ford Coppola, the flamboyant American filmmaker who is known for the *Godfather* trilogy, is taking time now at age fifty-eight to talk to insects. One evening as he and Fred Ferretti of *Gourmet* magazine ate dinner on the veranda of Coppola's Napa Valley farmhouse, a mosquito landed on a spot of spaghetti sauce on the tablecloth. Coppola noticed it but continued to eat quietly, letting the insect take its fill. Later over dessert Coppola told Ferretti, "We don't kill the bugs here. We kill nothing here.... I talk to the bugs, the grasshoppers, the spiders." When Ferretti asks, "About what?" Coppola replies, "I am always trying to understand the scope of our existence, knowing that we are allowed to see only a piece of it. That's what I want to write about and to film. And I will."[20]

Coppola is intuitively following Mantis. He has turned toward the insects to learn how to see the macrocosm in the microcosm and free his imagination to find the correspondences that exist between human beings and the natural world.

He is not alone. Many are turning now, answering an inner call, and all must eventually undertake the inner and outer journey to the bug-infested worlds that surround us. We must participate in their powers and try to sense the larger patterns of communication that sustain all life streams. We must also welcome, if not seek, personal contact with the small ones, understanding that new meaning must be lived and grounded in our physical life before it can be truly known.

The time has come for humans and the insects to turn toward each other, as Thomas Berry wrote in the introduction. We need to understand "the language of the sting" and cross the vast distance between false power and the power at the heart of creation where the insects await our return. The voice of the infinite, carried by butterfly and bee, beetle and fly, cockroach and spider, and countless others, bids us to remember their true names and open ourselves to their teachings.

What insect do you rejoice in where you come from?

Postscript

When I first sent copies of this manuscript to publishers for their consideration, a large praying mantis appeared over my front door. As I lifted my hand toward him, he climbed down the door and onto my hand. I photographed him, aware of the significance of his visit that morning and knowing that the book was traveling under Mantis's auspices. His picture is on the cover.

Joanne E. Lauck
and Shanti

As an environmental educator, professional writer, former counselor, and certified wildlife rehabilitator, Joanne Lauck brings an eclectic group of skills to her work. She received her Bachelor of Science degree in psychology from Grand Valley State University (Grand Rapids, Michigan) and received a Masters of Science degree in experimental psychology from Eastern Michigan University with continuing courses in transpersonal psychology from the Institute of Transpersonal Counseling in Menlo Park, California. She has a California teaching credential in psychology.

Since 1984, she has collected and explored animal and insect dreams, using the material to write articles that have subsequently been published in a variety of publications including Susan McElroy's book *Animals As Guides for the Soul: Stories of Life Changing Encounters* (Ballantine 1998), *Dream Network Journal*, *la Joie* magazine, *Earthlight*, *Tracks*, *Native Rescue Magazine*, *The Opossum Newsletter*, *The Monarch Newsletter*, and *It's a Wild, Wild Life*.

Ms. Lauck has presented at numerous conferences including the 8th International Conference of Shamanism and Alternative Modes of Healing in San Rafael, California, at the 6th International "Animals 'n Us" conference in Montreal, Quebec, and at Whole Life Expositions. In addition she has developed and implemented an elective course for grade school children in a San Jose Unified Public school called "Thinking Like a Bug," which is now four years old.

She resides in San Jose, California, and when she is not working with her brother in their graphic design studio, she spends her time teaching, lecturing, and writing about the human-animal bond.

Endnotes

Preface

1. Quinn, p. 17.
2. Nelson, p. 52.
3. van der Post, 1975, p. 24.

PART I: REDRAWING THE CIRCLE

Chapter 1: Coming Home

1. Carroll, p. 43.
2. Rood, 1971, p. 3.
3. Millman, p. 184.
4. Disch, p. 175.
5. Ibid., p. 184.
6. "Out of the Night."
7. Ibid.
8. Graeber, p. 36.
9. People vs. Pests.
10. de Waal, p. 8.
11. Welcome to Cockroach World.
12. Ibid.
13. Pringle, p. 1.
14. McLaughlin.
15. Berry, p. 1.
16. Hillman, 1988, p. 44.

Chapter 2: Clearing the Lens

1. Whitmont, p. 162.
2. Carson, 1962, p. 12.
3. Ibid.
4. Ibid. Introduction by Vice President Al Gore, 1994, p. xxii.
5. Keen, p. 23.
6. Lévy-Brühl, p. 24.
7. Keller, p. 206.
8. Wilson, 1991, p. 8.
9. Carson, 1956, p. 45.
10. Kellert, 1993, p. 4.
11. Swan, 1993, p. 182.
12. van der Post, pp. 6-7.

Chapter 3: Mirrors of Identity

1. Eckhart, p. 14.
2. van der Post, 1975, p. 24.
3. Margulis, p. 15.
4. Suzuki, p. 109.
5. Halifax, 1982, p. 78.
6. Stevens, p. 34.
7. Webster, p. 5.
8. Sonam, pp. 7-9.
9. Blassingame, p. 51.
10. Ibid. p. 56
11. Graham, p. 7.
12. Snell, p. 68.
13. Keen, p. 11.
14. Fleming, p. 79.
15. Kelly, 1983, p. 13.

PART II: REDEEMING PEST SPECIES

Chapter 4: My Lord Who Hums

1. Hubbell, p. 28.
2. Kennedy, p. 74.
3. Ibid.
4. Oldroyd, p. 75.
5. Lee.
6. Shuttlesworth, p. 11.
7. Called campaign diseases, typhoid and dysentery were common among troops in active service and, until recently, caused more deaths than wounds. By the end of the nineteenth century, again as overall sanitation improved, they were largely eliminated in Europe and North America but still claim lives in countries where inadequate sanitation practices and overcrowding permit the parasite to be transmitted easily.

   Nearly all recent outbreaks of typhoid in developed countries have been traced to immigrants, refugees, or returning visitors from infected places—not flies. Sometimes the microorganism responsible for the symptoms we call typhoid is carried in the bowels of a healthy person. If such a carrier is employed in the kitchen of a large institution, he or she may spread the disease. Public health officials make tracking and immunizing human carriers an ongoing priority.
8. Most doctors agree that dirty hands are a common means of transmitting illnesses including diarrhea and other intestinal problems. Findings released at a recent infectious-disease conference held by the American Society for

Microbiology showed that out of a sample of 6,300 men and women using public toilets in five cities only74% of women washed their hands afterward and only 61% of men (although a phone survey of over a thousand adults indicated that 94% say they always wash after using the toilet).

9. Margulis, p. 15.
10. Keen, p. 97.
11. Ibid.

Chapter 5: A Compassionate Response to Flies

1. Brussat, p. 204.
2. Butterfield, pp. 20-25.
3. Fox, 1995, p. 72.
4. Morgan, p. 70.
5. Rilke, pp. 121-122.
6. Perera, p. 13.
7. Dossey.
8. Ackerly, p. 626.
9. Ibid.

Chapter 6: Divine Genius

1. Hubbell, p. 158.
2. "Good Riddance to Roaches."
3. Pringle, 1971. p. 14.
4. Ibid.
5. Kerby, p. 11.
6. Ibid. p. 31
7. Ibid.
8. Gordon, p. 88.
9. Frishman.
10. Fox, 1996, p. 24.
11. Pest, p. 27A.
12. Gordon, p. 85.
13. Swan, 1992. p. 115.
14. Ibid.
15. Ibid.
16. Spangler, pp 70-73.
17. Weber, p. 23.
18. Entomologist Gilbert Waldbauer reports in his recent book *Insects through the Seasons* that in cockroaches, when a piece of paper that has been walked on by virgin female cockroaches is put in the cage of celibate male cockroaches, they go into a sexual frenzy. And different pheromones facilitate other behavior or physiological functions involved with sex and reproduction. For instance, some male cockroaches exude an aphrodisiac pheromone on their bodies called seducin that the females nibble on during copulation.
19. Jordan, p. 125.
20. Kerby, p.31.
21. Cooper, Gail, p. 172.
22. Gordon, p. 165.
23. Cooper, Gail, p. 175.
24. Hall, p. 41.
25. Ritchie, pp. 53-56.

Chapter 7: The War on Sawgimay

1. Nollman, p. 63.
2. Waldbauer, p. 202.
3. White, p. 80.
4. Ibid.
5. Goldman, p. 169.
6. Conrad, p. 1B.
7. Ibid. p. 4B.
8. Parker, p. 2A.
9. Harrison, p. 5.
10. An account from that time period makes no attempt to disguise its racist underpinnings. "With weapons in hand to destroy the lower forms of life that made men ill, Europeans could move into the tropical lands and supplant the lower forms of human life who were then in possession. The battle against disease and the battle for civilization were demonstrably one." (Harrison p. 4).

    Believing the great difference in susceptibility was due to racial differences, the conquerors defined their mission to overtake other lands in heroic terms. The self-deception went something like this: The White race would eliminate tropical diseases which they perceived as the cohorts of "barbarism." Since diseases like malaria "permitted backward, slothful races of man to ride best what were obviously the richest lands on earth, and so letting their riches go to waste,"

(Harrison p. 5) the Caucasian race would change all that.

12. Genetically, the Black race is relatively resistant to one form of malaria, a fact demonstrated among Black Americans. The sickling trait, an inherited abnormality of hemoglobin (the "working" constituent of red blood cells) also produces an inherited immunity to a malignant form of malaria indigenous to sub-Saharan Africa. Children with a sickle cell gene from both parents are likely to develop severe and often fatal anemia, but those with genes from only one parent have a mixture of normal and abnormal hemoglobin that usually doesn't affect them and gives them resistance to the malignant malarial parasite. They can get malaria, but the parasite doesn't thrive. The child survives the attack to eventually become an adult protected by an acquired immunity. So in a small agricultural community, the sickling trait has allowed the survival of the group.

    Like the sickle cell gene, two other abnormal hemoglobins (hemoglobin C and E) also protect populations in Africa and Asia, and an enzyme defect connected to the red blood cell metabolis protects malaria patients from death while occasionally producing a mild medical condition called hemolytic anemia.
13. Lappé, 1994, p. 9.
14. Shuttlesworth, p. 29.
15. Nollman, 1987, pp. 112, 114.
16. Boyd, p. 133.

Chapter 8: Lords of the Sun

1. Evans, Arthur, p. 9.
2. Martin, p. 108.
3. Thoreau, pp. 439-440.
4. Albrecht, p. xxi.
5. Ibid., p. xx.
6. Calahan, in *Secrets of the Soil.*
7. Ausubel, p. 53.
8. Graham, p. 249.
9. Evans, Arthur, p. 151.
10. As an insider, entomologist Robert Van den Bosch could see the corruption in his profession first hand. In 1978 he wrote *The Pesticide Conspiracy*, a book whose thesis is that corruption lies everywhere in the pest-control field. In his book, he blasted The Entomological Society of America which he saw as having been bought by the chemical companies and wined and dined by the "pesticide Mafia."(Graham, p. 289) Other books since then continue to detail how chemical companies (and now biotechnology companies) have manipulated scientific studies, the EPA, and the public to keep harmful products on the market.
11. An increasing number of plant hybrids on the market have been genetically altered to withstand great amounts of herbicides. In fact, these hybrid plants will allow farmers to freely apply a herbicide to their fields so that the hybrid plants are the only green things left alive. Some of the problems inherent in herbicide-resistant plants—like exchanging this herbicide-resistant gene with their wild relatives—have already been identified. Advocates, however, are quick to assure farmers and the public that solving that problem is simply a matter of perfecting the technique. But this focus on finding a more successful technique does not address the real issue. What will these liberal doses of herbicides do, for example, to the already damaged soil and to other species that have not been genetically manipulated to survive the onslaught? And if farmers can triple the amount of herbicides used on crops because we've manipulated their genetic make-up, it means we will put even more chemicals into the environment. And these are not environmentally friendly or human friendly chemicals—some, like the bromoxynil have been linked to cancer and birth defects. The entire line

of research is an untenable development, and yet seeds altered in this way are already in the fields and the focus of fifty-seven percent of all biotechnology research. In fact, to counter opposition and gain acceptance of GE plants, the USDA is trying to implement national organic standards that allow genetic engineering and other ways of producing food long disavowed by organic gardeners and those who buy organic meat and produce.

Critics of GM crops also argue that these plants may have unanticipated allergens or toxins, or have their nutritional value reduced in unexpected ways. They point out that those who shuffle genes around also risk removing or inactivating substances presumed to be undesirable in food, but which may actually have an unknown but essential quality such as a natural cancer-inhibiting ability. Critics say danger is also present when modified virus genes are inserted into crop plants. Dr. Joseph Cummins, professor emeritus of genetics at the University of Western Ontario is one concerned scientist who warns "it has been shown in the laboratory that modified viruses could cause famine by destroying crops or cause human and animal diseases of tremendous power. ("Eminent Scientists Comment on the Dangers of Genetically Engineered Foods," http://www.geocities.com/Athens/1527/scientists.html.) Finally, added to the real risks of genetically altered plants touched on in this section, the check to insure quality in one giant agrochemical conglomerate have already failed.

12. Hillman, 1995, p. 39.
13. Evans, Arthur, p. 11.
14. Ausubel, 1997, p. 60.
15. An example of a new approach is seen in the work of soybean farmers who call themselves The Practical Farmers of Iowa. This group began a tillage system and planting techniques that have completely replaced a carcinogenic herbicide. And Jim Bender, author of *Future Harvest*, outlines how he turned his acres of corn and soybeans into a chemical-free farming operation by staggering planting times, using crop rotation and diversification, and reintegrating livestock.

Readers interested in exploring visionary solutions are encouraged to read Kenny Ausubel's *Restoring the Earth*, which fosters a sense of optimism that we will find a way to heal our soils, plants, and animals and grow healthy food. *Secrets of the Soil* by Tompkins and Bird also offers solutions—both innovative and nontraditional—for reversing serious agricultural problems. Read too Janine Benyus' *Biomimicry: Innovation Inspired by Nature*. Benyus introduces us to "biomimics"—people learning from (not about) nature, including those who are trying to grow food plants the way nature grows plants. In her chapter on agriculture, she recounts clearly and concisely the mounting problems with current practices and believes revamping the way we grow food is the most important idea in her book—and the most radical:

"[Following nature] agriculture in an area would take its cue from the vegetation that grew there before settlement. Using human foods planted in the patterns of natural plant communities, agriculture would imitate as closely as possible the structure and function of a mature natural ecosystem." (Janine M. Benyus. *Biomimicry: Innovation Inspired by Nature*. NY: William Morrow & Co., 1997, p. 13.)

And this approach is already in practice. Wes Jackson, a California university professor who returned to Kansas, his home state, and started a school that focused on sustainable living practices, shows how a prairie-based, chemical free form of agricul-

ture will benefit farmers and their ecosystems as well as the consumer.
16. Findhorn Community, pp 166-168.
17. Wright, in "Perelandra," *Secrets of the Soil*, p. 309.
18. Wright, 1983, p. 85.
19. Ibid., p. 86.
20. Ryan, p. 14.

Chapter 9: First Born

1. Walker, 1988, p. 414.
2. Nollman, 1990, pp. 181-182.
3. Ibid., p. 182.
4. Lake-Thom, p. 132.
5. Vitebsky, p. 68.
6. Ibid.
7. Helfman, p. 34.
8. Rudolph Steiner saw striking similarities between the ant heap and the human being—with every part (ant or human cell) communicating and cooperating with every other part. He also attributed much of the natural world's ability to renew itself to the production of formic acid—produced in nature especially by stinging insects. Probing nature and the cosmos with his spiritual vision, Steiner saw in formic acid and bee venom the chemicals essential in our process of incarnation, and, as it exists in nature, the physical basis for the Earth's soul to be able to unite with the physical earth. Those interested are encouraged to read his thought-provoking lectures on the subject (Rudolph Steiner. "Nine Lectures on Bees: Given to Workmen at the Goetheanum. Marna Pease and Carol Alexander Mier, trans., Lecture 9, Dornach, Switzerland, Dec. 1923, NY: Anthroposophic Press, 1947, pp. 92-103.)
9. Holldobler, p. 227.
10. Wilson, 1975, p. 549.
11. Goodwin, p. 182.
12. London.
13. Russell, p. 143.
14. Chui, pp. 1F-2F.
15. Hall, Rebecca, pp. 81-82.
16. Shapiro, p. 53.
17. Wilson, 1991, p. 4.

Chapter 10: Birds of the Muses

1. Alternative medicine practitioner Andrew Weiss likes to tell the story of a 64-year-old man suffering from rheumatoid arthritis since he was in his 30s, who was stung by a bumblebee. Within a few days after the sting, the swelling in his joints subsided and a few weeks later the chronic body-wide pain vanished as well.
2. Teale, 1940, p. 18.
3. van der Post, 1962, p. 9.
4. Aïvanhov, 1984, p. 110.
5. Adam, pp. 81-88.
6. Davis, 1978, p. 101.
7. Nahmad, p.45.
8. Steiger, pp. 115-116.
9. Gunther Hauk in a private correspondence, March, 1998.
10. Studies by the Thailand government are underway right now to investigate if honey (used in many of their herbal medicines) is contaminated when bees make it from pollen taken from plants with a foreign insecticide gene inserted into them.
11. Sharon Callahan in a private correspondence, February, 1998.
12. Boyd, pp. 114-115.
13. Fernandes, p. 127.
14. "Chiquinho of the Bees: The Boy Who Talks to Animals."
15. Hoy, p. 40.
16. Schul, p. 190.

Chapter 11: All Our Relations

1. "Explorations: The Ties that Bind," p. 1.
2. For further information about microbes and their role as the building blocks of life see Margulis.
3. While seed patenting began in the 1950s (and has its own battlefields with the advent of GM or genetically modified seeds), in the 1980s the field expanded to the patenting of other life-forms and natural products. Critical to its expansion was a U.S.

Supreme Court ruling that allowed a patent on an oil-eating bacterium developed in 1972 by Ananda Chakrabarty—even though Chakrabarty admitted that he hadn't invented anything, but had merely moved some genes around.The ruling opened Pandora's box. Bioprospectors are hunting for heat-resistant and other microbes all over the world, hoping to profit from them, and an increasing number of companies are filing for biotechnology patents.

4. Chief Seattle, p. 317.
5. Instead of making the companies prove their harmlessness, for instance, our regulatory agencies tell us we must prove that they are harmful. And while testing is being done (even when preliminary evidence of harm is reported) the company with the patented GM organism is allowed to continue marketing its product until there is definitive evidence of harm. Those alarmed by the number of GM plants and animals on the market without adequate testing believe that to disrupt the genetic blueprint of an organism by randomly integrating genetic material into its DNA from an unrelated species is a procedure with totally unpredictable consequences—and may even invite a major health crisis.
6. As a fuel additive, ethanol boosts the octane rating of gas and as a pure fuel it burns cleanly; when blended with gasoline, it increases overall volatility and causes increased evaporation of smog-causing chemicals.
7. Keen, p. 100.
8. Zweig, p. xx.
9. Preston, pp. 1E, 11E.
10. Bacteriophages are target specific and do not prompt general allergic reactions. The problem with them in the past is that they did not always work and so were discarded in favor of antibiotics. Although new studies have shown that they may be effective and doctors are trying them on staphylococcus, pneumococcus, or salmonella, so far they haven't shown much promise.
11. A virus at its core is pure information encoded in molecules of DNA (or RNA). Until the moment a virus enters a cell, it seems more dead than alive. It can not reproduce or move on its own. Once in a living cell, however, the virus's tightly bundled genes begin to unfold and drift through the cellular fluid, ordering the cell's reproductive machinery to clone it. Eventually thousands of these viral clones bloat the cell, slipping out through the cell's membrane or multiplying until the swollen cell ruptures. If unchecked, the viruses eventually destroy enough cells to kill the host.

    Using an electron microscope we can now see viruses and the intricately shaped molecules called antigens that project from their surface. The shapes of these projections are critical to a virus's ability to function. Each strain of virus has its own unique configuration of surface molecules that work like keys in a lock. Viruses gain entry into hosts and cause infections by precisely fitting the molecules on their surfaces to those on the membranes of targeted cells. The major defense of our bodies against all viral diseases is our immune system. When white blood cells known as helper T cells spot these antigens, they mobilize the body's defenses to attack the viruses.
12. The scientists who insist that viruses are not just threats explain that under certain circumstances, their abilities to target specific cells, slip inside, and carry new genetic information make them heroes. In the treatment of cystic fibrosis, for instance, which occurs when children are born with a defect in the gene that helps lung cells prevent mucus buildup, a replacement gene has been spliced into a cold virus. The idea is that when the patient

inhales the virus through a nasal spray, the virus will take the new gene into the lung cells. Although experimental at this point, the technique promises to help thousands of people who suffer from this fatal illness.

13. A DNA vaccine in the works relies on swapping plasmids (the tiny rings of DNA that let bacteria acquire and pass on useful genes). The scientists working on it think that a plasmid with the right genes stitched into it can be used to create a mock infection that in turn will trigger a real and surprisingly effective immune response (an operating premise of homeopathic remedies). Advocates assure critics that since the disease causing portion of the virus is deleted, the virus can't run wild and reproduce itself—although, for the record, certain well-established people in the medical profession think it might do just that. And some experiments conducted by Agriculture Canada (K. Kliener. *New Scientist*. August 16, 1997) show that the deleted characteristic re-appears one in eight times in the next generation.
14. John Webster, a parasitologist at Simon Fraser University in British Columbia, has studied soil-dwelling microscopic worms called nematodes for twenty years. Nematodes carry a kind of bacteria that can only live inside them—not in the soil—and, interestingly, can only reproduce inside insects. When the worm finds an insect, it burrows into it and releases these bacteria. The bacteria starts to feed on the insect and reproduce at a rapid rate. In turn, the nematode starts feeding on the multiplying bacteria, and then it too starts reproducing. During this frenzied feeding and reproducing activity, the bacteria produce chemicals that kill competing bacteria. These chemicals are being isolated, including xenorxide which appears to attack human lung, breast, prostate, and colon cancer cells, while apparently leaving healthy cells alone. They also appear to be highly effective against bacteria now immune to the penicillin family of antibiotics.
15. Marc Micozzi. head of the National Museum of Health and Medicine in Washington, D.C.
16. Myss, p. 35.
17. Goodman, p. 196.
18. Ibid., p. 199.
19. "Outbreak the Real Risks," p. 3F.
20. Keen, p. 99.
21. Those interested can find a detailed explanation in Myss' book *Anatomy of the Spirit*. (See note #22).
22. Myss, p.108
23. Boone, p. 115.
24. Ibid.
25. Ibid., p. 117.
26. Kinkead, p. 13.
27. Ibid., p. 15.

PART III: EIGHT-LEGGED PEOPLE

Chapter 12: Spinner of Fate

1. Tanizaki, p. 95.
2. "Breakthroughs: Of Sex, Somersaults, and Death," p. 34.
3. Lake-Thom, p. 13.
4. Weigle, p. 5.
5. Moffett, 1991, p. 45.
6. Abram, p. 19.
7. Estés, p. 91.
8. Time-Life Editors, pp. 66-67.
9. Swan, 1992, p. 132.
10. Russell, p. 143.
11. McIlroy.
12. Compton, p. 238.
13. Cox.
14. Ibid.
15. Ibid.

Chapter 13: Insect Initiators

1. Stokes, p. 6.
2. Johari, p.58.
3. Rilke, p. 121.
4. Ibid., pp. 121-122.
5. Cassettari, p. 3.

6. Matthews, Marti Lynn, p. 146.
7. "Journey Into Mystery."
8. Alcock, p. 72.
9. Angier, p. 97.
10. Saunders, p. 108.
11. Eliot, pp. 173-174.
12. Noble, 1982, p. 100.
13. Taylor, Ronald, p. 200.
14. van der Post, 1984, pp. 236-237.
15. Hall, Rebecca, p. 40.
16. Hearne, p. 68.
17. Ibid.
18. Ibid.
19. Kowalski, p. 109.

PART IV: REJOICING IN INSECTS

Chapter 14: Nation of Winged Peoples

1. Spears.
2. Riegner, p. 33.
3. In 1982 Norie Huddle got the idea that by looking more closely at the actual process of insect metamorphosis, that is, at what specifically happens inside a chrysalis, we might gain insight into our own process of transformation. After three years and conversations with three entomologists, Huddle understood the fundamentals of this miraculous change from caterpillar to winged creature. In 1988 the information and her original idea came together one memorable day, and she wrote the book *Butterfly* (or as she tells it, "it wrote itself in an hour and then asked to be illustrated.") Later that same day she met the artist Charlene Madland who then collaborated with her to create the book's beautiful illustrations. *Butterfly* (Huddle Books) is "the first Official Game Piece" in Huddle's *The Best Game on Earth*, a "consciously redesigned game of life" aimed at bringing about universal peace, health, prosperity and justice on Earth—the "Butterfly Era of Human Civilization." For more information contact Huddle by email at: nhuddle@intrepid.net or write her at: P.O. Box 444, Bakerton, WV 25410.
4. Houston, p. 119.
5. Charbonneau-Lassay, p. 346.
6. Guggenheim, p. 188.
7. Ibid.
8. Ibid., p. 189.
9. Fisher, p. 114.
10. Hall, Mitchell, p. 62.
11. Peck, p. 2.
12. Lee, John, p. 88.
13. Jean Collins, Personal Correspondence, November 1997.
14. Anderson, Sherry Ruth, p. 27.
15. Ibid.
16. Werner, 1965.
17. Brown, 1992, p. 40.
18. Smith, Penelope, 1993, p. 329.
19. "Guarding a Forest Giant," *San Francisco Examiner*, February 12, 1998, p. 1A, 20A.
20. For more information about the spiritual nature of butterflies and their connections to plants and nature spirits, see Rudolph Steiner's *Man as Symphony of the Creative Word* (Sussex: Rudolf Steiner Press, 1991) and Karl Konig's lectures on bio-dynamic agriculture presented in *Earth and Man* (Wyoming, RI: Bio-Dynamic Literature, 1982).

Chapter 15: Finding Our Affinities

1. McVay, p.1.
2. Hillman, P. 11.
3. Nabokov, pp. 23-24.
4. Hope, p. 58.
5. Matthews, Patrick, p. 30.
6. Evans, Arthur, p. 170.
7. Nixon, p. 31.
8. Eisner, p. 48.
9. Evans, Howard Ensign, p. 50.
10. Gould, pp. 12-20.
11. Eisner, p.46.
12. Elkins, p. 55.
13. Lovell, p. 7.
14. Blassingame, p. 51.
15. Ibid.
16. Ackerman, p. 83.
17. McVay, p. 11.
18. Benyus, p.4.

19. Laughlin, p. 5.
20. Holland, p. 9.
21. Gywnn Popovac. Personal correspondence, November, 1997.
22. Hope, p. 47.
23. Wilson, 1992, p. 7.
24. Houston, 1996.
25. Robert Garmston and Bruce Wellman ("Adaptive Schools in a Quantum Universe," *Educational Leadership*, April 1995, pp. 6-12) discuss the "new sciences" looking for insights into new approaches to school improvement by asking "Which Butterfly Wings Should Schools Be Blowing On?" (p. 8).
26. Hope, p. 15.
27. Ibid.
28. van der Post, 1974, p. 300.

Chapter 16: Strange Angels

1. Brown, 1992, p. 46.
2. Jung, p. 94.
3. Tanner, 1988, pp. 140-141.
4. Eagle, pp. 249-250.
5. Rothschild, p. 149.
6. Fabre, 1949, p. xiv.
7. Taylor, Jeremy, p. 2.
8. Gywnn Popovac in a private correspondence, November, 1997.
9. Longgood, 1985, p. 54.
10. Moss, p. 25.
11. Ibid.
12. Mails, p. 95.
13. Halifax, 1993, p. 179.
14. Noble, 1991, p. 117.
15. Ibid.
16. Hillman, 1988, p. 70.
17. Lauck, p. 26.
18. Marcia Lauck in a private correspondence, March 1997.
19. Lauck, p. 27.

Chapter 17: Following Mantis

1. Levertov, p. 43.
2. Andrews, p. 343.
3. Charbonneau-Lassay, p. 357.
4. Ibid., p. 356.
5. Cheneviere, p. 97.
6. Fabré, p. 143.
7. Goethe, p. 70.
8. van der Post, 1962, p. 9.
9. Berebaum, p. 317.
10. van der Post, 1962, p. 9.
11. van der Post, 1957, p. 21.
12. Sharon Callahan. Personal Correspondence, February 1998.
13. van der Post, 1975, p. 20.
14. Ibid., p. 22
15. Ibid.
16. van der Post, 1962, p. 21.
17. van der Post, 1975, p. 34.
18. Ibid.
19. Rumi, p. 268.
20. Ferretti, p. 63.

Bibliography

Aardema, Verna. *More Tales From The Story Hat*. NY: Coward-McCann Inc., 1966, pp. 7-11, 16-21.

Abram, David. *The Spell of the Sensuous: Perception and Language in a More-Than-Human World*. NY: Pantheon Books, 1996.

Ackerly, J. R. My *Father and Myself*. (1968, p. 174) in Keith Thomas, *Religion and the Decline of Magic: Studies in Popular Beliefs in Sixteenth- and Seventeenth-Century England* Hammondsworth: Penguin University Books, 1973.

Ackerman, Jennifer. "Parasites: Looking for a Free Lunch," *National Geographic*, October 1997, pp. 72-90.

Adam, Frank. "Quantum Honeybees," *Discover*, November 1981, pp. 81-88.

Adams, Jean, ed. *Insect Potpourri: Adventures in Entomology*, Gainesville, FL: The Sandhill Crane Press, 1992.

Aisling, Irwin. "Gene Crops May Lose Power to Kill Pests," *The Electronic Telegraph*, March 5, 1997. in http://www.geocities.com/Athens/1527/scientists.html.

Aïvanhov, Omraam Mikhaël. *The Key to the Problem of Existence.* Fréjus, France: Prosveta, 1985.

—. *Sexual force or the Winged Dragon.* Fréjus, France: Editions Prosveta, 1984.

Aizenstat, Stephen. "Jungian Psychology and the World Unconscious," *Ecopsychology: Restoring the Earth Healing the Mind*, Theodore Roszak, Mary E. Gomes, and Allen D. Kanner, eds., San Francisco: Sierra Club, 1995, pp. 92-100.

Albrecht, William A. in "Introduction," *Secrets of the Soil: New Age Solutions for Restoring Our Planet.* Peter Tompkins and Christopher Bird. NY: Harper & Row: 1989, p. xxi.

Alcock, John. in *A Desert Garden: Love and Death Among the Insects*. NY: W.W. Norton & Co., 1997.

Allen, Louis A. *Time Before Morning: Art and Myth of the Australian Aborigines*. NY: Thomas Y Crowell, 1975.

"Altered Arthropods: Powerful Weapons or Spineless Menaces?" *San Jose Mercury News*, January 2, 1996, p 1F, 2F.

"America's Dirty Secret," *San Jose Mercury News*, September 17, 1996, p. 6A.

Anderson, Curt. "EPA Backs 7 Plant Pesticides," *Associated Press*, Sept. 27, 1997.

Anderson, Sherry Ruth. *Noetic Science Review*, Autumn, 1995, pp. 25-27.

Andrews, Ted V. *ANIMAL SPEAK: The Spiritual & Magical Powers of Creatures Great & Small*. St. Paul, MN: Llewellyn publications, 1994.

"Antibiotic-resistant Bacteria Arrives in U.S.," *San Jose Mercury News*, August 22, 1997, p. 11A.

Angier, Natalie. *The Beauty of the Beastly: New Views of the Nature of Life*. Boston: Houghton Mifflin Company, 1995.

Arritt, Susan. "Antcologists Unearth Secrets of the Ancients," *New Mexico Magazine*, Feb. 1995, pp. 36-38.

Ausubel, Kenny. "Restorative Farming and the Real Green Revolution," in http://www.light-party.com/Environment/Restore.html, p. 1.

—. *Restoring the Earth: Visionary Solutions From the Bioneers.* Tiburon, CA: H.J. Kramer, 1997.

Barasch, Marc Ian. *The Healing Path: A Soul Approach to Illness*. NY: G.P Putnam's Sons, 1993.

Barber, Theodore Zenophon. *The Human Nature of Birds: A Scientific Discovery with Startling Implications.* NY: St. Martin's Press, 1993.

Bardens, Dennis. *Psychic Animals: A Fascinating Investigation of Paranormal Behavior*, NY: Henry Holt & Co., 1987.

Beals, Carleton. *Nomads and Empire Builders: Native Peoples and Cultures of South America*. NY: Chilton Company, 1961.

Beard, Paul. *Living On*. London: George Allen & Unwin, 1980.

Benson, E.F. "Caterpillars," in *100 Creepy Little Creature Stories*, Robert Weinberg, Stefan Dziemianowicz and Martin H. Greenberg, eds. NY: Barnes & Noble, 1994, pp. 78-85.

Benyus, Janine M. *Biomimicry: Innovation Inspired by Nature.* NY: William Morrow and Co., 1997.
Berebaum, May. *Bugs in the System: Insects and Their Impact on Human Affairs.* Reading, MA: Addison-Wesley, 1995.
Berliner, Nancy Zeng. *Chinese Folk Art: The Small Skills of Carving Insects.* Boston: Little, Brown and Co., 1986.
Berry, Thomas. *The Dream of the Earth.* San Francisco: Sierra Club Books, 1988.
Bigham, Joe. "Biological Pest Control Not Simple," *Associated Press.* Sept. 24, 1997.
"Biology: One Microbe's Meat," *Discover,* March, 1995, p. 20.
"Biotechnology: Bacterial Cement," *Discover,* April, 1997, p. 22.
Blassingame, Wyatt. *The Little Killers.*NY: G.P. Putnam, 1975.
"Bloodcurdling Therapies" *American Health,* October 1994, p. 13.
Bly, Robert, trans. *Selected Poems of Rainer Maria Rilke.* NY: Harper & Row, 1981.
Boone, J. Allen. *Kinship With All Life.* San Francisco, Harper & Row, Inc., 1954.
Boyd, Doug. *Rolling Thunder,* NY: Dell Publishing Co, 1974.
"Bravo for Bacteria," Aug. 15, 1995, *San Jose Mercury News,* pp. 1F, 3F.
"Breakthroughs: Biology: Deliberate Resistance," *Discover,* April, 1996, pp. 21-22.
"Breakthroughs: Of Sex, Somersaults, and Death," *Discover,* November, 1995, p. 34.
Brennan, Barbara Ann. *Light Emerging: The Journey of Personal Healing.* NY: Bantam Books, 1993.
Brown, Joseph Epes. *Animals of Soul: Sacred Animals of the Oglala Sioux.* Rockport, MA: Element, Inc., 1992.
—. *The Sacred Pipe.* Norman, OK: University of Oklahoma Press, 1953.
Brussat, Frederic and Mary Ann. *Spiritual Literacy: Reading the Sacred in Everyday Life.* NY: Scribner, 1996.
"Building Better Mosquitoes," *Discover,* September, 1996, p. 16.
Burton, Robert. *Venomous Animals.* NY: Crescent Books, 1977.
Butler, J.F. 1994. Personal Communication from J.H. Byrd. University of Florida Book of Insect Records. Department of Entomology & Nematology, May, 1994.
Butterfield, Stephen T. "The Face of Maitreya" *The Sun,* Feb., 1989, pp. 20-25.
"Butterfly Man," *People,* 2/26/98, pp. 131-132.
Caldwell, Mark. "The Dream Vaccine," *Discover,* Sept., 1997, pp. 85-88.
Calahan, Philip S. *The Soul of the Ghost Moth.* Old Greenwich, CT: The Devin-Adair Co., 1981.
—. in *Secrets of the Soil,* Tompkins and Bird op site., p. 273.
Campbell, Joseph. *The Hero with a Thousand Faces.* Princeton, NJ: Princeton University Press, 1949, 1968.
"Carbon Dioxide May be Weapon Against Termites," *San Jose Mercury News,* February 3, 1998, 1F.
Carroll, Lewis. *Through the Looking Glass and What Alice Found There.* NY: Clarkson N. Potter, Inc. 1972.
Carson, Rachel. *Silent Spring.* NY: Houghton Mifflin, 1962, 1994.
—. *The Sense of Wonder.* NY and Evanston: Harper & Row, 1956.
Charbonneau-Lassay, Louis. *The Bestiary of Christ.* D.M. Dooling, trans., NY: Parabola Books, 1991.
Cassettari, Stephen. *Pebbles on the Road: A Collection of Zen Stories and Paintings.* Auckland, New Zealand: HarperCollins, 1992.
"Chemical-Eating 'Bug' to the Rescue," *USA Today,* June 1990, p 3.
Cheneviere, Alain. *Vanishing Tribes: Primitive Man on Earth.* Garden City, NY: Doubleday and Co., Inc. 1987, p. 97.
Cherry, Ron H. "Insects in the Mythology of Native Americans," *American Entomologist* (39), 1993, pp. 16-21.
Chief Seattle. in *The Extended Circle: A Commonplace Book of Animal Rights.* ed. Jon Wynne-Tyson. NY: Paragon House, 1989, p. 317.
Chimenti, Elisa. *Tales and Legends of Morocco.* Arnon Benamy, trans., NY: Ivan Obolensky Inc., 1965.

"Chiquinho of the Bees: The Boy Who Talks to Animals," ("Chiquinho da Abelha, O Menino Que Fala Com Os Bichos,") Feb. 1980.
Chui, Glennda. "An Invasion Underfoot," *San Jose Mercury News*, Oct. 28, 1997, pp. F1-F2.
Clausen, Lucy W. *Insect Fact and Folklore.* NY: The Macmillan Company, 1954.
Chauvin, R. *The World of Ants*. NY: Hill & Wang, 1971.
Cody, John. *Wings of Paradise: The Great Saturniid Moths*. Chapel Hill & London: The University of North Carolina Press, 1996.
Cole, Joanna. *Cockroaches*. NY: William Morrow and Company, 1971.
Combs, Alan and Mark Holland. *Synchronicity: Science, Myth, and the Trickster.* NY: Paragon House, 1990.
Compton, John. *The Spider.* NY: Nick Lyons Book, 1987.
Conrad, Katherine. "Swat Teams Rev Up Battle Against Larvae," *San Jose Mercury News*. Feb. 5, 1995, pp. 1B, 4B.
Conrotto, Eugene L. *Miwok means People: The Life and Fate of the Native Inhabitants of the California Gold Rush Country.* Fresno, CA: Valley Publishers, 1973.
Cooper, Gail. *Animal People.* Boston: Houghton Mifflin, 1983.
Cooper, J. C. *Symbolic & Mythological Animals.* London: HarperCollins, 1992.
Costello, Peter. *The Magic Zoo: The Natural History of Fabulous Animals.* NY: Saint Martin's Press, 1979.
—. *The World of Ants*. NY: Hill & Wang, 1971. p. 122.
Cotterell, Arthur. *The Macmillan Illustrated Encyclopedia of Myths and Legends.* NY: Macmillan Publishing Co., 1989, p. 114.
Cox, Christie, Unpublished manuscript.
"Cuba Accuses U.S. of Biological Attack," Geneva, *Associated Press*, October, 1997.
Davis, Bennett. "A Living Battery: Power Lunch," *Discover*, October, 1997, pp. 32-34.
Davis, Flora. *Eloquent Animals: A Study in Animal Communications: How Chimps Lie, Whales Sing, and Slime Molds Pass the Message Along.* Berkley: Berkley Publishing Corp., 1978.
de Waal, Frans. "Simian Sympathy" *Natural History.* March 1996, p. 8.
Deer, John (Fire) and Richard Erdoes. *Lame Deer Seeker of Visions*. NewYork: Washington Square Press, (Simon and Schuster), 1972.
Desowitz, Robert S. *The Malaria Capers: More Tales of Parasites and People, Research and Reality.* NY: W.W. Norton & Company, 1991.
Dillard, Annie. *Pilgrim at Tinker Creek.* NY: Harper & Row, 1974.
Disch, Thomas. "The Roaches," in *Strangeness: A Collection of Curious Tales*. Thomas M. Disch and Charles Naylor, eds., NY: Charles Scribner's Sons, 1977, pp. 175-184.
Donnelly, Kathleen. "Germ Warfare," *West Magazine*, Sept. 28, 1997, pp. 10-19.
Dossey, Larry. *Recovering the Soul—. A Scientific and Spiritual Search.* NY: Bantam, 1989.
Eagle, Brooke Medicine. *Buffalo Woman Comes Singing.* NY: Ballantine Books, 1991.
Eberhard, Wolfram, ed., *Folktales of China.* Chicago: University of Chicago Press, 1965.
Eckhart, Meister. in *Meditations with Meister Eckhart.* Matthew Fox, trans. Santa Fe, NM: Bear & Co., 1983.
Edman, John D. "Biting the Hand that Feeds You," *Natural History*, July 1991, pp. 8-10.
Eisner, Thomas. In "Uncovering the Chemistry of Love and War," *National Wildlife*, August-September 1990.
Eliot, Alexander (with contributions by Mircea Eliade, Joseph Campbell, Detlef I. Lauf, and Emil Bührer). *Myths.* NY: McGraw-Hill Book Co., 1976, pp. 173-4.
Elkins, James. *The Object Stares Back: On the Nature of Seeing.* NY: Simon & Schuster, 1996.
"Eminent Scientists Comment on the Dangers of Genetically Engineered Foods," http://www.geocities.com/Athens/1527/scientists.html.
Engelmann, Larry. "Legends of the fall: Vietnam War: 20 years later," *San Jose Mercury News*, April 30, 1995, 1C, 5C.
Estés, Clarissa Pinkola. *Women Who Run With the Wolves: Myths and Stories of the Wild Woman Archetype.* NY: Ballantine Books, 1995.
Evans, Arthur V. and Charles L. Bellamy. *An Inordinate Fondness for Beetles.* NY: Henry Holt & Company, Inc., 1996.

Evans, Howard Ensign. *Life on a Little Known Planet.* Washington D.C.: Smithsonian Institution Press, 1988.
—. *The Pleasures of Entomology.* Washington D.C.: Smithsonian Institution Press, 1985
—. "To the Ant, and Beyond, with Edward O. Wilson," *Orion Nature Quarterly*, Summer 1986, pp. 46-55.
Ewald, Paul W. "On Darwin, Snow, and Deadly Diseases," *Natural History*, June 1994, pp. 42-45.
"Explorations: The Ties that Bind," *Scientific American*, March, 1997, in http://www.sciam.com/explorations/032797funding/031797stix.html, p. 1.
Fabre, Henri J. *The Insect World of J. Henri Fabre*, Alexander Teixeira de Mattos, trans., Boston: Beacon Press, 1949.
Fagan, John. "Importation of Ciba-Geigy's Bt Maize is Scientifically Indefensible," http://www.netlink.de/gen/BTCorn.htm.
Fernandes, Alvaro. *Como Usar o Magnetismo e a Hipnose.* Rio de Janeiro: Ediouro S.A., 1987.
Ferretti, Fred. "Master of Movies and Wine," *Gourmet*, April 1998, pp. 60-63.
Fewkes, J. Walter. "The Butterfly and Hopi Myth and Ritual," *American Anthropologist* (12), 1910, pp. 576-594.
(The) Findhorn Community. *The Findhorn Garden: Pioneering a New Vision of Man and Nature in Cooperation.* NY: Harper & Row, 1975.
Fisher, Helen M. *From Erin With Love.* San Ramon, CA: Swallowtail Publishing, 1995.
Fisher-Nagel, Heiderose and Andreas. *The Housefly.* Minneapolis: Carolrhoda, 1990.
Flakus, Greg. *Living With Killer Bees: The Story of the Africanized Bee Invasion.* Oakland, CA: Quick American Archives, 1993.
Fleming, Pat, Joanna Macy, Arne Naess, and John Seed. *Thinking Like A Mountain: Towards a Council of All Beings.* Philadelphia: New Society Publishers, 1988.
Fox, Maggie. "Weevils go haywire in biocontrol project," *Reuter Information Service*, 1997.
Fox, Matthew and Rupert Sheldrake. *Natural Grace.* NY: Doubleday, 1996.
Fox, Michael. In "Matthew Fox," *Listening to the Land: Conversations About Nature, Culture, and Eros.* Jensen Derrick, ed., San Francisco: Sierra Club Books, 1995. pp. 67-77.
Frank, Adam. "Quantum Honeybees," *Discover*, November, 1997, pp. 81-88.
Fredrickson, James K. and Tullis C. Onstott. "Microbes Deep Inside the Earth," *Scientific America*, October, 1996. in http://www.sciam.com/1096issue/1096onstott.html.
Freedman, David H. "The Butterfly Solution" *Discover*, April, 1997, pp. 47-53.
Frishman, Austin M. and Arthur P. Schwartz. *The Cockroach Combat Manual.* NY: Wm Morrow and Co. 1980.
Furth, David. "Beetlemania!" *Wings: Essays on Invertebrate Conservation*, Summer, 1991. pp. 3-13.
Gadsby, Patricia. "Why Mosquitoes Suck," *Discover*, August 1997, pp. 42-45.
Garfield, Patricia. *The Healing Power of Dreams.* NY: Simon & Schuster, 1991.
"Garlic and Mosquitoes #1002," in *Ivanhoe's Medical Breakthroughs*, Ivanhoe Broadcast News, Inc., 1997.
Garmston, Robert and Bruce Wellman. "Adaptive Schools in a Quantum Universe," *Educational Leadership*, April 1995, pp. 6-12.
Geographica, "The Stowaway Beetle That Ate Wall Street," *National Geographic*, April 1995.
Gillett, J.D. *The Mosquito: Its Life, Activities, and Impact on Human Affairs.* NY: Doubleday & Co., 1972.
Gimbutas, Maria. *Language of the Goddess.* San Francisco: Harper & Row, 1989.
Goethe. "The Holy Longing," in *News of the Universe: Poems of Twofold Consciousness.* Robert Bly, trans., ed., San Francisco, Sierra Club Books, 1980, p. 70.
Goleman, Daniel. *Emotional Intelligence: Why It Can Matter More than IQ.* NY: Bantam Books, 1995.
"Good Riddance to Roaches," *San Jose Mercury News.* September 11, 1992, pp. 1F-2F
Goodwin, Brian. *How the Leopard Changed Its Spots: The Evolution of Complexity.* NY: Simon & Schuster, 1994.
Gordon, David George. *The Compleat Cockroach: A Comprehensive Guide to the Most Despised (and Least Understood) Creature on Earth.* Berkeley, CA: Ten Speed Press, 1996.

Gould, Stephen Jay. "Of Mice and Mosquitoes" *Natural History*, July, 1991, pp. 12-20.
Graeber, M. J. "Big Bugs From Outer Space," *Fate*, February 1998, pp. 36-39.
Graham, Jr., Frank. *The Dragon Hunters*. NY: Truman Talley Books, E. P. Dutton, Inc., 1984.
Gray, Leslie. "Shamanic Counseling and Ecopsychology," in *Ecopsychology: Restoring the Earth Healing the Mind*. Theodore Roszak, Mary E. Gomes, and Allen D. Kanner, eds., San Francisco: Sierra Club, 1995.
Griffin, Donald R. *Animal Minds*. Chicago: The University of Chicago Press, 1992.
—. *Animal Thinking*. Cambridge, MA: Harvard University Press, 1984.
Grinnell, G.B. "The Butterfly and Spider Among the Blackfeet," *American Anthropologist* (1), 1899, pp. 194-196.
Grissell, Eric. "City Toads and Country Bugs," in *Insect Potpourri: Adventures in Entomology*, Jean Adams, ed., Gainesville, FL: The Sandhill Crane Press, Inc., 1992, pp. 248-251.
Guggenheim, Bill and Judy. *Hello From Heaven*. New York: Bantam, 1996.
Halifax, Joan. *Shaman, the Wounded Healer*. NY: Thames and Hudson, 1982.
—. *The Fruitful Darkness: Reconnecting with the Body of the Earth*. NY: HarperCollins, 1993.
Hall, Manly P. *The Secret Teachings of All Ages: An Encyclopedia Outline of Masonic Hermetic Qabbalistic and Rosicrucian Philosophy*. Los Angeles: The Philosophical Research Society, Inc., 1975.
Hall, Mitchell. "Some Animal Tales," *Orion Nature Quarterly*. Spring 1990, pp. 62-64.
Hall, Rebecca. *Animals are Equal: Humans and Animals – the Psychic Connection*. London: Century Hutchinson Ltd., 1980.
Hanson, Jeanne K. *The Beastly Book: 100 of the World's Most Dangerous Creatures*.NY: Prentice Hall, 1993.
Harbrecht, Doug. "Rescuing Rare Beauties," *National Wildlife*, (29) (5) Aug./Sept. 1991, pp. 4-8.
Harman, Willis. "Biology Revisioned," *Ions: Noetic Sciences Review*, (41) Spring 1997, pp. 12-17,39-42.
—. interviewed by Sarah van Gelder. "Transformation of Business," *In Context*, (41), 1994, pp. 52-55.
Harrison, Gordon. *Mosquitoes, Malaria and Man: A History of the Hostilities Since 1880*. NY, E.P. Dutton, 1978.
Haug, Kathia. "Sethian Love Stories Communicating with Animals," *Reality Change* 9, October, 1985, pp. 48-49.
Hausman, Gerald. *Sitting on the Blue-Eyed Bear: Navajo Myths and Legends*. Westport, CT: Lawrence Hill & Co., 1975.
Hawkes, Nigel, ed. "Ladybirds Harmed in Transgenic Crop Test: Cross-pollination Reported by Scottish Crop Reserach Institute," http://www.geocities.com/Athens/1527/scientists.html.
Hearn, Lafcadio. *Kotto: Being Japanese Curios, with Sundry Cobwebs*. NY: The Macmillan Co., 1927, pp. 57-61, pp. 137-169.
—. *KWAIDAN Stories and Studies of Strange Things*. Rutland, VT: Charles E. Tuttle Co., 1971, pp. 117-118. (Originally published Boston: Houghton Mifflin Co., 1904.)
Hearne, Vicki. *Animal Happiness*. NY: HarperCollins, 1994.
Heinrich, Bernd. "Some Like it Cold," *Natural History*, Feb., 1994, pp. 42-48.
Helfman, Elizabeth S. *The Bushman and Their Stories*. NY: The Seabury Press, 1971.
"Hidden Unity," *Discover*, January, 1998, pp.46-47.
Hillyard, Paul. *The Book of the Spider: From Arachnophobia to the Love of Spiders*. NY: Random House, 1994.
Hillman, James. "Going Bugs" *Spring:A Journal of Archetype and Culture*. Dallas: Spring Publications, 1988. pp. 40-72.
—. *Kinds of Power: A Guide to its Intelligent Uses*. NY: Currency Doubleday, 1995.
—. *The Soul's Code*. NY: Random House, 1996.
—. and Margot McLean. *Dream Animals*. San Francisco: Chronicle Books, 1997.
Hitchcock, Stephen W. "Insects and Indians of the Americas" *Bulletin of the Entomology Society of America* (8), 1962, pp 181-187.
Hogan, Linda, Theresa Corrigan, and Stephanie Hoppe, eds. *And a Deer's Ear, Eagle's Song, and Bear's Grace: Animals and Women*. Pittsburgh: Cleis Press Inc., 1990, pp. 52-53.

Hogue, Charles. "Cultural Entomology" *Annual Review of Entomology* (32), 1987.
Holland, Gail Bernice. "Changing Society Through the Arts," Gail Bernice Holland. *Institute of Noetic Sciences Connections*, September., 1997, pp. 9-11.
Holldobler, Bert and Wilson, Edward O. *The Ants*. Cambridge MA: Harvard University Press, 1990.
Hope, David B. *A Sense of the Morning: Inspiring Reflections on Nature and Discovery*. NY: Fireside, 1988.
Houston, Jean. *A Mythic Life: Learning to Live Our Greater Story*. San Francisco: HarperCollins, 1996.
Howell, Bill. "California Invertebrates Rejected," *The Monarch Newsletter*, Dec. 1994.
Hoy, Michael J."Amazing Boy Talks to Animals —. And They Obey His Commands," *MANCHETE Revista Semanal*, 1452, February 16, 1980. p. 40-41.
Hubbard, Barbara Marx. *Conscious Evolution: Awakening the Power of Our Social Potential*. Novato, CA: New World Library, 1998.
Hubbell, Sue. *Broadsides from the Other Orders: A Book of Bugs*. NY: Random House, 1993.
Huddle, Norie. *Butterfly*. NY: Huddle Books, 1990.
Inouye, David W. "Portrait of a Mountain Meadow," *Orion Nature Quarterly*, Summer 1986, pp.18-25.
"Itsy-bitsy Spiders Alarm Japan," *San Jose Mercury News*, December 2, 1995, pp. 1A, 18A.
James, Mary. *Shoebag*. NY: Scholastic Inc., 1990.
Jangchub, General Kelsang. "Are You My Mother?" *Yoga Journal*, October 1995. p. 151.
Johnson, Buffie. *Lady of the Beasts*. NY: Harper & Row, Inc. 1988.
Jordan, William. *Divorce Among the Gulls: An Uncommon look at Human Nature*. San Francisco: North Point Press, 1991.
Johari, Harish. *The Monkeys and the Mango Tree: Teaching Stories of the Saints and Sadhus of India*. Rochester, VT: Inner Traditions, 1998
"Journey Into Mystery," NY: Marvel Comics Group, 1987.
Jung, Carl. *Synchronicity, A Causal Connecting Principle*. Princeton, NJ: Princeton University Press, 1973, p. 94.
Kanigher, Bob. "FireDance," *Weird Mystery Tales* (4) (19), NY: National Periodical Publications, 1975.
Keen, Sam. *Faces of the Enemy: Reflections of the Hostile Imagination*. San Francisco: Harper & Row, 1986.
Keller, Evelyn Fox. *A Feeling for the Organism: The Life and Work of Barbara McClintock*. NY: W. H. Freeman & Co., 1983.
Kellert, Stephen R., "Editorial" *Wings: Essays on Invertebrate Conservation*, Winter 1991, p. 13.
—. and Edward O. Wilson, eds. *The Biophilia Hypothesis*. Washington D.C.: Island Press, 1993.
Kelly, Peter. "Understanding Through Empathy," *Orion Nature Quarterly*, Winter 1983, pp. 12-16.
Kennedy, Des. *Nature's Outcasts:A New Look at Living Things We love to Hate*. Pownal, VT: Storey Communications, 1993.
Kerby, Mona. *Cockroaches*. NY: Franklin Watts, 1989.
"The Killing Effect of Destroying Nature," *The Straits Times*, Life! Section, September 10, 1997.
Killion, Michael J. and S. Bradleigh Vinson. "Ants with Attitude," *Wildlife Conservation*, January/February, 1995, pp. 44-51, 73.
Kinkead, Eugene. "Profile of Roman Vishniac," in *The New Yorker*, July 2, 1955, pp. 28-29.
Klausnitzer, Bernhard. *Insects: Their Biology and Cultural History*. New York: Universe Books, 1987.
Klinkenborg, Verlyn. "Biotechnology and the Future of Agricultre," *New York Times*, December 8, 1997.
Kluger, Jeffrey. "Mr. Natural," *Time*, May 12, 1997, pp. 68-75.
Knutson, Roger M. *Furtive Fauna*. New York: Penguin Books, 1992.
Koehler, Philip G. and Richard S. Patterson. "Cockroaches," in *Insect Potporrri: Adventures in Entomology*. Jean Adams, ed., Gainesville, FL: The Sandhill Crane Press, 1992, pp. 147-149.
Kowalski, Gary. *The Souls of Animals*. Walpole, NH: Stillpoint, 1991,

Kraus, Sibella. "The Flight of the Honey Bees," *San Francisco Chronicle*, June 23, 1993, pp. 4-5.
Kritsky, Gene. "Beetle Gods, King Bees & Other Insects of Ancient Egypt," *KMT*, pp.32-39.
Lake-Thom, Bobby. *Spirits of the Earth: A Guide to Native American Nature Symbols, Stories, and Ceremonies*. New York: Plume Trade Paperback, 1997.
Laland, Stephanie. *Peaceful Kingdom: Random Acts of Kindness by Animals*. Berkeley, CA: Conari Press, 1997.
Lame Deer, John (Fire) and Richard Eredoes. *Lame Deer Seeker of Visions*. New York:Simon and Schuster, 1972.
Lappé, Marc. *Evolutionary Medicine: Rethinking the Origins of Disease*. San Francisco, Sierra Club. 1994.
—. *Broken Code: The Exploitation of DNA*. San Francisco: Sierra Club. 1984.
—. and Britt Bailey. " Genetically engineered Cotton in Jeopardy," *Center for Ethics and Toxics*, Sept. 10, 1997.
Lauck, Marcia. "Dreamtime & Natural Phenomena: The Release of Transformative Energy into Collective Consciousness," *Dream Network: A Journal Exploring Dreams & Myths*, (14) (4), 1995, pp. 25-27.
Laughlin, Robin Kittrell. *Backyard Bugs*. San Francisco: Chronicle Books, 1996.
Leach, Maria. *How the People Sang the Mountains Up*. New York: The Viking Press, 1967.
Lee, John. *The Flying Boy: Healing the Wounded Man*. Austin, TX: New Men's Press, 1987.
Lee, Stan. "Bugs," in Peter Parker, *The Spectacular Spider-Man*. (1) (86), New York: Marvels Comic Group, 1984.
Lehane, Brendan. *The Compleat Flea*. NY: The Viking Press, Inc. 1969.
Lessen, Don. "Dr Ant" *International Wildlife*, Jan-Feb 1991, pp. 28-34.
Levertov, Denise. "Come Into Animal Presence," in *We Animals: Poems of Our World*. Nadya Aisenberg, ed. San Francisco: Sierra Club Books, 1989, p. 43.
Levine, Stephen. *Healing into Life and Death*. NY: Doubleday, 1987.
Lévy-Brühl, Lucien. in *Man and his Symbols*. Carl Jung. NYk: Doubleday & Co., 1964.
Limburg, Peter R. *Termites*. NY: Hawthorn Books, 1974.
Line, Les. ed., *The Audubon Society Book of Insects*, NY: Harry N. Abrams, Inc., 1983.
Locke, Raymond Friday. *Sweet Salt: Navajo Folktales and Mythology*. Santa Monica: Roundtable Publishing Company, 1990.
London, Scott. "The Science of Leadership, interviewing Margaret Wheatley," Online Noetic Science Institute, onnjoel@libertynet.org, December 16, 1997.
Longgood, William. *The Queen Must Die: And Other Affairs of Bees and Men*, NY: W.W. Norton & Co., 1985.
Lovell, Gary J. "A Sixth Sense or Coincidence?" *Y.E.S. Quarterly* (9) (2), April/June, 1992, pp. 6-7.
Macy, Joanna. *World as Lover, World as Self*. Berkeley, CA: Parallax Press, 1991.
Madigan, Michael T. and Barry L. Marrs. "Extremophiles," *Scientific American*, April, 1997. In http://www.sciam.com/0497issue/0497marrs.html.
Mails, Thomas E. *Plains Indians: Dog Soldiers, Bear Men and Buffalo Women*. NY: Bonanza Books, 1985.
Malone, Fred. *Bees Don't Get Arthritis: The Healing Powers of Bee Stings, Honey, Pollen and Propolis*. NY: E.P. Dutton, 1979.
Margulis, Lynn and Dorian Sagan. *Microcosmos: Four Billion Years of Microbial Evolution*. NY: Summit Books, 1986.
Marriott, David. "Monarch Poachers Profit on California Overwintering Sites," *The Monarch Newsletter*, December 1995, p. 3.
Marsh, Richard. "The Beetle," in *Victorian Villainies*. Graham Greene and Hugh Greene, eds. Middlesex, England: Viking, 1984.
Martin, Laura C. *Wildlife Folklore*. Old Saybrook, CT: The Globe Pequot Press, 1994.
Matthews, Marti Lynn. *Pain: the Challenge and the Gift*. Walpole, NH: Stillpoint Publishing, 1991.
Matthews, Patrick. *The Pursuit of Moths and Butterflies: An Anthology*. London: Chatto & Windus, 1957.

Matthews, Richard. *Nightmares of Nature*. London: HarperCollins, 1995.
McClellan, Michael G. "If We Could Talk With the Animals: Elephants and Musical Performance During the French Revolution," in *Cruising the Performative: Interventions into the Representation of Ethnicity, Nationality, and Sexuality* (13), Sue-Ellen Case, Philip Brett, and Susan Leigh Faster, eds., Bloomington, IN: Indiana University Press, 1995.
McIlroy, A. J. "Water firm risks hosepipe ban to bail out spiders," *Electronic Telegraph*, Monday May 5, 1997.
McLaughlin, John. In "Biological Pest Control Not Simple," Joe Bigham, *Associated Press*, Sept. 24, 1997.
McVay, Scott. "Prelude: A Siamese Connexion with a Plurality of Other Mortals," in *The Biophilia Hypothesis*. Stephen Kellert and Edward O. Wilson, eds., Washington D.C., Island Press, 1993, pp. 3-19.
"Medicine: Worms and Wonder Drugs," *Discover*, March 1997, p. 23.
Mercatante, Anthony S. *Zoo of the Gods: Animals in Myth, Legend, & Fable*. San Francisco: Harper & Row, 1974.
Miller, Neil Z. "Vaccines and Natural Health" *Mothering*, Spring, 1994, pp. 44-52.
Millman, Lawrence. *A Kayak Full of Ghosts: Eskimo Tales*. Santa Barbara, CA: Capra Press, 1987.
Moffett, Mark. "Why I Like Jumping Spiders," *International Wildlife*, May/June 1995, pp.30-36.
—. "All Eyes on Jumping Spiders," National Geographic, Sept., 1991, pp. 43-63.
Moore, Daphne. *The Bee Book*. NY: Universe Books, 1976.
Moore, Robert. ed. *A Blue Fire: Selected Writings by James Hillman*. NY: Harper & Row, 1989.
—. Care of the Soul: A Guide for Cultivating Depth and Sacredness in Everyday Life. NY: Harper Collins, 1992.
Morgan, Marlo. *Mutant Message DownUnder*. Lees Summit, MO: MM CO.,1991.
Morton, Miriam, ed. *A Harvest of Russian Children's Literature*. Berkeley and Los Angeles: University of California Press, 1967.
Moss, Richard. *The Black Butterfly: An Invitation to Radical Aliveness*. Berkeley, CA: Celestial Arts, 1986.
Myss, Caroline. *Anatomy of the Spirit: The Seven Stages of Power and Healing*. NY: Harmony Books, 1996.
Nabokov, Vladimir. "Butterflies" in *The New Yorker*, June 1948 originally from *The Pursuit of Moths and Butterflies An Anthology*. Patrick Matthews. London: Chatto & Windus, 1957, pp. 17-24.
Nagel, Ronald L. "Malaria's Genetic Billiards Game" *National History*, July, 1991, pp. 59-61.
Nahmad, Clair. *Magical Animals: Folklore and Legends from a Yorkshire Wisewoman*. London: Pavilion Books Limited, 1996.
National Wildlife Association.*The Unhuggables: The Truth About Snakes, Slugs, Skunks, Spiders, and Other Animals That Are Hard to Love*. Washington D.C.: National Wildlife Federation, 1988.
Navarro, Mireya. "Beetle Brigade," *NY Times*, In San Jose Mercury News, May 5, 1997. p 15A.
Nelson, Richard. *The Island Within*. San Francisco, North Point Press, 1989.
Neto, Eraldo Medeiros Costa. "Ethnotaxonomy and Use of Bees in Northeastern Brazil," *The Food Insects Newsletter*, (9) (3) November, 1996, p. 1.
"New El Niño Nuisance: Ants, Ants, Ants, Ants," *San Jose Mercury News*, February 22, 1998, 1A, 18A.
Nichol, John. *Bites & Stings: The World of Venomous Animals*. NY: Facts on File, 1989, pp. 151-152.
Nielsen, Lewis T. "Mosquitoes Unlimited" *Natural History*, July, 1991, pp. 4-5.
Nixon, Will. "Letting Nature Shape Childhood," *The Amicus Journal*, Fall, 1997, pp. 31-34.
Noble, Vickie. *Motherpeace: A Way to the Goddess through Myth, Art, and Tarot*. San Francisco: Harper & Row, 1982.
—. *Shakti Woman: Feeling Our Fire, Healing Our World*. San Francisco: HarperSanFrancisco, 1991
Nollman, Jim. *Spiritual Ecology: A Guide to Reconnecting With Nature*. NY: Bantam Books, 1990.

—. *Dolphin Dreamtime.* NY: Bantam New Age Books, 1987.

Oldroyd, Harold. *The Natural History of Flies.* In *Nature's Outcasts: A New Look at Living Things We love to Hate.* Des Kennedy. Pownal, VT: Storey Communications, Inc. 1993.

Osborne, Harold. *South American Mythology.* Feltham, Middlesex, England: The Hamlyn Publishing Group Ltd., 1968.

"Outbreak the Real Risks," *San Jose Mercury News,* March 28, 1995, pp. 1F, 3F.

"Out of the Night," *Journey Into Mystery,* December (8), NY: Marvel Comic Group, 1973.

Parker, Laura. "Bugs taking Big Bite Out of Paradise," *Washington Post.* in *San Jose Mercury News,* July 8, 1991, p. 2A.

Patent, Dorothy Hinshaw. *Mosquitoes.* NY: Holiday House, 1986

Patrick, William. *Spirals.* Boston: Houghton Miffin ? 1983.

Payne, Mark. *Superhealth in a Toxic World.* London: HarperCollins, 1992.

Pearlman, Edith. "Coda: An Inordinate Fondness," *Orion Nature Quarterly,* Autumn 1995, p. 72.

Peck, M. Scott. *The Road Less Traveled.* in *Daybook: A Weekly Contemplative Journal,* Grass Valley, CA: Iona Center, January, 14-Feb.10, 1991.

"Penicillin is Losing Its Punch: Rise in Resistant Germs Shock Researchers," *San Jose Mercury News,* August 24, 1995. p. 10A.

"People vs. Pests: Fighting Pests Without Pesticides: Introduction to Pests," http:www.gsn.orglweb/models/cb/aast/index3.htm

Perera, Sylvia Brinton. *Descent to the Goddess: A Way of Initiation for Women.* Toronto: Inner City Books, 1981.

Pest, Talmadge. Associated Press, Tokyo. "Pest Control: Cockroaches Equipped with Tiny Electronic Backpacks Scuttle around Japanese Lab at Scientists' Commands." *San Jose Mercury News,* 10 January, 1997, p. 27A.

Peterson, Brenda. "Animal Allies," *Orion Nature Quarterly,* Spring 1993, pp.47-50.

Petrenko, Alexa. "I'd Rather Eat A Bug," *Dimensions,* (VII), July/August, 1992, pp. 38-39.

"Pill Takes Bite Out of Disease-bearing Mosquitoes," *San Jose Mercury News,* Dec. 9, 1997, p. F1.

Pressly, W. L. "The Praying Mantis in Surrealist Art," *Art Bulletin* (55), 1973, pp. 600-615.

Preston, Robert. in "Killer Viruses: Concern Proves Quite Contagious," *San Jose Mercury News,* February, 28, 1995, pp. 1E, 11E.

Pringle, Laurence. *Cockroaches: Here, There, and Everywhere.* NY: Thomas Y. Crowell Co., 1971.

—. *Scorpion Man: Exploring the World of Scorpions.* NY: Charles Scribner's Sons, 1994.

Puzzanghera, Jim. "Why *E. coli* is so Hard to Corral," *San Jose Mercury News,* November 15, 1996, p. 1A, 13A.

Quammen, David. *Natural Acts: A Sidelong View of Science and Nature.* NY: Thomas Y. Crowell Co., 1971.

Quinn, Daniel. *Providence: The Story of a Fifty-Year Vision Quest.* NY: Bantam Books, 1994.

Reichard, Gladys A. *Navaho Religion: A Study of Symbolism.* Princeton, NJ: Princeton University Press. 1950.

Riegner, Mark. "Blossoms and Butterflies: A New Look at Metamorphosis," *Orion Nature Quarterly,* Summer, 1986, pp. 30-39.

Rigby, Byron. "Klebsiella Bacteria Destroy Fields," *Living Now,* Jan-Feb, 1997, p.7.

Riley, Norman. *Butterflies and Moths.* NY: Viking Press, 1965.

Rilke, Maria. "A Man Watching," in *News of the Universe: Poems of Twofold Consciousness.* Robert Bly, ed. San Francisco, Sierra Club Books, 1980, pp. 121-122.

Ritchie, Elisavietta."The Cockroach Hovered Like a Dirigible," in *And a Deer's Ear, Eagle's Song, and Bear's Grace: Animals and Women.* Theresa Corrigan and Stephanie Hoppe eds., Pittsburgh: Cleis Press Inc., 1990, pp. 53-56.

Rood, Ronald. *Animals Nobody Loves.* Brattleboro, VT: The Stephen Greene Press, 1971.

—. *It's Going to Sting Me: A Coward's Guide to the Great Outdoors.* NY: Simon and Schuster, 1976.

Ross, Gary Noel. "Winged Victory," *Wildlife Conservation,* July/August, 1994, pp. 60-65.

Rothschild, Miriam. *Butterfly Cooing Like a Dove.* NY: Bantam Doubleday Dell Publishing Group, 1991.

Roszak, Theodore, Mary E. Gomes and Allen D. Kanner, ed. *Ecopsychology: Restoring the Earth, Healing the Mind.* San Francisco: Sierra Club Books, 1995.

Rumi, Jelaluddin. "The Name," in *News of the Universe: Poems of Twofold Consciousness*, Robert Bly, ed., San Francisco: Sierra Club Books, 1980, p. 268.

—. in *News of the Universe: Poems of Twofold Consciousness*. Robert Bly, trans., ed., San Francisco, Sierra Club Books, 1980, p. 268.

Russell, Peter. *The Global Brain Awakens: Our Next Evolutionary Leap*. Palo Alto, CA: Global Brain Inc., 1995.

Ryan, Lisa Gail. *Insect Musicians and Cricket Champions: A Cultural History of Singing Insects in China and Japan*. San Francisco, CA: China Books & Periodicals, Inc., 1996.

Saunders, Nicholas J. *Animal Spirits*. Boston: Little Brown & Co., 1995.

Schul, Bill. *Life Song: In Harmony With All Creation.*Walpole, NH: Stillpoint Publishing, 1994.

Schultz, George F. *Vietnamese Legends*. Rutland, VT: Charles E. Tuttle Co., 1965.

Shapiro, Robert and Julie Rapkin. *Awakening to the Animal Kingdom.* San Rafael, CA: Cassandra Press, 1988.

Sharp, David. Cambridge *Natural History*, p.119-120.

Shealy, C. Norman and Caroline M. Myss. *The Creation of Health: The Emotional, Psychological, and Spiritual Responses that Promote Health and Healing.* Walpole, New Hampshire: Stillpoint Publishing, 1993.

Sheldrake, Rupert. *Seven Experiments That Could Change the World: A Do-It Yourself Guide to Revolutionary Science*, NY: Riverhead Books, The Berkley Publishing Group, 1995.

Shell, Ellen Ruppel. "Resurgence of a Deadly Disease," *The Atlantic Monthly*, August 1997, pp. 45-60.

Shepard, Paul. *Thinking Animals: Animals and the Development of Human Intelligence.* NY: Viking Press, 1976.

Sherlock, Philip. *Anansi, the Spider Man.* NY: Thomas Y. Crowell, 1954.

Shiva, Vandana. *Biopiracy: The Plunder of Nature and Knowledge*. Boston, MA: South End Press, 1997.

Shuttlesworth, Dorothy. *The Story of Flies*. NY: Doubleday, 1970.

Smith, Christopher. "Park Deal: Some Call It Biopiracy," *The Salt Lake Tribune*, November 9, 1997.

Smith, Penelope. *Animal Talk*. Point Reyes, CA: Pegasus Publications, 1989.

—. *Animals...Our Return to Wholeness.* Point Reyes, CA: Pegasus Publications, 1993.

Snell, Marilyn Berlin. "Little Big Top: Maria Fernanda Cardosa Reinvents the Flea Circus," *Utne Reader*, May-June, 1996. pp. 67-71.

Sob, Zong. *Folk Tales from Korea.* Elizabeth, NJ: Holly International Corp. 1970, 1982.

Sonam, Ruth, trans. and ed. *The 37 Practices of Bodhisattvas: An Oral Teaching by Geshe Sonam Rinchen.* Ithaca, NY: SnowLion Publications, 1997, pp 8-9.

Spangler, David. "Decrystallizing the New Age," *New Age Journal*. Jan/Feb, 1997, pp. 70-73, 136.

Spears, Robert. "Gypsy Myths: News, Information, Alternatives & Opinion About Coexisting with the Gypsy Moth," http://www.erols.com/rjspear/gyp_welcom.htm.

Speck, F. G. *Naskapi: the Savage Hunters of the Labrador Peninsula*. Norman, OK: University of Oklahoma Press, 1935.

Steiger, Sherry Hansen and Brad. *Mysteries of Animal Intelligence.* NY: Tom Doherty Associates, 1995.

Steingraber, Sandra. *Living Downstream: An Ecologist Looks at Cancer and the Environment*. Reading, MA: Addison-Wesley, 1997.

Stevens, J. R. *Sacred Legends of the Sandy Lake Cree*. Toronto: McClelland and Stewart Ltd., 1971.

Stokes, John. "Finding Our Place On Earth Again," *Wingspan: Journal of the Male Spirit*. Summer 1990. p.1, 6-7.

Stolzenburg, William. "Silent Sirens," *Nature Conservancy*, May/June, 1992, pp. 8-13.

Suzuki, David and Peter Knudtson. *Wisdom of the Elders: Honoring Sacred Native Visions of Nature*. NY: Bantam Books, 1992.

Swan, James A. *Nature as Teacher and Healer: How to Reawaken Your Connection With Nature*. NY: Villard Books, 1992.

—. in *Voices on the Threshold of Tomorrow: 145 Views of the New Millennium*. Georg Feuerstein and Trisha Lamb Feuerstein, eds. Wheaton, IL: Quest Books, 1993.

Swift, W. Bradford. "Down the Garden Path: How Ten Thousand Years of Agriculture Has Failed Us: An Interview with Daniel Quinn," *The Sun* (7) Dec., 1997, pp. 7-12.

Tanizaki, Junichirø. "The Tattoo," *Modern Japanese Stories*, Ivan Morris, ed., Rutland, VT: Charles E. Tuttle Company, 1962. pp. 95.

Tanner, Wilda B. *The Mystical Magical Marvelous World of Dreams*. Tahlequah, OK: Sparrow Hawk Press, 1988.

Taubes, Gary. "Malarial Dreams," *Discover*, March 1998, pp.109-116.

Taylor, Jeremy. *Dream Work Techniques for Discovering the Creative Power in Dreams.* NY: Paulist Press, 1983.

Taylor, Ronald. *Butterflies in My Stomach or: Insects in Human Nutrition.* Santa Barbara, CA: Woodbridge Press, 1975.

Teale, Edwin Way. *The Golden Throng.* Binghamton: Vail-Ballou Press, Inc. 1940.

—. *Grassroot Jungles*. NY: Dodd, Mead & Co., 1966.

"The Year In Science: Ebola Tamed—. for Now," *Discover*, January, 1996, pp. 16-18.

Thomas, Keith. *Religion and the Decline of Magic: Studies in Popular Beliefs in Sixteenth and Seventeenth Century England*. Hammondsworth: Penguin University Books, 1973, p. 626, quoting J.R. Ackerly, *My Father and Myself* (1968) p. 174.

Thoreau, Henry David. *Walden*. NewYork: Thomas Y. Crowell, 1961.

Thurmon, Howard. *The Search for Common Ground*. NY: Harper & Row, 1971.

Time-Life Editors. *The Spirit World.* New Jersey: Time-Life Books, 1992.

Tompkins, Peter and Christopher Bird. *Secrets of the Soil.* NY: Harper & Row. 1989.

—. *The Secret Life of Nature: Living in Harmony with the Hidden World of Nature Spirits from Fairies to Quarks*. San Francisco, CA: HarperSanFrancisco, 1997.

Toner, Mike. "Spin Doctor," *Discover*. May, 1992, pp. 32-36.

Tyler, Hamilton A. *Pueblo Gods and Myths*. Norman, OK: University of Oklahoma Press, 1964.

Underwood, Anne. "The Witness Was a Maggot," *National Wildlife*, Spring, 1993.

van der Post, Laurens. "The Creative Pattern in Primitve Africa," *Eranos Lectures* 5, Dallas: Spring Publications, 1957.

—. "Creative Patterns of Renewal," *Pendle Hill Pamphlet* (121), Chester, PA: John Spencer, Inc., 1962.

—. *A Far-Off Place*. NY:William Morrow and Company, 1974.

—. *A Mantis Carol.* Covelo, CA: Island Press. 1975.

—. "Wilderness: A Way of Truth," in *Wilderness The Way Ahead*. Vance Martin and Mary Inglis, eds., Forres, Scotland: The Findhorn Press, 1984, pp. 231-237.

"Viruses: On the Edge of Life, On the Edge of Death," *National Georgraphic*, July, 1994, pp. 57-86.

Vitebsky, Piers. *The Shaman*. Boston: Little Brown and Co., 1995.

von Franz, Marie-Louise. *Alchemy: An Introduction to the Symbolism and the Psychology.* Toronto: Inner City Books, 1980.

—. *The Psychological Meaning of Redemption Motifs in Fairytales.* Toronto: Inner City Books, 1980.

The Visual Encyclopedia of Science Fiction. NY: Harmony Books, 1977.

Waldbauer, Gilbert. *Insects Through the Seasons*. Cambridge, MA: Harvard University Press, 1996

Walker, Barbara. *The Woman's Dictionary of Symbols and Sacred Objects.* NY: Harper & Row, 1988.

—. *The Woman's Encyclopedia of Myths and Secrets*. San Francisco: Harper & Row, 1983.

Wallace, Alfred Russel. in "Malarial Fit," Robert H. Mohlenbrock. *Natural History*, July, 1991, p. 53.

—. in "Alfred Russel Wallace & the Birth of Biogeography: The Living Evidence for Evolution," *Pacific Discovery*, Spring, 1993.

Walterscheid, Ellen. "Ill wind," *The Sciences*, March/April 1998, pp10-11.
Warren, Thomas D. and Daniel Kaufman. *Dolphin Conferences, Elephant Midwives and Other Astonishing Facts About Animals*. Los Angeles, CA: Jeremy P. Tarcher, Inc. 1990.
Waters, Frank. *Book of the Hopi*. NY: Ballantine Books, 1963.
Weber, Christin Lore. *A Cry in the Desert: The Awakening of Byron Katie*. Barstow, CA: The Work Foundation, 1996.
Webster, David. *Let's find Out About Mosquitoes*. NY: Franklin Watts, 1974.
Weigle, Marta. *Spiders and Spinsters: Women and Mythology*. Albuquerque: University of New Mexico Press, 1982.
Weil, Andrew. *Health and Healing: Understanding Conventional and Alternative Medicine*. Boston: Houghton Mifflin Co., 1983.
"Welcome to Cockroach World," http:www.np.com/yucky/roaches/index.html.
Werner, Alfred and Josef Bijok. *Butterflies and Moths*. Norman Riley, ed. NY: Viking Press, 1965.
Wexler, Mark. "How to Feed a Visiting Monarch," *National Wildlife*, 1994, pp. 15-21.
Wheeler, W. M. *Ants:Their Structure, Development and Behavior*. NY: Columbia University Press, 1910.
White, Steward Edward. *The Forest*, 1903. quoted in Sue Hubbel. *Broadsides from the Other Orders: A Book of Bugs*. NY: Random House, 1993.
Whitmont, Edward C. *The Symbolic Quest: Basic Concepts of Analytical Psychology*. Princeton, NJ: Princeton University Press, 1969.
Whynott, Douglas. *Orion Nature Quarterly*.
Williams, Greer. *The Plague Killers*. NY: Charles Scribner's Sons, 1969.
Willow, Sara. in "Stories of Animal Companions," *SageWoman* Readers, SageWoman, Spring 1995, pp. 26-27.
Wilson, Edward O. *Sociobiology: The New Synthesis*. Cambridge, MA: Harvard University Press, 1975.
—. *On Human Nature*. Cambridge, Mass.: Harvard University Press, 1978.
—. *Biophilia*. Cambridge: Harvard University Press, 1984
—. "Ants," *Wings: Essays on Invertebrate Conservation*, Fall, 1991. pp. 4-13
—. *The Diversity of Life*. Cambridge, MA: Harvard University Press, 1992.
—. *The Naturalist*. Washington D.C.: Island Press, 1994.
Wolkomir, Joyce and Richard. "High-wire Artists," *National Wildlife*, February-March, 1991, pp.52-59.
—. "Uncovering the Chemistry of Love and War," *National Wildlife*, August-September, 1990, pp. 44-50.
Wolkomir, Richard. "The Bug We Love To Hate," *National Wildlife*, December-January, 1993, pp. 34-37.
Woodfin, Max. "Bugging Out: How Monsanto's Biotech Miracle Undermines Organic Farming," *Sierra*, January-February, 1997, p. 27.
Wright, Barton. *Hopi Kachinas: The Complete Guide to Collecting Kachina Dolls*. Flagstaff, AZ: Northland Publishing, 1977.
Wright, Machaelle Small. *Behaving As If the God In All Life Mattered: A New Age Ecology*. Jeffersonton, VA:Perelandra, Ltd., 1983.
—. *Perelandra Garden Workbook*. Jeffersonton, VA, Perelandra, LTD, 1987.
—. in "Perelandra," *Secrets of the Soil*, p. 309.
Yager, David and Mike May. "Coming In on a Wing and an Ear," *Natural History*, January, 1993, pp. 29-32.
Zimmer, Carl. "A Secret History of Life on Land," *Discover*, February 1998, pp. 76-83.
Zweig, Connie and Jeremiah Abrams, eds. *Meeting the Shadow:The Hidden Power of the Dark Side of Human Nature*. NY: Jeremy P. Tarcher, 1991.

Index

N

The New Millennium Library

Building a New Paradigm
for a New World
in a New Millennium

Volume I— Theology and Science
The God Hypothesis: ET Life and Its Implications for Science and Religion
Dr. Joe Lewels
0-926524-40-2 (Wild Flower Press)

Volume II—Sociology
At the Threshold: UFOs, Science and the New Age
Dr. Charles F. Emmons
0-926524-42-9 (Wild Flower Press)

Volume III—Medicine and Healing
Plant Spirit Medicine: Healing with the Power of Plants
Eliot Cowan
0-926524--09-7 (Swan•Raven & Co.)

Volume IV—Anthropology and Cryptozoology
The Psychic Sasquatch: And Their UFO Connection
Jack Lapseritis, M.S.
0-926524-17-8 (Wild Flower Press)

Volume V—Ecology
The Voice of the Infinite in the Small: Revisioning the Insect-Human Connection
Joanne E. Lauck, M.S.
0-926524-49-6 (Swan•Raven & Co.)

Volume VI—Psychology
Soul Samples: Personal Exploration in Reincarnation and UFO Experience
Dr. R. Leo Sprinkle
0-926524-47-x (forthcoming - Wild Flower Press)

Order from your favorite bookstore

To receive more information on books
by Blue Water Publishing,
(Imprints: Swan•Raven & Co. and Wild Flower Press)

or to contact the author,
please write to

P.O. Box 190
Mill Spring, NC 28756

or call 800•366•0264

Swan•Raven & Co.

Place Stamp Here

Swan•Raven & Co.
P.O. Box 190
Mill Spring, NC 28756

Contact us at—

Swan•Raven & Co., P.O. Box 190, Mill Spring, NC 28756

Email: BlueWaterP@aol.com

Web Site: www.5thworld.com *or* www.bluewaterp.com

The Swan and the Raven traditionally carry the Sacred Message between the Otherworld and our world. Swan•Raven & Company publishes books whose themes explore this Sacred Message.

If you would like to receive our latest catalog and be placed on our mailing list, please send the enclosed card.

Name ______________________ Date __________

Address ______________________________

City ______________ State ________ Zip __________

Country ______________________